THE BIOSPHERE

Edited by **Natarajan Ishwaran**

INTECHWEB.ORG

The Biosphere
Edited by Natarajan Ishwaran

Published by InTech
Janeza Trdine 9, 51000 Rijeka, Croatia

Copyright © 2012 InTech
All chapters are Open Access distributed under the Creative Commons Attribution 3.0 license, which allows users to download, copy and build upon published articles even for commercial purposes, as long as the author and publisher are properly credited, which ensures maximum dissemination and a wider impact of our publications. After this work has been published by InTech, authors have the right to republish it, in whole or part, in any publication of which they are the author, and to make other personal use of the work. Any republication, referencing or personal use of the work must explicitly identify the original source.

As for readers, this license allows users to download, copy and build upon published chapters even for commercial purposes, as long as the author and publisher are properly credited, which ensures maximum dissemination and a wider impact of our publications.

Notice
Statements and opinions expressed in the chapters are these of the individual contributors and not necessarily those of the editors or publisher. No responsibility is accepted for the accuracy of information contained in the published chapters. The publisher assumes no responsibility for any damage or injury to persons or property arising out of the use of any materials, instructions, methods or ideas contained in the book.

Publishing Process Manager Ivona Lovric
Technical Editor Teodora Smiljanic
Cover Designer InTech Design Team

First published March, 2012
Printed in Croatia

A free online edition of this book is available at www.intechopen.com
Additional hard copies can be obtained from orders@intechweb.org

The Biosphere, Edited by Natarajan Ishwaran
 p. cm.
ISBN 978-953-51-0292-2

INTECH
open science | open minds

free online editions of InTech
Books and Journals can be found at
www.intechopen.com

Contents

Preface IX

Part 1 Perspectives on Origins and Evolution 1

Chapter 1 **Early Biosphere: Origin and Evolution 3**
Vladimir F. Levchenko, Alexander B. Kazansky,
Marat A. Sabirov and Eugenia M. Semenova

Chapter 2 **The Search for the Origin of Life:
Some Remarks About the Emergence
of Biological Structures 33**
Mirko Di Bernardo

Chapter 3 **The Biosphere: A Thermodynamic Imperative 51**
Karo Michaelian

Part 2 Ecosystems and Resource Utilization 61

Chapter 4 **Yukon Taiga – Past, Present and Future 63**
Rodney Arthur Savidge

Chapter 5 **Trace Metals and Radionuclides
in Austrian Forest Ecosystems 93**
Stefan Smidt, Robert Jandl, Heidi Bauer, Alfred Fürst,
Franz Mutsch, Harald Zechmeister and Claudia Seidel

Chapter 6 **Plant Extracts: A Potential Tool for Controlling
Animal Parasitic Nematodes 119**
Pedro Mendoza de Gives, María Eugenia López Arellano,
Enrique Liébano Hernández and Liliana Aguilar Marcelino

Chapter 7 **Understanding Linkages Between
Public Participation and Management
of Protected Areas – Case Study of Serbia 131**
Jelena Tomićević, Ivana Bjedov, Margaret A. Shannon
and Dragica Obratov-Petković

Chapter 8 **The Materially Finite Global Economy
Metered in a Unified Physical Currency   143**
John E. Coulter

Part 3 **Biosphere Reserves   173**

Chapter 9 **Biosphere Reserves   175**
Murat Özyavuz

Chapter 10 **The Importance of Biosphere Reserve in Nature Protection
and the Situation in Turkey   193**
Sibel Saricam and Umit Erdem

Chapter 11 **Wealth of Flora and Vegetation in the La Campana-Peñuelas
Biosphere Reserve, Valparaiso Region, Chile   215**
Enrique Hauenstein

Chapter 12 **Investigation of Reproductive Birds in Hara Biosphere
Reserve, Threats and Management Strategies   245**
Elnaz Neinavaz, Elham Khalilzadeh Shirazi,
Besat Emami and Yasaman Dilmaghani

Chapter 13 **Livelihoods and Local Ecological Knowledge in Cat Tien
Biosphere Reserve, Vietnam: Opportunities and Challenges
for Biodiversity Conservation   261**
Dinh Thanh Sang, Hyakumura Kimihiko and Ogata Kazuo

Chapter 14 **UNESCO Biosphere Reserves
Towards Common Intellectual Ground   285**
Richard C. Mitchell, Bradley May, Samantha Purdy and Crystal Vella

Preface

Biosphere is the living envelope that encompasses both the abiotic and the biotic components of planet earth. It is the total sum of all world's ecosystems. Since the Apollo missions of the late 1960s began to introduce humanity to "earthrise" visions via photographs taken from the lunar orbit, scientists, artists and the public have increasingly warmed to the idea that planet earth is a wholesome living being. This holistic view has encouraged the use of biosphere as an object of study in a range of disciplines spread across the natural and social sciences as well as the humanities. This "Biosphere" book presents a collection of articles resulting from the work of individuals and teams from all geographic regions of the world on several dimensions of understanding of this important idea.

The book is divided into three sections as follows:

- Perspectives on Origins and Evolution
- Ecosystems and Resources Utilization, and
- Biosphere Reserves

The three papers that address origins and evolution of the biosphere explore a number of stimulating ideas and hypotheses. Collectively they shed light on processes and mechanisms that probably contributed towards the emergence of life from non-living chemical substances and reactions. One of them (Michaelian Karo) attempts to explain the thermodynamic basis for that emergence. All three papers clearly stress and illustrate the increasing role of new sciences addressing complexity and emergence for our overall understanding of the origins and evolution of the biosphere.

The five papers that constitute the section on ecosystems and utilization of resources cover a wide range of changes that are now evident within ecosystems and are attributable to a mix of natural and anthropogenic influences. Dr. Rodney Savidge's tracing of the changes in the Yukon Taiga is an illustration of how a combination of climatic and human influenced changes is impacting a critical ecosystem and the risk that some of those changes may be altering the physiognomy and composition of the Taiga irreversibly. On the other end Dr. Coulter's attempt to use energy based accounting to question the assumptions and workings of the current economic and financial models of resource use and consumption is an indicator of the drastic shifts in lifestyles and resources extraction patterns we may have to re-engineer if we are to alter current trajectories of change and their impacts on the biosphere. Other papers in this section include examples of the use of biological resources for treatment of ailments and illnesses, trace metals in forest ecosystem, the discovery of a new reptilian species in a unique cave ecosystem and the role of people's participation in the management of a protected area.

The third section, entitled "Biosphere reserves" includes a collection of papers on research, conservation and socio-economics and their relevance to people resident in and around places in the world designated by UNESCO (United Nations Educational Scientific and Cultural Organization) as biosphere reserves under that Organization's Man and the Biosphere (MAB) Programme. To-date 580 biosphere reserves have been designated in 114 countries.

As shown by the authors of the first two papers from Turkey, the concept of biosphere reserve originated to make conservation of nature and natural resources have a better scientific basis and to be more inclusive of local community interests. Papers from the Chilean biosphere reserve of La Campana-Penuelas and the Iranian Hara biosphere reserve in this section clearly illustrate the important role assigned to research and conservation in these internationally designated places.

But the concept of biosphere reserve and its applications have undergone considerable changes during the 40 years (1971-2011) of the MAB Programme of UNESCO. What originated as a concept that aimed to make protected area management more scientifically based and people friendly is now a concept that aims to significantly contribute towards sustainable development objectives at global, national and local levels. Dinh Thanh Sang et al's paper on the opportunities and challenges to biodiversity conservation in the Cat Tien National Park of Vietnam is a clear illustration of the complexity of problems and issues at the human-nature interface that biosphere reserve managers face. Biosphere reserve management is intended to generate lessons and experiences that enable reflections and participatory deliberations for "theorizing sustainability through transdisciplinarity", as so elegantly demonstrated by Richard C. Mitchell and his colleagues from the Brock University in Canada in their paper.

As humanity becomes increasingly aware of the possibility that it may have tinkered unwisely with natural processes dealing with climate change and evolution and the extinction of species, populations, communities and ecosystems, it must work towards attaining a new compact with biosphere processes and functions in order to assure its own future survival. The 21st century that is in now its second decade will probably mark the beginning of a new moment in human history when the depth and breadth of our efforts to live with the consciousness and knowledge that we are an integral part of the biosphere will express itself via new life styles and cultures that are mutually beneficial for the wellbeing of ourselves and the planet. I hope that contributions in this book on the "Biosphere" will stimulate thoughts, reflections and further research and study to strengthen our movement towards establishing a new partnership with fellow species and natural and modified ecosystems to create sustainable biosphere futures.

Dr. Natarajan Ishwaran
Division of Ecological and Earth Sciences, UNESCO, Paris,
France

Part 1

Perspectives on Origins and Evolution

Early Biosphere: Origin and Evolution

Vladimir F. Levchenko, Alexander B. Kazansky,
Marat A. Sabirov and Eugenia M. Semenova
*Laboratory of Evolution Modeling, Institute of Evolutionary Physiology and Biochemistry
of the Russian Academy of Sciences, St. Petersburg
Russian Federation*

1. Introduction

We argue that the process of the origin and evolution of living organisms is indivisible from the process of the origin and evolution of biosphere, the global planetary system. The most general features of biosphere evolution as a directed process are formulated in the principle of "bio-actualism". It is demonstrated, that this principle, stated for Phanerozoic eon, hold for early biosphere up to the period of biosphere coming into being. To support this thesis we considered an episode of pre-Cambrian history – Vendian phytoplanktonic crisis.

Traditionally, the problem of the origin of life on Earth is the study of how biological life arose from inorganic matter and primary living organisms spread around the planet. Some philosophers and scientists such as Helmholtz and Arrhenius proposed the hypothesis of so-called "panspermia" and place the origin of life outside the Earth somewhere in cosmos. We suggest that such approaches are naive in view of modern progress of life sciences and astrophysics. It is clear that occasional appearance of some living forms (organisms) on any planet doesn't mean they will survive, settle and evolve there. These conceptions ignore the problem of longtime existence and evolution of the earliest forms of life on the Earth. Moreover, they overlook the problem of the origin of the biosphere – unique milieu for early organisms survivorship and reproduction. Therefore we suppose that origin of earthly life and the origin of the biosphere are aspects of a whole indivisible process (Levchenko, 2010, 2011). Hence, we consider the appearance of such conditions on the early planet, which guaranteed the origin and survivorship of organic life. We suppose also that primary organisms should be incorporated in natural geological processes, accelerating them and transforming surroundings in directions favorable for the creation of higher forms of life.

The abiogenesis (or biopoiesis) hypothesis claims that life emerged from non-living matter in early terrestrial conditions. This is the traditional approach to the question of life's origin. Main questions of abiogenesis hypothesis are reduced to an origin of biological membranes via emergence of ionic asymmetry of the cells and chiral asymmetry of biomolecules, and to the problem of the appearance of matrix synthesis and polymer molecules which are able to store hereditary information. Below we will give a brief review of the different hypotheses. In order to find out the regularities of the pre-biosphere development, features of processes of evolution and development for different biological systems were considered and the embryosphere hypothesis (Levchenko, 1993, 2002, 2011) proposed. According to it, the appearance of different sub-systems of primary organisms could happen independently from each other within an united functional system called an embryosphere. This is a system with a self-regulated process of that allows the interchange of substances between its different parts under the influence of external energy flows.

Many difficulties in the explanation of the origin of primary life can be surmounted if we suppose that surface of early Earth contained a large quantity of organic matter, for example, hydrocarbons. This assumption looks very plausible, accounting for the results of investigations of other bodies of Solar system, e.g. Titan (one of the big satellites of Saturn – Niemann et al., 2005). It is very likely also for at least earliest phases of planetary evolution that water and the hydrocarbon complexes (admittedly slightly like by composition to oldest oils) could be present simultaneously in considerable quantities on some planets of small and medium size. Early Earth had apparently appropriate surface temperature, water and some hydrocarbons in the liquid phase. "Water-oil" emulsion could be a fine matrix for the origin and long evolution of the initial forms of life on our planet (Levchenko, 2010). Such emulsion could be an appropriate material for forming of liquid structures of the pre-cell envelops and moreover, it could be a suitable food and chemical energy source for initial forms of life.

The second half of this chapter is devoted to the description of the earliest stages of biosphere development, in particular Riphean and Vendian periods of the history of Earth. Many known regularities of biological evolution already became apparent during these periods. Among these regularities one can note those that describe usual stages of biotic crises. Moreover, the different systemic principles as well as some conditions and mechanisms, necessary for the origin of biosphere and living organisms are also discussed.

2. What is Biosphere? Modern approaches

Though Eduard Seuss had coined the term **biosphere** more than hundred years ago, it is Vladimir Vernadsky's concept of the biosphere, formulated in 1926, that is accepted today (Vernadsky, 1989). Biosphere is a specific envelope of the Earth, comprising totality of all living organisms and that part of planet matter which is in constant material exchange with these organisms. Biosphere includes **geospheres** i.e. lower part of the atmosphere, hydrosphere and upper levels of lithosphere. Virtually, it is a spherical layer 6-12 km thick. In the view of Vernadsky, life is the geological force. Indeed all geological features at Earth's surface are bio-influenced. The planetary influence of living matter becomes more extensive with time. The number and rate of chemical elements transformed and the spectrum of chemical reactions engendered by living matter are increasing, so that more parts of Earth are incorporated into the biosphere. Life, as Vernadsky viewed it, is a cosmic phenomenon which is to be understood by the same universal laws that applied to the physical world. All his life Vernadsky was dreaming about creating the universal biosphere science, comprising not only biogeochemical processes, but also processes, peculiar to **noosphere** (part of planet where biogeochemical processes may depend on human activity).

Biosphere according to Vernadsky is a self-regulating system, including both living and non-living constituents. The work of living matter in the biosphere is manifested in two main forms: (a) chemical (biochemical) and (b) mechanical. Vernadsky made a detailed analysis of different forms of biochemical and mechanical transformations on environment activity of life and realized that there is no force on the face of the Earth more powerful in its results than the totality of living organisms. No phenomena in the biosphere are separated from life and biogeochemical cycles. To analyze these processes Vernadsky introduced the notions of "**living matter**" of the biosphere – the sum of its living organisms, "**inert matter**" – non-living substance and "**bioinert matter**", which is an organic composition of living organisms with not-living substance. The last concept is of special significance in the context of self-modification and indirect interactions in ecosystems. Soil is an example of bioinert matter. Great many living forms permeate soil and organize it. Biological activity of organisms constantly modifies this environment, and thus modifies organisms themselves, forming self-organized and self-modifying system. Vernadsky noticed, that bioinert matter have unusual physical properties. But indeed, soil being an open system demonstrates some properties of living tissue. Soil at the same time is an environment as well as an organic constituent part of the biosphere.

Vernadsky's ideas are very relevant now in the light of progress of the Earth system science and the interest in the role of life in global climate regulation. James Lovelock in his Gaia hypothesis virtually concentrated on this aspect of biosphere's self-regulation (Lovelock, 1988). Gaia hypothesis suggests that not only do organisms affect their environment; they do so in a way that regulates the biosphere's global climate to conditions that are suitable for life. Gaia phenomena include the regulation of local climate by marine algae that influence the formation of clouds over the oceans, global temperature regulation by biotic enhancement of rock weathering, the maintenance of constant marine salinity and nitrogen/phosphorus ratios by the aquatic biota, etc. Lovelock reflected about Earth "**geophysiology**": the temperature, alkalinity, acidity, and reactive gases that are modulated by life. According to his Gaia hypothesis (Lovelock, 1988), with the appearance of life, our planet had got self-regulatory, homeostatic properties, inherent to living biological organism. Particularly, aspects of surface temperature and chemistry are being self-supported at constant, comfortable levels for life forms for more than 3.6 Gya (i.e. billions of years – giga-years ago).

Biosphere or Gaia, which are almost synonyms nowadays, can be represented as a very specific, dynamic, constantly self-modifying, multilevel evolving system, composed of living, inert and bioinert components, which dynamically exchange. Wholeness of this system is provided by the dynamic interrelatedness of all components, participating in metabolic processes, local trophic cycles, local and global biogeochemical cycles. Living biological components fundamentally differ from inert non-living components. The former are active and goal-seeking systems, demonstrating wholeness and relative autonomy. The later are passive, submit to influence and do not demonstrate autonomy and wholeness; their dynamics is controlled by physical forces and activity of biological organisms. But, bioinert matter with high concentration of symbiotically related biological organisms of different taxa, like a vegetable mold demonstrates intermediate properties and that is true for the whole biosphere. Biosphere functioning, which includes processes, proceeding in a small time-scales does not differ in principle from the process of biosphere evolution, noticeable in geological time-scales. Even in the case of microevolution we cannot apply the model of Markov's chain, stochastic process, in which the role of previous history is reduced to the parameters of current state of the system. First of all, this state is unmanageable, because it should also include for example non-coding sequences of genome (e.g. transposons, etc.), which are potential source of innovation. Biosphere functioning as well as its evolution is a creative, developmental process. In the process of biosphere evolution all biotic components participate in the process of constant self-modification, extinction (destruction) and emergence (collective, mutual production). The abiotic components are open systems nevertheless; they are passive participators in biosphere processes and are subjected to biogenic and physical transformations. Biosphere is a dynamic and emergent system, and its components demonstrate "mutual co-construction". The only absolutely autonomous, self-constructive, self-modifying and self-constitutive system on Earth is the whole biosphere. Apart from biosphere as a whole, all biological organisms and super-organismic systems are half-open, semi-autonomous and more or less sustainable components of biosphere.

Though biological systems have some form of autonomy, they urgently depend on each other directly, through trophic and other classical ecological relationships, but also indirectly, via environment. Organisms interact with all components, which also interact with each other, and the nature of this interaction determines the selection pressure which organisms experience. Species populations evolve, it is possible that selection may occur on their environment-altering traits so that traits are favored, which change the global environment in some beneficial way for living organisms. Such reflections stimulate modeling experiments, demonstrating possibility of Gaia effect (Lovelock, 1988). In his work "The Self-Organizing Universe", E. Jantsch (1980) suggested, that Lovelock's "Gaia" is a multilevel autopoiesis, viz., a cyclically organized network of productions (syntheses, transformations, and destructions) of components, that recursively regenerate and support this very complicated network. "**Autopoiesis**", a neologism meaning "self-production" was coined by Chilean neurophysiologists and system theorists H. Maturana and F. Varela (1987) in the seventies of the 20th century. It is the name of a minimal formal model of life (autonomic biological cell), interacting with its environment holistically by adaptive change of structure ("structural coupling") for the sake of conserving the autopoietic organization. Structurally determined reactions on reciprocal perturbations bring about the co-evolution of the system and its environment. Autopoiesis is now accepted as a theoretical basis of contemporary cybernetics (the second – order cybernetics, von Foerster, 1974), sociology, management and robotics. Jantsch subordinated autopoiesis to the notion of specific self-organization, dissipative regenerative systems and virtually ignored phenomenological aspects of the theory. That led him to classify living cells, organisms, populations, ecosystems as autopoietic systems. The philosophy of radical constructivism and phenomenology of included autopoietic observer are excluded from consideration under the assumptions that Jantsch made.

The last two decades were marked by great progress in the development and critical analysis of Gaia theory – geophysiology as well as of the phenomenological aspects of autopoiesis. In the following sections we describe different phases of origin of life and earliest stages of its evolution in context of biosphere development.

3. Traditional approaches to the origin of life

The appearance and expansion of life on Earth is a versatile, very complex and long-term process. At present, there is no complete theory, exact scientific basis of the origin of life. However we can specify several clear-cut requirements for the emergence of life. First, it is accepted, that a double-layer membrane presence is necessary for maintenance of biological processes and thus for life origin (Baross & Hoffman, 1985; Deamer et al., 2002). Such membrane structures could form **phase-isolated system** which is crucial for substance interchange processes (Fox & Dose, 1977; Rutten, 1971). On the other hand the origin of life exactly requires mechanisms of heredity. According to many hypotheses, such mechanisms could arise on molecular basis of RNA, DNA and (or) proteins. The appearance of RNA and DNA is connected with the problem of rise of the nucleotides as well as with the problem of their abiotic polymerization to linear (non-branchy, non-cyclic) molecules. Also it is important to note that any considerations of origin of life have to give explanations of emergence of ionic asymmetry, chiral asymmetry and metabolism – see below section 3.2.

Of course, it is very difficult to cover the whole spectrum of hypotheses concerning of the origin of life; therefore we mention only some of them, which in our opinion, are basic ones. The majority of hypotheses of **abiogenesis (or biopoiesis)** concentrates only on specific aspects of life origin and cannot account for the whole picture. On the other hand, many holistic hypotheses which make efforts to describe the whole process are weak in details. As we have already mentioned, in this section we shall restrict our consideration to the hypotheses of chemical and successive biological evolution on Earth. It is worth noting here that biospheric approach leads to additional assumptions for the question of life's origin (section 4).

3.1 Hypothetical phases of the origin of life in the view of traditional approaches

The emergence of life is a process, which can be subdivided into several successive phases: 1) appearance of specific conditions on the Earth (or in space) in which synthesis of complex organic compounds could be possible; 2) formation of compartmentalized structures (membrane structures, liposomes and perhaps others) with half-permeable membrane which was able to selectively absorb and excrete different substances; appearance of the primordial substances interchange; 3) synthesis of chemical polymers (RNA, DNA) with specific properties within these structures; such self-replicable polymers could be relatively stable inside the compartment and they were able to keep the information about features of compartmentalized structures by means of variation of their own structure; 4) development of certain correspondence between features of compartmentalized structures and features of these polymers. Such correspondence is likely to have enhanced the stability of the whole system. Improving of the internal biochemical machinery: emerging of specialized catalysts (proteins); origin of genetic code: their structure came into correspondence with the structure of the polymers.

These stages, as they described above suggests that the transition from complex organic substances to biochemistry progressed like a chain of a successive events. In accordance with this point of view compartmentalized structures were able to be selective in absorption of the monomers and their subsequent polymerization. Nevertheless, there is no reason to reject the possibility of parallel (i.e. relatively simultaneous) realization of some of the stages.

3.2 Candidates for initial matrix. Membrane structures forming. Ionic and chiral asymmetry

There are several hypotheses concerning a problem of initial matrix or abiotic synthesis of the primordial informational polymer (most likely it was an RNA molecule – see Gilbert,

1986; Joyce & Orgel, 1993). Primordial informational polymer should be able to self-replicate (otherwise it cannot persist for a long time). Self-replicating, in its turn is based on two properties: preserving information about its own structure and catalytic activity, targeted on the same set of monomers. DNA, protein and RNA, in this respect, are the unique molecules. It was revealed for several small RNA that they are able to catalyze certain reactions of protein matrix synthesis (Cech, 1986). Because of their enzymatic activity they are called **rybozymes**. At present time, many scientists assume that RNA predated the appearance of DNA and proteins: the "RNA-World" hypothesis (Gilbert, 1986). In order to explain emergence of DNA and proteins many hypotheses invoke the idea of functional specialization: DNA has emerged as a special structure to secure storage of information, while proteins have specialized on just enzymatic functions.

A film of hydrocarbons mixture, which was apparently present on the early Earth, may have been a basis for membrane structures formation. Polyaromatic hydrocarbons are the probable candidates as templates for matrix synthesis of RNA. Hydrocarbons are reliable templates for RNA synthesis especially because they are widespread in the Universe and particularly, in the Solar System (section 5). "PAH World" (where PAH means polyaromatic hydrocarbons) is a novel chemical structural model for the formation of prebiotic informational oligomeric material, which could provide for the rise of the "RNA-world", now widely accepted (Platts, 2005). Under assumptions of this model, the primary binding of nucleotides may have been realized by virtue of photochemical and/or geohydrothermal derivations at the edges of PAH complexes (Platts, 2005).

RNA may have been synthesized using some clay minerals (for example, montmorillonite) as template (Ferris & Ertem, 1992; Cairns-Smith, 2005). The authors pay attention to some properties of clay minerals such as: (i) a significant diversity of surface structures which make them potentially storages of information ("crystal genes") and (ii) positive electric charge on the surface of some of these minerals (which in turn could facilitate interaction with RNA or RNA-like molecules carrying negative charge). Therefore the so called "genetic takeover" by RNA could arise on the basis of a primitive crystal gene which was functionally substituted and gradually displaced by it (Cairns-Smith, 2005). Another hypothesis suggests that ferro-sulphuric minerals, notably greigite (Fe_3S_4), due to its magnetic properties could be appropriate informational matrix for the first biological systems (Mitra-Delmotte & Mitra, 2010). There are also some extravagant hypotheses, concerning the origin of primary RNA and (or) DNA in cosmos, e.g. in cores of some comets. For example, Kaimakov (1977) proposed a mechanism of such synthesis which is based on particularities of sublimation of the water ice with solved nucleotides in vacuum.

Concerning of life origin Oparin (1966) wrote about hypothetical "coacervates" which have gel envelops, consisted of proteins and a gelatinous medium; mechanisms for the origin of lipid double-layer membranes, as found in real cells, have also been proposed later (Bangham, 1981; Fox & Dose, 1977; Rutten, 1971). Is it possible that membrane structures (we recognize this term following Bangham, 1981) have arisen independently from RNA and/or simultaneously with RNA (or RNA-protein complexes)? There are several simple physical mechanisms through which it could occur: 1) disturbance of hydrophobic films on the water surface due to the wave action (e.g. Goldarce, 1958; Rutten, 1971); 2) spontaneous self-assembling from dispersion in aqueous phase; what is more, this process may progress in an autocatalytic manner (Luisi et al., 1999; Monnard & Deamer, 2002); 3) forming from marine aerosols; the aerosols could be formed, for example, when gas bubbles from entrails of Earth reach the water surface, covered with an hydrophobic film (Dobson et al., 2000; Yakovenko & Tverdislov, 2003).

The first membranes may have consisted more simple hydrophobic substances than modern membranes which contain predominantly phospholipids (Deamer et al., 2002); it is important because in the absence of specialized transport proteins permeability of the membranes is too low. The primordial membrane vesicles may consist of fatty acids (Chen & Walde, 2010; Deamer et al., 2002). It is interesting that such clay mineral as montmorillonite (a candidate for initial matrix for RNA synthesis – see above) is able to accelerate the spontaneous conversion of fatty acid micelles into vesicles (Ferris & Ertem, 1992; Hanczyc et al., 2003). It is not clear whether clay minerals were responsible for forming membrane structures or if they emerged in aqueous medium? All in all, it seems that this process was likely to have occurred in the boundary between different phases (Chaykovsky, 2003; Fox & Dose, 1977; Rutten, 1971). In case of mineral-dependent mechanism it could be solid-aqueous phase whereas wave- and aerosol-dependent mechanism may apply to aqueous-gaseous phases. Ancient shallow-water hydrothermal vents might be a source both of gases and minerals; hence this source was likely to have provided for the generation of membrane structures on the early Earth.

Also, there are several hypotheses concerning the rise of ionic asymmetry between external environment and internal space of primary membrane vesicles. It is known that modern cells are able to maintain significantly bigger concentrations of potassium and calcium in their interior compared to the concentrations of these ions in the environment. For sodium and magnesium we can see an inverse pattern – in contrast to a cell, concentration of these ions are often relatively high in environment (for example, intercellular liquid of higher animals where ionic content resembles sea water). This asymmetry is important for biochemical processes in living cells; hence, conceptions of origin of life should account for this (Natochin, 2010). Membranes have the potential to maintain concentration gradients of ions, creating in that way a source of energy that is necessary to support transport processes across the membrane boundary and drive them (Deamer et al., 2002). Intensive water evaporation may provide concentration of prebiotic substances in primordial "dilute soup" (Monnard & Deamer, 2002). It was revealed, that the cooling of surface of water by means of evaporation can lead to ionic thermo-diffusion and provide increasing of concentration of calcium and potassium ions near to the surface of oceanic water. It is widely regarded that ions such as Na, K, Ca and Mg play very important roles in the energy processes of a cell. However, divalent cations have well expressed tendency to bind to the anionic head groups of amphiphilic (i.e. which have hydrophilic and hydrophobic parts) molecules, strongly inhibiting their ability to form stable membranes that can prevent self-assembly (Monnard et al., 2002). Nevertheless the vesicles could be formed as a film-coated microspray which forms when gas bubbles reach water surface covered with amphiphilic film and burst. If such single-layered microspray fall on the same amphiphilic film it may be covered with second ampiphilic layer due to hydrophobic interactions between molecules. This may lead to formation of a double-layer membrane structures with relatively rich interior with calcium and potassium as have been shown in experiments (Yakovenko & Tverdislov, 2003).

There is also another kind of asymmetry which should be considered in the context of the origin of life. As was discovered by Louis Pasteur in 1848 there are two types of molecules:

L-molecules (rotate polarized light leftwards) and D-molecules (rotate polarized light rightwards). At present time it is well-documented evidence that all living organisms include only L-amino acids in their proteins and only D-sugars in their nucleic acids. This chiral asymmetry of life (on the contrary a mixture of L- and D-isomers in equal quantities exist in the abiotic environment and have no optical activity on polarized light). Several mechanisms for emergence of such kind of asymmetry have been proposed based on experimental data. At first it was experimentally revealed that intensive evaporation of solution of both L- and D-amino acids leads to increasing of L-forms near the water surface (Yakovenko & Tverdislov, 2003). Therefore authors assume that existing fractioning of amino acids could be established by virtue of natural factors. Complex prebiotic chemical systems could emerge in conditions where L-forms of amino acids prevail over D-forms. Subsequently, "fixation" of L-amino acids in primary proteins could have arisen. However this hypothesis doesn't explain how RNA could be predated to proteins ("RNA-World" hypothesis – see above and Gilbert, 1986).

Another hypothesis claims that amino acid homochirality could be created by homochirality of RNA which in turn was inherited from mineral matrix structure (Bailey, 1998). Such considerations seem to be plausible although symmetry breaking processes have also been observed along crystallization (Viedma, 2001). There are a number of hypotheses on the problem of chiral asymmetry of life but we cannot describe all of them here. At the present time this question remains unsolved.

3.3 "Metabolism first" or "Replication first"?

We agree that the question of the origin of life is the question of the origin of the matrix synthesis (Chaykovsky, 2003; Gilbert, 1986; Kenyon & Steinman, 1969; Yakovenko & Tverdislov, 2003), and this is one of the central problems in the origin of life. However the question arises: what appeared first – "metabolism" (substances interchange) or replication? "Metabolism first" hypotheses try to offer plausible mechanisms on how membrane structures emerged or how prebiotic catalytic cycles evolved, whereas "replication first" hypotheses are focused mainly on formation of molecular replicating mechanisms (Mulkidjanian & Galperin, 2009). Obviously such division of hypotheses is rather conditional and depends on definitions of metabolism. Note, that "metabolism-first" hypotheses also could be subdivided in two classes: those which propose an existence of compartmentalized structures as a necessary condition for "metabolism" and those which consider primary "metabolism" as catalytic chemical cycles in surroundings. However such considerations seem to ignore the fact that metabolism is the property of living organisms only. If we define metabolism as selective chemical substances interchange between prebiotic system and its environment (like in the case of Oparin's coacervates) then we have to assume that compartmentalized structure predates "metabolism".

For "replication first" hypothesis appearance of membrane structure to allow for self--replicating RNA (or hypothetical RNA-like molecules) seems necessary. For example "Zinc World" hypothesis, where zinc sulfide (ZnS) is described as catalyst, considers some "standard" sequence of events: RNA → proteins → membranes (Mulkidjanian & Galperin 2009). In contrast Egel (2009) supposes that hydrophobic peptides could form primordial protein-dominated membranes before replicating RNA appear. In this case the sequence of events is: membranes + proteins → RNA.

"Metabolism first" and "replication first" hypotheses often resemble the well-known "egg--or-chicken" paradox. However there is no reason to oppose them to each other. The possibility of relatively independent emergences of membrane structures and matrix (apparently RNA) molecules should be considered: emergence of membrane structures and appearance of matrix synthesis could be parallel.

3.4 The deficiencies of traditional approaches

Most of the theoretical approaches to the problem of appearance of life in our planet include some abiotic mechanisms for the origin of primary organisms. In order to explain these mechanisms different chemical and (or) physical processes are being described. There are, for example, suggestions about the origin of the cell membrane, hypothesis concerning coacervates or other secluded structures, assumptions concerning mechanisms of genetic memory, energy processes etc (Fox & Dose, 1977; Kenyon & Steinman, 1969; Oparin, 1966; Ponnamperuma, 1972; Rutten, 1971). All of them consider various hypothetical ways for abiotic creation of different systems of organisms. The first principal defect of all these theoretical constructions is that the biosphere is described as some mechanical sum of all organisms but not as a united functional system which provides suitable conditions for the existence of its own components. These hypotheses are interesting because they sometimes use very exotic mechanisms to explain the origin of life. But all of them have several unavoidable defects because they usually require assumptions that are quite unlikely: 1) very peculiar conditions in localities where the life could arise and, thus, very low probability of origin of initial organisms; 2) very peculiar but relatively stable conditions for development of primary life; 3) very long period (at least 1 Gya) when such conditions have to exist. Moreover, it is not clear; 4) what is the food which was consumed by primary organisms during the period when initial life evolved; and 5) how the appearance of several trophic levels in primary biosphere could be explained.

4. The biosphere approach and the origin of life

There is also another point of view concerning the origin of the life and its evolution on Earth. It is based on the assumption that all living organisms on the planet depend on one another i.e. any life outside biosphere is impossible (Capra, 1995; Lovelock, 1991; Ludwig von Bertalanffy, 1962; Maturana & Varela, 1980; Vernadsky, 1989). In this view, the biosphere (or – in western tradition – Gaia) is considered a separate system of the highest structural level of life-organization. This system controls and directs evolution of living organisms and earthly ecosystems. Such assertion is named the **pan-biospheric** paradigm (Starobogatov & Levchenko, 1993; Levchenko, 2011). The problems of both biological evolution as well as the origin of life can be described within the framework of this paradigm. The above assumption leads us to several conclusions. Firstly, biological evolution has to be described in the context of the ecosystems evolution. This is named the **ecocentric** conception of evolution; according to it, the relationships between evolutionary processes on different levels of biological organization are inter-coordinated. In other words, the evolution on the species level and above (genus, familia, etc) can't take place without evolutionary changes of corresponding ecosystems and the biosphere. Some morphological variations can be restricted by necessity of preserving geochemical functions (Vernadsky, 1989). Known exceptions of micro-evolutionary processes are connected usually with neutral (ecologically-neutral) drift (Eigen, 1971). Only the biosphere is a relatively independent "whole" and live system, although it is, of course, dependent on geological and cosmic processes, in particular, in the interplanetary medium (Levchenko, 1997, 2002, 2011).

Secondly, to explain the development of the pre-biosphere environment, the **embryosphere hypothesis** – (see below and Levchenko, 1993, 2011) is proposed. According to it the appearance of different sub-systems of primary organisms could happen independently from each other (see above) within a unified functional system – the embryosphere. This is a system with self-regulated processes of interchange of substances between its different parts under the influence of external energy flows. Embryosphere isn't another name for any hypothetic pre-biospheres, because it has some functional properties of living systems. There are various regularities in its appearance and existence in the past, and analogies between these regularities and processes of evolution and development could be described.

4.1 Hypothesis of embryosphere

Vladimir Levchenko (1993) identified several evolutionary principles that can be applied to organisms, ecosystems and the biosphere. In particular, there are: 1) principle of evolution of functions; it can be formulated as the intensification of processes providing important functions of the biosystem (e.g. organisms) during evolution; 2) principle of increasing of multi-functionality of separate sub-systems of organisms or ecosystems during evolution; 3) principle of superstructure (or over-basis): new functions do not replace previous ones but they subordinate old functions and are "superimposed" on them (see in detail Orbeli, 1979). These principles can be applied to the development of embryos as well. Comparing all these features of evolution and development, one can argue that the pre-biosphere is a system, which is self-preserved, and which is similar to primitive organism without generative organs. In other words, the pre-biosphere in this view is weakly differentiated system, which develops as embryo by means of successive differentiations. The pre-biosphere with above particularities was named by us **embryosphere** (Levchenko, 1993, 2002). Consecutive differentiation leads to complication of the embryosphere. Known in paleontology microfossils could be non-independent separate primary organisms but they are similar functionally to cell organelles. The embryosphere hypothesis is based also on the following assumptions: 1) the interchange of substances between the embryosphere parts occurs under influence of external energy flows; 2) embryosphere is a unified system with a closed circulation of substances; 3) the origin of different vital (prebiotic) processes could happen independently in different parts and regions of the embryosphere; 4) embryosphere is the system which is self-preserved and self-instructed (Eigen, 1971); in particular, this means, the system is able to switch between some branches of processes depending on external conditions in order to maintain important vital functions. Sustainability is one of the main requirement for existence of the primordial community, so the evolutionary process should not disturb this sustainability in any case (Zavarzin, 1993). In other words, the embryosphere is able to adapt its interactions with surroundings in order to survive.

This approach, which explains the origin of living organisms within embryosphere by means of successive coordinated differentiations, implies that the Earth had initially sufficient quantity of mobile substances (liquids, gases) for embryosphere functioning. The organic substances must have also been sufficient in order that primary life could evolve a long time, at least, hundreds of millions of years or more. The biosphere approach implies the existence of "biosphere memory". It is being reflected factually as irreversible results of previous evolutionary changes. At first, they create different evolutionary constraints at the organism level, for example, morphogenetic ones. Later, changes in biosphere surroundings modify factors which control directions of global evolutionary process (in other words, canalize it). The idea of auto- or self-regulation or even auto-canalization of the life evolution is important at this juncture (Chaykovsky, 2003; Levchenko, 1997; Levchenko & Starobogatov, 1994; Zherikhin, 1987) and it conforms to the general systems theory of Ludwig von Bertalanffy (1962) and approach of Lovelock (1991) as well as Capra (1995). One more else known fact about the biosphere evolution concerns the growth of the energy flow, which is used by the biosphere, during its evolution (Levchenko, 1999, 2002, 2011). We assume this regularity is fulfilled for the earliest biosphere and embryosphere too.

4.2 Bio-actualism principle

In our earlier works (Levchenko, 1999, 2002, 2011; Levchenko & Starobogatov, 1994; Starobogatov & Levchenko, 1993) we discussed some regularities which characterize different stages of biospheric evolution: 1) gradual expansion of the life to new, before lifeless conditions and areas; 2) complication of the structure of the biosphere, the appearance of new circulations and ecosystems; as regular consequence of that – the gradual increasing of biodiversity up to some optimal level for the current conditions; 3) gradual (although sometimes irregular) inclusion of more and more matter of the planet into the living processes of the biosphere; 4) gradual increasing of energy flow through the biosphere during its evolution; 5) relative steadiness of all circulations of the biosphere; the disturbance of balance between circulations leads to a biospheric crisis; 6) auto-regulation (auto-canalization) of the biosphere evolution: every evolutionary step leads to the creation of new specific conditions on the planet which constrain the availability of directions for the subsequent steps. Moreover, trends known in evolutionary physiology and used in the formulation of embryosphere hypothesis could also be taken into consideration (see above). At last, we have not to forget the Darwinian principles of mutability, competition and natural selection, which promote the raising of the most optimal variants during evolution (Levchenko, 2010, 2011; Levchenko & Kotolupov, 2010). For historical reasons these principles have been discussed for organisms but they can as valid for ecosystems and even for non-living components of nature.

We assume that all these regularities and principles can be applied not only for the last stages of the biosphere evolution but for all ones including most ancient stages and also for the case of the embryoshere. Using an analogy in geology, i.e. the **actualism principle**, proposed by Charles Lyell in 1830, we propose a **bio-actualism principle**. Of course, it is a subset of the more general actualism principle but we would like to emphasize here just the biological aspects of planetary changes. Sections 7 of this is devoted to the earliest known crises of the biosphere on the border between the Riphean and Vendian periods; the description of this exceptional event is given in a bio-actualism context.

4.3 The origin of primary organisms in the biosphere approach

When we talk about life we mean firstly living organisms (although ecosystems and biosphere are living ones too). The problem of the origin of primary organisms within a biospheric framework proceeds through the following phases (Levchenko, 2002): **1)** mixing, and the increasing complexity of the chemical components and structures (membranes including lipid membranes) of different parts of the embryosphere and **2)** self-organization of self-sustained reactions, which are preserving these structures (we suppose they are so called **cooperons** – Levchenko, 2011; Levchenko & Kotolupov, 2010). **3)** the successive structural and functional differentiation of the embryosphere and **4)** the appearance and using of the mechanisms of molecular memory (on the basis of RNA and DNA) to provide self-preserving functioning; perhaps as a consequence of natural selection of some functional modules among multitude of different self-sustained structures (Eigen, 1971; Kenyon & Steinman, 1969).

What could be the embryosphere system in this point of view? How could the life arise in the Solar system just in our planet? In order to come near to the answers to these questions, one can consider the conditions in some other planets of the Solar system.

5. Hydrocarbons in cosmos and in small planets of the Solar system

Astrophysical observations have revealed the presence of organic matter as well as carbonaceous compounds in the gas and solid state, refractory and icy, and ubiquitous in our and distant galaxies (Ehrenfreund & Cami, 2010). Complex **polyaromatic hydrocarbons (PAHs)** in the mid-infrared spectra were detected in two infrared galaxies (Yan et al., 2005). According to mathematical models, abiotic synthesis could occur simultaneously with the formation of protoplanet bodies in early stage of evolution of the Solar System from Nebula (Snytnikov, 2006).

Explanation of the origin of primary life can be made much easier if the presence of a large quantity of hydrocarbons at the Earth's surface is assumed. This assumption isn't too fantastic or improbable based on the results of different investigations of other bodies of the Solar system. For example the thermodynamical calculations demonstrate that abiotic synthesis of PAHs through the Fischer-Tropsch reaction and quenching of $CO + H_2$ was possible on the ancient Mars (Zolotov & Shock, 1999). PAHs were detected in the martian meteorite ALH 84001 in 1996 (McKay et al., 1996; Becker et al., 1999), although some scientists (Stephan et al., 2003) argue for their terrestrial origins.

Many satellites of the biggest bodies in the Solar systems (Jupiter and Saturn) have much hydrocarbons on their surfaces and atmospheres (Woolfson, 2000). In particular, Titan – the satellite of Saturn - has even oceans and seas of liquid hydrocarbon – methane. There are other, high-molecular hydrocarbons which are dissolved in these seas. The diameter of Titan is about 5150 Km; it is more than the diameter of Mercury (about 4880 Km) and less than in the case of Mars (about 6800 Km). The investigations of Cassini orbiter in 2005 have demonstrated, that Titan has dense atmosphere (pressure could amount to 1.6 Kg/cm^2), which consists of nitrogen mainly (more than 95%), methane, other gaseous hydrocarbons and ammonia. The temperature of the planet is quite low: it is approximately minus 160 degrees by Celsius. Therefore, the water on Titan (its quantity is also large) is only in "stone" solid form (Woolfson, 2000). The seas in the planet consist of liquid methane mainly, the color of the sky is yellow-orange. You can find beautiful reconstructions of Titan landscape made on the basis of the Cassini mission data (see one of such reconstructions by M. Zawistowski (Canada) on http://www.astrogalaxy.ru/271.html).

Is it possible that water and hydrocarbons are present simultaneously in considerable quantities on other, smaller planets, which are alike Earth? It is quite probable for at least the earliest stages of planetary evolution in the case of planets of Earth's group and satellites of big planets. This point of view isn't contrary to planetology (Kenyon & Steinman, 1969; Rauchfuss, 2008; Woolfson, 2000). Moreover, we know one contemporary planet of the Solar system with oceans and seas of liquid water on its surface. This is, of course, Earth, which has appropriate temperature of its surface for that. As water and hydrocarbons can be found on ancient planets together, then the evident question arises: could liquid water and liquid medley of hydrocarbons exist on the Earth surface somewhere jointly? If to remember that fossilized Precambrian and even Archean oils (the origin of which are not explained obviously by biological processes) are present in Earth then one can suppose the ancient Earth had on its surface somewhere medleys or emulsions of some liquid hydrocarbons (which dissolved into more heavy ones, probably PAHs) and water. The temperature of ancient Earth was apparently suitable in order for some of hydrocarbons (or ancient oils) to exist in liquid form. In these conditions hydrocarbons are insoluble in water and their thickness small, so the hydrocarbons films ("oil" films) on the surface of ancient seas seem quite possible. Furthermore, during evolution ancient hydrocarbons have decomposed and much carbon exists in other forms on the Earth's surface today – mainly as living substance.

If so, then water-"oil" emulsion could be fine surroundings for the origin and long evolution of the initial forms of life on our planet. Such emulsion could create liquid secluded structures (Fox & Dose, 1977; Kenyon & Steinman, 1969; Rutten, 1971) and, moreover, it could be suitable source of organic matter and chemical energy (i.e., food) for initial forms of life (Levchenko, 2010).

6. The joint origin of the biosphere and life on the Earth

All modern biological membranes consist mainly of phospholipids. However protocells may have been self-assembled from components which are different from normal ones in biochemistry of modern organisms (Ehrenfreund et al., 2006). Taking into consideration that organic substances such as hydrocarbons are widespread in space they could be convenient raw materials for membrane structures. Amphiphilic molecule has an electric charge in certain domain (so called hydrophilic domain) whereas other part of the molecule have no electric charge. Therefore such molecules can form a film with directional characteristics on the water surface.

We already have noted evidence that fatty acids could form membranes (section 3.2). One can consider fatty acids as probable derivatives of hydrocarbons. Then amphiphilic film (hydrocarbons with fatty acids) on the water surface can hypothetically play a role as a starting point for the origin of membrane structures (Rutten, 1971; Yakovenko & Tverdislov, 2003). Assemblies based on aromatic hydrocarbons may be the most abundant flexible and stable organic materials on the ancient Earth and some scientists discuss their possible integration into minimal life forms (Ehrenfreund et al., 2006). It is interesting to note that polyaromatic hydrocarbons, substantial quantities of which were detected in space, are not the most complicated compounds. According to data from recent observations there are also buckyballs (or fullerenes) C_{60} found in cosmic space, they comprise up to 0.1-0.6 % of carbon in the interstellar medium (Sellgren, et al., 2010). It seems that the Universe is plentiful of complex carbonaceous organic compounds, and chemical evolution of different complex substances is real cosmic phenomenon. So we can make a conclusion that hydrocarbon molecules could play a crucial role in the abiotic evolution of organic matter in space and on Earth, that in turn have provided origin of life on our planet.

One of the hypotheses concerning physical mechanisms of origin of pre-biological structures was proposed by L.V.Yakovenko and V.A.Tverdislov (2003). It describes the formation of marine aerosols with simple one-layer membrane; these aerosols are enriched by potassium and calcium. Moreover, it can explain the rise of the ionic asymmetry between external environment and internal space of pre-biological structure (section 3.2). Let's try to develop this hypothesis using our approaches presented in this chapter. We argue that liquid water and a film of liquid hydrocarbons on its surface create appropriate conditions for abiotic emergence of the asymmetric conditions that characterize the membranes. For this to happen the following conditions have to be met: 1) presence of large quantities of water (with dissolved inorganic and organic substances) and substantial quantities of hydrocarbons on the planet surface; 2) relatively stable surface temperatures, which allow for both water and hydrocarbons to exist in a liquid state (at that, hydrocarbons form a dispersed film on the water surface); 3) intensive water evaporation and, as consequence, appearance of considerable thermal gradient nearby the surface (this leads to higher concentration there of more massive ions Ca and K in contrast to Mg and Na ones; 4) the presence of sources of gas bubbles. They may appear as result of water choppiness or can be produced in shallow-water hydrothermal fissures. When gas bubbles come out on the water surface with the hydrocarbons film, they could originate specific microsprays which are secluded aerosol structures with simple membrane.

Such aerosols could be snatched by wind and raised in high parts of atmosphere where they could remain up to 2 years. They have higher internal concentration of potassium and calcium compared to sea water (see point 3). When such aerosols come back to natural water reservoirs without vigorous hydrocarbon film on the surface, aerosols could begin to be appropriate candidates to the role of pre-biological structures because they have secluded structure, both a membrane and high internal concentration of K and Ca ions. Taking into consideration that hydrocarbons are good solvents of many lipids and in particular phospholipids, it is not difficult to suggest that hydrocarbon membranes can concentrate lipids as well as other amphiphilic molecules from environment into their own composition. Of course, such chemical evolution is possible before and after aerosol stage(s). When concentration of amphiphilic molecules amount to some critical value, it can promote the appearance of aerosols with double-layer membranes (section 3.2 and Fox & Dose, 1977; Rutten, 1971). That is quite important in the context of the origin of life because all modern cells have phospholipid double-layer membranes.

In summary one can come to the following statements about the joint origins of the biosphere and life on the Earth:

1. The origin of the embryosphere was a necessary phase in the origin of the biosphere as well as life. In our opinion the conception of the biosphere has to be included to any definition of contemporary life. These statements are a conclusion of the pan-biospheric paradigm (see above) concerning of that any biological organism cannot exist outside the surroundings of the biosphere or embryosphere. This approach

conflicts with so called panspermic hypothesis of Arrenius (Ponnamperuma, 1972) because it supposes that the early lifeless Earth may be appropriate in order that "cosmic sperm" can settle down there. We don't disclaim entirely the panspermic hypothesis but suggest there are serious limitations for the environment where this hypothesis may be applicable.

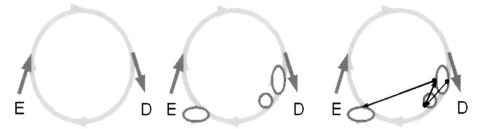

Fig. 1. Stages of evolution of embryosphere. On the left: the simplest natural circulation: the external energy (E) creates a circulation of substances and then is being dissipated as thermal energy (D). In the middle: the appearance of the micro-circulation events on the energy gradients of main circulation; this is possible when the micro-circulation processes accelerate some parts of main circulation. On the right: the appearance of immediate interactions (for example, by way of chemical catalysis) between different cycles of micro-circulations; the development of the controlling net of embryosphere.

2. The embryoshere hypothesis tries to explain how conditions, which are necessary for life, could be created without the contribution of life during early stages of the Earth's history. We argue that the origin of the embryosphere was connected with organization and evolution of the autocatalytic circulations of substances in early planet (in particular, in pristine "bouillon") under influence of external energy. The large-scale circulations can contain "micro-circulations"; moreover, some micro-circulations can use sometimes factors from other circulations. The competition for resources between different circulations results in the emergence of some of them that are able to be self-preserved. This tendency is conjugate with the organization of complicated nets of autocatalytic reactions. In cybernetic context these nets are the controlling mechanisms which provide the existence of the embryosphere – see Fig. 1.

3. The self-preserving embryosphere is connected with the development of the system of compensative, adaptive reactions in response to conditions of changes; different branches of the circulation processes are being switched on when surroundings vary. The mechanism can probably be the same as described for the physical evolution of the biosphere: when external "interruptions" of energy flow weaken some of existing processes, new compensative processes are originated (Levchenko, 2002). One can describe that in the terminology for mechanisms of auto-canalization and self--instruction (Eigen, 1971) for the chemical evolution of the planet (Fox & Dose, 1977; Kenyon & Steinman, 1969; Rutten, 1971). Self-instruction mechanism implies anticipatory behavior of the system (Rosen, 1991); in fact this is the possibility to change features of functioning using current internal instructions.

4. Every circulation, which is being self-preserved, supports its own dynamic structure (so called "cooperon" – see Levchenko, 2011; Levchenko & Kotolupov, 2010). New

circulations, which arise during evolution of the embryosphere, create new dynamic structures. These processes lead to gradual formation of such relatively stable environment on the planet, which is more suitable for the origin of pre-life and life. If the energy flow through embryosphere grows along time then preconditions of origin of different trophic levels of embryosphere appear.
5. The origin of primary organisms, within a framework of this biospheric conception, was a result of several continuous events and evolutionary processes described above in sections 3 and 4. Developed embryosphere sustains necessary conditions for that, the pre-biological processes promote the development of embryosphere.

Many of the self-supporting processes in biological cells of today (for example, the processes of energy transformation based on ADP <-> ATP transmutation cycles) could arise in some fragments of the embryosphere before the origin of independent organisms. In other words, some important processes in modern organisms may in general traits repeat ancient processes in embryosphere. The water-"oil" emulsion which very probably existed in large quantities on the surface of early Earth, was a suitable environment for the development of different physical and chemical processes in the embryosphere and later – in primitive biosphere. When hydrocarbons on the surface of the planet were largely exhausted, then life found the mechanism of photosynthesis (Levchenko, 2010, 2011).

7. The development of early biosphere. The biospheric crisis on the border between Riphean and Vendian

After the appearance of living organisms (apparently at the beginning of Archean (3.5 – 4 Gya), the further development of the biosphere (at the beginning, populated only by bacterial forms of life) was proceeding along directions which increased complexity of communities structure, appearance of new connections of trophic chains, and at the same time, changes of living surroundings as well as appearance of new habitats. After the long period of high volcanic activity, which ended approximately 1350 Mya (i.e. millions of years mega-years ago), the beginning of Proterozoic upper division – Riphean - of rather stable conditions were set on the planet. Variations in the changes in parameters narrowed in comparison to earlier times. It could mean the establishment or a strengthening of the role of either self-regulating mechanisms in biosphere or an ecosystem of higher complexity. These mechanisms "control" the processes occurring in a geospheres within some framework. The most dynamic and changing part of biosphere – living organisms – probably played an important role in such processes. In order to understand a role of biological communities in the history of Earth we have to remember that majority of global planetary processes are involved in different circulations. The changes of them lead to global modification of planetary conditions. The role of life is very important for biospheric circulations, and now there is considerable interlacing of biological and geological ones; hence, organisms are indeed a geological force (Vernadsky, 1989). As we have already noted above, biological (or, in early stages of the planet history, pre-biological) circulations could carry out a role of micro-circulation, which superimposed on natural abiotic circulations (Levchenko, 2010). This micro-circulations use available energy sources of surroundings.

Let's imagine the following model: surroundings, in which some cycles of exchange by substances and energy flows exist, favor the existence of a particular biological community.

When circulation in the system "biological community-environment" is being settled in some equilibrium state then biota is using all available energy resources, and it doesn't produce any superfluous production. Such system can exist factually eternally. But sometimes (due to changes in the external condition, availability of resources or due to modifications in biota) this balance can be disturbed that results in the community becomes unable to support the current level of circulation. The weakening of circulation provokes further disintegration and, finally can lead to a collapse of the system. Good examples of such sequences of events with positive feed-back can be found in a description of the cretaceous crisis (Zherikhin, 1978). Of course, any crisis events have important role in various biospheric and biocoenosis changes. We shall discuss briefly below the first known global biospheric crises, i.e the Vendian crisis.

7.1 Climate in Precambrian. Development of early life

The climate of the Earth changed all the time. Over long periods of time, the climate fluctuated not only periodically but also had irreversible changes which, took place in Precambrian (Semikhatov & Chumakov, 2004). The intensive volcanism, an abundance of carbonic acid gas and high temperatures limited the development of communities of organisms on the planet. But from the middle Riphean, about 1 Gya, periods of cooling started (some authors name them as glaciations). During those periods average temperatures dropped affecting environmental conditions. Semikhatov and Chumakov (2004) write about three periods in the Earth history. The first was a period without exposed glaciations (3.5 – 2.9 Gya) when there were no great cooling periods. After that the period from 2.9 to 1 Gya was an episodic known glacial period. And from 1 Gya till present time, the authors mark as a period of repeated glaciations.

Biological communities of the Precambrian period have been presented in paleontology basically by cyanobacteria which formed so-called bacterial floor-mats and maritime planktonic communities. First multicellular organisms also started to appear in Proterozoic (2.7 Gya) but were rare. Among other major events in Proterozoic it is necessary to note appearance of developed photosynthesizing communities (2 – 2.5 Gya) and eukaryotes. The first eukaryotes have most likely appeared 2.7 Gya, but they become to be widespread much later (Fig.2). Approximately 620 – 650 Mya (the end of Riphean) there was a sharp impoverishment of fossil microorganisms which a number of researchers attribute to the extinction of a majority of dominating forms. This important event was named Vendian phytoplanktonic crisis. After that deposits contain a lot of fossil remains of living communities again. These biological communities consisted already from eukaryotes and rather numerous multicellular organisms. Especially many multicellular animals appear in the unique and unusual fauna of Ediakaran (upper division of Vendian). In accordance with certain estimations (Fedonkin, 2006), structure of Ediakaran fauna comprised 67 % coelenterates, 25 % worm-like organisms and about 5 % arthropods. However several years after it was found out that the majority of the animals of Ediakaran fauna have no modern analogues. Most likely they were not predecessors of animals who appeared in Cambrian. Nevertheless they maintained suitable conditions for the further development of the Cambrian fauna. Also we should note the appearance of the first seaweeds with limy covers (Rozanov, 1986) and various other animals with limy covers and tubes (Fedonkin, 2006; Rozanov, 1986).

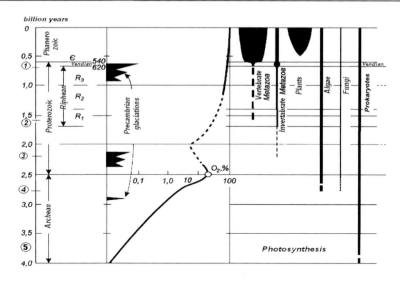

Fig. 2. The main events in early biosphere history (by Rozanov, 1986). In center: changes of the atmospheric oxygen concentration. On the right: times of appearance of some main taxa are given by full lines; dotted lines give possible presence of the group, but are unconfirmed or indirectly confirmed. In circles left: 1 – Vendian phytoplanktonic crisis; 2 – appearance of first hypothetic pre-vertebrate organisms (by hypothesis of Rozanov, 1986); 3 – developed photosynthesizing communities, appearance of first multicellular animals (invertebrate); 4 – in the sediments of this period first eukaryotes have been found; appearance of first plants and fungi; 5 – period of appearance of prokaryotes and probably mechanism of photosynthesis. The "flags" mean periodic glaciations, which took place in Precambrian.

Since Cambrian, *Sponges, Mollusks, Polychaetes, Bryozoans* and other elements of fauna start to be found; they have been usual for all Phanerozoic and continue today (Rozanov, 1986). Besides the remains of early Cambrian inhabitants, whose taxonomic status is definitely clear, there were also some groups of remains which can't be assigned to any known taxa. *Coelenterates* (jellyfishes, corals), though still rare, are known since early Cambrian. Regardless of scientific explanations of evolutionary events, there are some indirect evidence revealing changes in structure of biota on the Vendian and Cambrian border (Rozanov, 1986). There was an increase in variety of forms and abundance. This increase of ancient biodiversity has begun in Vendian and continued in early Cambrian (550 – 570 Mya). Ediakaran fauna disappeared as suddenly as it arose, making it difficult to connect it with the Cambrian "explosion". But for a relatively brief period between 650 – 620 Mya there was substantial impoverishment of plankton (Semikhatov & Chumakov, 2004), known as the Vendian phytoplanktonic crisis. This extinction of Proterozoic communities was the first ever documented and caused considerable changes in biosphere.

7.2 The first mass extinction, its possible reasons and consequences

In order to describe the reasons of Vendian Phytoplanktonic Crisis it is necessary to take into account earlier stages of biosphere development, before this first known mass extinction took place. Burzin (1987) viewed the crisis as irreversible changes of a system at the period from the beginning of Riphean (about 1500 Mya) to the Vendian (620 Mya). The author assumes that a hypothetical consumer has appeared in the food webs. Such consumers might eat away plankton organisms, and this might be the main internal reason of the crisis.

Let's see how it might be. Size classes of organisms demonstrate growth from early Riphean to Vendian (Veis, 1993). At the beginning of Riphean, organisms with the size of cells about 16 microns prevailed; at this time the large hypothetical consumer appears. During successive evolutionary changes the prey size reached 16-60 microns. More and more large--sized consumers were appearing and further, thus, forcing a prey to adapt and become more and more large-sized as well, till it has reached the size of 200-300 microns by the end of Riphean. It is the top size of prokaryotic cell. In principle instead of this strategy of avoiding of a predator (by increase of the cell size) there is also another one: formation of thorns and offshoots. But it is impossible for prokaryotes with their rigid cellular membrane that doesn't permit to exploit such way. Thus, cyanobacteria, being the basis of phytoplanktonic communities, cannot adapt to "pressure" of predators neither using larger sizes inaccessible for them, nor using either thorns. Therefore they have been factually by degrees almost eaten away that is proved by impoverishment of fossiled remains of a phytoplankton during this period (620-630 Mya). According to Burzin the elimination of cyanophyta by larger their consumers is the main reason of this crisis. However, except the internal reasons, crisis could be caused by external ones, in particular by decrease of orogenesis level (at which emission of carbonic acid grew). The intensity of this process probably synchronized with so called **galactic year** which has duration about 200 Mya (Marochnik, 1982; Woolfson, 2000). Known last decreases of orogenesis level was from Cretaceous till present, from Carboniferous till Permian as well as the end of Vendian (Ronov, 1976). It is logically to assume that the previous minimums and accordingly crises were around 800 Mya, and a little earlier 1 Gya, etc.

Further development of Vendian crisis, based on extinction of phytoplankton organisms was connected with decrease of food for predators of these organisms. That leads to their extinction too and the crisis reaches the bottom. Seemingly, in the end of Riphean it was a situation when reducers of dead biomass were insufficient. It reminds upper Carbonian (when a big amount of mortified biomass was accumulated), likely it was unfavorable for development of ancient cyanobacteria, and it might caused "second tide" of extinction.

Fossil samples demonstrate new increase of number of planktonic organisms about 620-570 Mya, but eukaryotes already dominate among them. As we have noted above the amassing of atmospheric oxygen could be connected with the insufficiency of reducers in biological communities. Mortified biomass of plankton is buried without aerobic oxidation. Of course, it led to disturbing of biospheric circulations as well as balance between biological and geological processes in biosphere. We can see relatively fast growth of the oxygen concentration along all Riphean (see Fig. 2) and establishment of almost constant level of it at the beginning of Cambrian. It could be related with so called great taxa radiation on the border of Vendian and Cambrian. The appearance of majority of new biological forms might had provided some stabilizing mechanisms. In this time many new groups of organisms has arisen; likely some of them are ancestors of many modern animal taxa, such as *Sponges, Mollusks, Polychaetes, Bryozoans* (Rozanov, 1986).

Some new biological forms, which appear as result of the crisis, give many different microtaxa. They started to carry out more differentiated functions in new ecosystems but they have narrower ecological niches than Precambrian forms. That could be interpreted as the increase of "rigidity" of biosphere system (Slavinsky, 2006). Rigid systems with highly specialized forms have a shortcoming: loss of even one element can sometimes destroy the whole system (Malinovsky, 1970; Rautian & Zherikhin, 1997). Moreover, specialized forms are not able to evolve because they are well-adapted to current environmental conditions. On the other hand, ecosystems with such forms have an advantage: they are able to use available resources more effectively under appropriate conditions. Last statements can be illustrated by mass appearance of different reducers and predators in Cambrian biosphere. Deep functional reorganization of ecosystems and formation of additional circulations of substances permit them to use vegetable biomass more effectively. Therefore quantity of buried organic matter decreases. All this gave the possibility to establish more complicated but durable structure of biosphere and to increase essentially the level of biological diversity. Such changes occurred at the beginning of Cambrian.

7.3 The role and some regularities of crises and mass extinctions

Throughout all the history of a planet the shape of ecosystems changed. Changes affect floral and faunal communities, both in land and sea. Being formed throughout millions of years, steady communities of living organisms used resources of environment. They closed the major biogeochemical circulations, thereby maintained a balance of ecosystems. But external conditions were varied for a long time and broke this fragile balance. Appearance of new species, variations of orogenesis intensity and climate changes, other external influences on the biosphere, in particular, modifications of relief of surface of Earth and drift of continents changed the environment and led to reorganizations in ecosystems. One of the "tools" of elimination of such resource inefficiency in ecosystems were mass extinctions. They led to change in dominating biota. There was a point in the history of the biosphere, when they led to disappearance of up to 96 % of all sea species and 70% of land species of terrestrial vertebrates. It occurred in the Permian and is considered the greatest mass extinction of all times (Burzin, 1987; Malinovsky, 1970).

It is necessary to distinguish background extinction which in contrast to biotic crisis occurs more or less permanently. Mass extinction represents a fast (in geological scale of time), catastrophic change in dominant taxa (Armand et al., 1999). Some of the scientists name mass extinctions "global ecological crises" (Leleshus, 1986), others – "biocoenotic crises" (Rasnitsyn, 1987) or "planetary ecosystem reorganizations" (Krasilov, 1986). In respect to biological crisis Najdin and coworkers (1986) offers the term "biocoenotic turn", underlining the fact that sharp changes occurred not only in biota, but also in the abiotic nature. All mass extinctions described in the literature developed under more or less similar scenarios. Phanerozoic crises, their possible reasons and consequences have been described in detail by paleontologists and ecologists. Among the various reasons we want to emphasize one that leads to the different parts of community becoming incompatible with one another. Owing to various critical changes in ecosystems and biological communities elimination of dominating, and then sub-dominating, species occurs. It in turn releases of ecological licenses ("vacant ecological niches" – Levchenko, 1997) and their occupation by more competitive and suitable species for new environmental conditions. Crises are an extremely important mechanism for the development of the biosphere, because they drive extinction of dominant species and allow rapid liberation of a great number of ecological niches. We assume that such events were favorable for species which weren't numerous in community earlier or were oppressed by dominants. These species may constantly be present in the ecosystems and could be referred as "cryptic biodiversity". One can reveal that soon (30 – 50 Mya) after each Phanerozoic crisis new macro taxa have emerged. This effect is observed in Cambrian also, just after Vendian extinction.

Many authors (Armand et al., 1999; Rautian & Zherikhin, 1997; Sepkoski, 1996; Zherikhin, 1984, 1997) try to find the role of mass extinctions. Symptoms of crises and laws of their transition were developed. The system of indicators (extinction level, intensity of extinction), allow to compare stages and amplitude of different mass extinctions to estimate their global value by comparison with background extinction level (Alexeev, 2009). Studying all mass extinctions allows one to assess the significance as a crucial mechanism for biosphere development, playing the key role in the regulation of evolutionary processes and opening up possibilities for new diversifications. Summarizing above considerations and using approaches from our own earlier works (Levchenko, 1997, 2011; Starobogatov & Levchenko, 1993) as well as those from other authors (Armand et al, 1999; Kalandadze & Rautian, 1993; Zherikhin, 1984, 1987) a succession of events which characterize known crises and mass extinctions can be proposed:

Beginning of crises: 1) decreasing of the orogenesis level and the excretion of carbonic acid gas from Earth entrails; 2) intensive withdrawing of carbonic acid from biosphere due to geological processes (formation of coal in upper Carbonian and carbonates in Cretaceous), resulting in the reduction of quantity carbon which is accessible for life, lowering of organic matter production, increasing of the oxygen concentration in atmosphere, and a more continental climate; 3) "evolutionary search" for more effective producers, appearance and dissemination of new forms of organisms with new adaptations (before beginning of crisis they are as a rule eliminated by selection).

Development of crises: 4) crash and destruction of some large "traditional" biocoenoses, extinction of many taxa, elimination of community dominant organisms, release of some quantity of carbon from biospheric circulation; 5) use of liberated territories as well as free carbon for development and broad dissemination of new biological forms, including forms of so called cryptic biodiversity, radiation of taxa; 6) rise of new biocoenoses with new producers, increasing of biodiversity of the biosphere (Levchenko, 2011).

As we know from the planetary history after Cambrian the periods of decreasing of orogenesis were taking turns with periods of its increasing (Ronov, 1976). That promoted the development of new taxa up to the next crisis. We suppose that earlier crises of biosphere could be also connected with periodical variations of orogenesis. Moreover, one can assume that appearance of new adaptations of living organisms could be caused by some crises events.

8. Biosphere as an autopoietic-like system

Applying the criterion of the autopoietic system, (Varela, 1979), some scientists concluded that Gaia is a real autopoietic system. It should be noted here, that Francisco Varela denied that Gaia is an autopoietic system. He proposed that it is an autonomous system, a system from a wider class without regeneration of elements and border. But, there are problems with this interpretation of biosphere as well with the definition of autopoiesis itself. First, let us pose two questions, concerning interpretation of biosphere as an autopoietic system. What can we say about the biosphere environment? How can we explain Gaia development (**epigenesis**) as a result of Gaia and cosmos co-evolution through structural coupling?

It is clear, that in many relations biosphere is a self-determining system, and its interior milieu is the main source of perturbations, responsible for its development and evolution. How can we coordinate this fact with the role of environment in autopoiesis? Strictly speaking, autopoietic systems are not evolving in traditional sense. Is it possible to combine evolutionary and autopoietic mechanisms? Classical, pure autopoietic model looks non-productive and non-constructive when applied to biosphere. The idea of the "multilevel autopoiesis" first suggested by E. Jantsch looks promising, but it should be perfected and corrected. Biosphere can be represented as Russian matryoshka (Kazansky, 2004) of nested structure of autopoietic-like systems of different organizational levels (be it hierarchy or "holarchy"). "Space" between levels is filled by "metasystems" or sympoietic systems (Dempster, 2000). In biosphere we see alternation, interchange of system levels with rigid autonomy, autopoietic organization just as living cell, organism or some modular organisms and system levels with loose, population-like or community-like organizations. The latter are organizationally half-opened, "ajar", sympoietic, as Beth Dempster refers to them. These levels play the role of buffer between rigid organizational levels and local environments. This flaky structure makes it possible to realize multilevel evolutionary process.

From the systems point of view the simplest form of life – unicellular organism is a form of relative dynamic autonomy. This autonomy is provided by circular, closed recursive organization of metabolic processes and by cellular half-permeable membrane. Circular closed production organization means, that it organizes permanent reproduction via metabolic process not only of its structure, elements, but also of itself. Well-known conceptual descriptive half-formal model of autopoiesis by H. Maturana and F. Varela was a prelude to formalization of biologically dynamic autonomy. From this position, chemical circular closed organization of autocatalytic hyper-cycles of M. Eigen (1971) could be a predecessor of biogenesis in the form of proto-cells. But, of course, problem of the origin of biological membrane is a special problem, related with the presence of special bio-matrices.

Conditions of pre-biotic Earth was favorable for the origin of life, which arose **polyphyletically** i.e. independently or concurrently (see above sections 3 and 4). Many parallel, different and relatively independent proto-populations occurred at different times and at different locations on Earth. Almost instantly very thin film of primitive unicellular bacterial communities covered the surface of our planet. These bacterial communities included a minimal standard set – primary producers, consumers and reducers. As a result, biosphere, as a specific global environment and a system appeared. Organisms need in biogenous elements and other materials and resources. Sustainable existence and development of this system is possible only if needs of all organisms are balanced and resources are recycled. So, biosphere became a sort of global ecosystem with relatively closed cycles of biogeneous elements. Evolution of biosphere is really a process of co-evolution of biota and abiotic elements of biosphere (Lovelock, 1988; Starobogatov & Levchenko, 1993). This process of biosphere evolution can be described as the **punctuated**

epigenesis (Kazansky, 2004), an conceptual attempt to refine well known model of punctuated equilibrium suggested by Gould (see below). In this conceptual model the periods of gradual development, occasionally interrupted by structural crises, inevitably lead to organizational crises, characterized by the perturbation of the autopoietic organization, which brings to the self-construction of a new, superior structural level of the system. Traditionally, biological evolution is understood in a Darwinian sense as the historical development in succession of replications and reproductions. That is why Lovelock spoke about "Gaia epigenesis" or individual development. Gaia is developing internally, without reproduction, through periodical organizational self-transformations ("transfigurations"). It looks as a new form of open-ended organizational non-darwinian evolution of the autopoietic system with darwinian-like processes in its parts. This evolution can be represented as a gradual process of self-harmonization of the system of global biogeochemical cycles. This process reaches its peak, then eventually fails and brings to destabilization of biosphere organization, its reconfiguration and construction of a new, more complex one. Crisis begins with rising of promising basic element of new type, which could emerge in a new symbiogenesis (Margulis, 1987; Margulis & Guerrero, 1996). Then we observe expansion of this new structure, generation of new organizational levels with the accumulation of actual biodiversity. But, all this ends with contradictions on micro-level, provoked by inevitable perturbation of balance on micro-and macro-level followed by global crisis and transformation of biosphere (Kazansky, 2004, 2010). Some general aspects of interaction between different levels of life organization are given in Levchenko (2011).

The Cambrian or more precisely, the Ordovican time-rock period (about 500 Mya) is an enigmatic turning point in biosphere development, separating long stage of biological stagnation and stage of explosive increase of biological diversity. First endeavors to look on evolution of biosphere as a self-organizing system were undertaken by physicists P. Back and K. Sneppen in pure mechanistic model of **self-organized criticality**, or model of "**sand pile**" (Back & Sneppen, 1993). According to this paradigm, systems consisting of many interacting constituents exhibit some characteristic behavior. According to Kauffman (1995), all biological systems exist in dynamic steady state on the border between a region of parametric state with chaotic behavior and a region with rigid, determined behavior. System, having stable focus "at the edge of chaos" demonstrates properties, characteristic for self-organizing systems. Besides, in this states the system is most labile and evolvable (high ability to evolve).

P. Back and K. Sneppen revealed, that in the state of self-organized criticality probability distribution of sizes of fluctuation in the system is described by power-law slopes with long tail. Such type of distribution makes all processes and structures fractal, scale-free. Most changes occur through catastrophic events rather than a graduate change. As a result, we have punctuations, large catastrophic events that affect the entire system. Small fluctuations can have global effect. The model of self-organized criticality fall into conception of "**punctuated equilibrium**" (Gould, 1994; Gould & Eldredge, 1977). According to the model, evolution of biota demonstrates alternation of durable quasy-ordered states and more short-term avalanche-like processes of destruction (crises) by power law statistics. Crises are interpreted as mass extinctions of species and whole taxa. Biodiversity recovers in the periods of stasis. But, this pure mechanistic model well fit only with the statistics of global ecological crises only after Cambrian period. Kauffman (1995) was mistaken, when he applied it to so-called Cambrian explosion (Kovalev & Kazansky, 1999, 2000). In this case we really have fundamental change of global organization of biosphere. Biosphere had cardinally different types of organization in the Archean, Proterosoic and Phanerozoic eons. In the beginning of the Phanerozoic eon there was a leap from simple, organizationally closed systems, formed by biological dominants - cyanobacterial mats to stochastic, open, complex systems, which gave rise to multi-level open-ended creative evolution. The continuous noticeable taxonomic diversification started only about 500 Mya. Two principally different, mutually exclusive strategies of biological systems were dominating at these two biosphere ages. Rigidly structured life-cycles with aggressive, expansionistic behaviour of soliton-like population waves were distinctive characteristic of biota before Phanerozoic eon. In contrast, in the beginning of Phanerozoic eon we can see the emerging and fantastic development of multi-level cyclical organization of life (Kovalev & Kazansky, 1999, 2000). That is why any universal mechanistic model, like punctuated equilibrium cannot be applicable to the whole history of biosphere. We should separate fundamental, organizational crises in the evolution of biosphere, which are associated with mass extinction of the whole taxonomic units, but which does not change the global structure of biosphere. Transition from Archean to the Proterozic related with increase of atmospheric oxygen or Phanerozoic explosion of diversity of multicellular organisms are examples of organizational crisis. Model of **punctuated epigenesis** is the endeavor to conceptualize the idea of biosphere evolution as interchanging of periods of development and periods of crises.

9. Conclusions

Biosphere is a planetary, global expression of life activity and from the other side it is the only favorable milieu for durable existence of biota. Any living organism has relatively autonomous organization of metabolic processes and at the same time, all living creatures are fundamentally dependent on each other via trophic, behavioral, sexual relationships. Any organism is the element of many systems – group (family, flock, school, population, etc.), community, ecosystem and biosphere as a global planetary system. All organisms are included in local and global biogeochemical cycles. Systemic status of biosphere and mechanisms of its development (or evolution) are still debatable questions. But, it is clear, that biosphere have durable periods of gradual development and relatively short periods of uneven change. In some of these periods, often associated with so called global crises, structure of biosphere can radically change, as in the case of gradual increase of oxygen in atmosphere at the transition from the Archean to the Proterozoic, or in the period of Phanerozic explosion of biological diversity. We believe that biospheric crises are processes regulating texture and functioning of the biosphere. The crisis which took place on the border of Riphean and Vendian (630 Mya) has resulted in considerable changes of the early biosphere. Great Cambrian taxa radiation and the rise of role of predators in communities were important specific consequences of the crisis. Nevertheless, this Vendian phytoplanktonic crisis developed by the same way as that is known for more later Phanerozoic crises. We tried to discover the main traits of biospheric crises and present them in this chapter.

The question "what is the origin of life: either origin of organisms or origin of biosphere?" isn't correct because only whole biosphere is independent unit of life among all known living forms (Lovelock, 1991). There are two important regularities of the evolution of life on the Earth. The first one demonstrates the autocanalization of physical evolution of the biosphere – increase of energy flow which is used by life. That defines many traits of biospheric development and leads to the hypothesis of embryosphere. Embryosphere developed since the time of primitive chemical processes on the surface of our planet since early Archean up to modern biosphere. The second regularity demonstrates non predetermination of phenotypical realization of biological evolution, demonstrated, in particular, by ecologically neutral changes of some biological forms in standing ecosystems (Levchenko, 2011). Hence, it is reasonable to assume non-predetermination of origin of strongly concrete pre-biological structures in embryosphere.

We also suppose that life on Earth has arisen under conditions when liquid hydrocarbons and water could meet each other and interact during long period of time. At that the hydrocarbons, which seemingly were present at early planet in big quantity, could be appropriate food for initial forms of life. We don't disclaim traditional (abiogenesis or biopoiesis) approaches to the problem of origin of life; they could be promising especially taking into consideration great achievements in the fields of biophysical and biochemical methods and experimental techniques. More recently, new experimental methods, for example such as artificial designing of a "minimal cell", mathematical and computer modeling have been elaborated. But we suggest that the phenomenon of life isn't connected only with life of separated organisms; we believe following Vernadsky that life without biosphere is impossible and doesn't exist. In other words to understand what is life it is necessary to explore the biosphere. It seems we are living at a time when all massive of amounts of biological data could be generalized within a framework of some unified theory. We hope that hypotheses and principles suggested herein by us in general will stimulate new insights on the problems of origin and evolution of life not only on our planet but also in the wider Solar system and perhaps in the Universe.

10. References

Alexeev, A.S. (2009). *Mass extinction in Phanerozoic* (doctoral thesis in Russian), Moscow. Available from http://www.evolbiol.ru/alekseev.htm

Armand, A.D.; Lyuri, D.I. & Zherikhin, V.V. (1999). *The anatomy of crises* (monograph in Russian), Nauka, Moscow

Back, P. & Sneppen, K. (1993). Punctuated equilibrium and criticality in a simple model of evolution. *Phys. Rev. Lett.*, Vol.71, No.24, pp. 4083–4086

Bailey, J. M. (1998) RNA-directed amino acid homochirality. *FASEB J.*, Vol.12, pp. 503–507

Bangham, A.D. (1981). *Liposomes: From Physical Studies to Therapeutic Applications*, G. Knight, (ed.), Elsevier, Amsterdam. pp. 1–17

Baross, J.A. & Hoffman, S.E. (1985). Submarine hydrothermal vents and associated gradient environments as sites for the origin and evolution of life. *Orig. Life.*, Vol.15, pp. 327–345

Becker, L.; Popp, B.; Rust, T. & Bada, J.L. (1999). The origin of organic matter in the Martian meteorite ALH84001. *Earth and Planetary Science Letters*, Vol.167, pp. 71-79

Burzin, M.B. (1987). Strategy of defence from predation, trophic and size structure of pelagic organisms and stages of Acritarchs morphological evolution in latter Precambrian and early Cambrian (in Russian), *Thesis from 3-rd USSR paleontological symposium of Precambrian and early Cambrian*, Petrozavodsk, pp. 12-14.

Cairns-Smith, A.G. (2005). Sketches for a Mineral Genetic Material. *Elements*, Vol.1, No.3, pp. 157-161

Capra, F. (1995). *The Web of Life*, Anchor Books, a Division of random House, Inc., New York

Cech, T.R. (1986). A model for the RNA-catalyzed replication of RNA. *Proc. Natl. Acad. Sci. USA*, Vol.83, No.12, pp. 4360-4363

Chaykovsky, Yu.V. (2003). *Evolution* (monograph in Russian), Cenological Researches. Vol.22, The Centre of Systems Research, IHST of Russian Acad. Sci., Moscow, 472 P

Chen, I.A. & Walde, P. (2010). From Self-Assembled Vesicles to Protocells. *Cold Spring Harb. Perspect. Biol.*, Vol.2. Available from http://sysbio.harvard.edu/csb/Chen_Lab/publications/publications.htm

Deamer, D.; Dworkin, J.P.; Sandford, S.A.; Bernstein, M.P. & Allamandola, L.J. (2002). The First Cell Membranes. *Astrobiology*, Vol.2, No.4, pp. 371-381

Dempster, B. (2000). Sympoietic and autopoietic systems: A new distinction for self-organizing systems, In: *Proceedings of the World Congress of the Systems Sciences and ISSS*. J. K. Allen & J. Will, (eds.), Toronto, Canada

Dobson, C.M.; Ellison, G.B.; Tuck, A.F. & Vaida, V. (2000). Atmospheric aerosols as prebiotic chemical reactors. *Proc. Natl. Acad. Sci. USA*, Vol.97, pp. 11864-11868

Egel, R. (2009). Peptide-dominated membranes preceding the genetic takeover by RNA: latest thinking on a classic controversy. *Bioessays*, Vol.10, pp. 1100-1109

Ehrenfreund, P. & Cami, J. (2010). Cosmic carbon chemistry: from the interstellar medium to the early Earth. *Cold Spring Harb. Perspect. Biol.*, Vol.2, No.12. Avalaible from http://cshperspectives.cshlp.org/content/2/12/a002097.full.pdf

Ehrenfreund, P.; Rasmussen, S.; Cleaves, J. & Chen, L. (2006). Experimentally tracing the key steps in the origin of life: The aromatic world. *Astrobiology*, Vol.6, No.3, pp. 490-520

Eigen, M. (1971). *Self-organization of Matter and the Evolution of Biological Macromolecules*. Springer-Verlag, Berlin, Heidelberg, New York.

Fedonkin, M.A. (2006). *Two annals of life: comparison experience (a paleobiology and genomics about early stages of evolution of biosphere)* (monograph in Russian), Syktyvkar Geoprint, Syktyvkar, Russia

Ferris, J. P. & Ertem, G. (1992). Oligomerization Reactions of Ribonucleotides on Montmorillonite: Reaction of the 5'-Phosphorimidazolide of Adenosine. *Science*, Vol.257, pp. 1387-1389

Fox, S.W. & Dose, R. (1977). *Molecular Evolution and the Origin of Life*, Dekker, NewYork.

Gilbert, W. (1986). The RNA world. *Nature*, Vol.319, p. 618

Goldacre, R.J. (1958). Surface films: their collapse on compression, the shapes and sizes of cells, and the origin of life. In: *Surface phenomena in biology and chemistry*, J.F. Danielli; K.G.A. Pankhurst & A.C. Riddiford, (eds.), Pergamon Press, New York, pp. 12-27

Gould, S.J. & Eldredge, N. (1977). Punctuated equilibria: the tempo and mode of evolution reconsidered. *Paleobiol.*, Vol.3, pp. 115-151

Gould, S.J. (1994). The evolution of Life on Earth. *Scientific American*, No.10, pp. 63-69

Hanczyc, M.M.; Fujikawa, S.M. & Szostak, J.W. (2003). Experimental Models of Primitive Cellular Compartments: Encapsulation, Growth, and Division. *Science*, Vol.302, No.5645, pp. 618–622

Jantsch, E. (1980). *The self-organizing universe*, Pergamon Press, New York

Joyce, G.F. & Orgel, L.E. (1993). Prospects for understanding the origin of the RNA world. *The RNA World*, R.F. Gesteland & J.F. Atkins, (eds.), New York, Cold Spring Harbor Laboratory Press, Cold Spring Harbor, pp. 1–25

Kaimakov, E.A. (1977). The Formation of Aminoacid Successions (in Russian). *Biofizika (Biophysics)*, Vol.22, No.6, pp. 1105 – 1106

Kalandadze, N.N. & Rautian, A.S. (1993) Simptomatology of ecological crises (in Russian). *Stratigraphy. Geological correlation*, Vol.1, No.5, pp. 3–8

Kauffman, S. (1995). *At home in the universe*, Oxford University Press, New York. 321 P

Kazansky, A.B. (2000). James Lovelock's Gaia Phenomena (in Russian, abstract in English). *Ecogaiasophic Almanac*, Issue 2, pp. 4–21

Kazansky, A.B. (2004). Planetary Bootstrap: A Prelude to Biosphere Phenomenology. *Proceedings of the 6-th International Conference on Computing Anticipatory Systems, AIP Conference Proceedings 718*, D.M. Dubois, (ed.), American Institute of Physics, Melville, New York, pp. 445–450

Kazansky, A.B. (2010). Bootstrapping of Life through Holonomy and Self-modification. *Computing Anticipatory Systems: Proceedings of the 9-th International Conference on Computing Anticipatory Systems, AIP Conference Proceedings*, D.M. Dubois, (ed.), American Institute of Physics, Melville, NewYork, Vol.1303. pp. 297–306

Kenyon, D.H. & Steinman, G. (1969). *Biochemical Predestination*, McGraw-Hill Book Company, New York, London.

Kovalev, O.V. & Kazansky, A.B. (1999). Two Faces of Gaia. *Gaia Circular*, Vol.2, issue 2, p. 9

Kovalev, O.V. & Kazansky, A.B. (2000). Biosphere evolution as a process of planetary organization types change (in Russian). *Proceedings of the 2-nd International Congress "Weak and Hyperweak Fields in Biology and Medicine"*, St. Petersburg, pp. 99 –102.

Krasilov, A.V. (1986). *Unsolved problems of evolution* (monograph in Russian), Vladivostok, Russia

Leleshus, V.L. (1986). About increasing of tabulates specialization growth during their evolution. Phanerozoic corals of the USSR. In: *Whole-union symposium on corals and reefes*, Nauka, Moscow, pp. 47–53

Levchenko, V.F. (1993). The Hypothesis of Embryosphere as a Consequence of Principles of Modeling (in Russian). *Proceedings of Russian Acad. Sci.*, No.2, pp. 317–320

Levchenko, V.F. (1997). Ecological Crises as Ordinary Evolutionary Events Canalised by the Biosphere, *International Journal of Computing Anticipatory Systems*, D.M. Dubois, (ed.), Vol.1, pp.105–117

Levchenko, V.F. (1999). Evolution of Life as Improvement of Management by Energy Flows. *International Journal of Computing Anticipatory Systems*, D.M. Dubois, (ed.), Vol. 5, pp.199–220.

Levchenko, V.F. (2002). Evolution and Origin of the Life: Some General Approaches. In: *Astrobiology in Russia (Proceedings of International Astrobiology Conference, Russian Astrobiology Center, NASA)*, St.Petersburg, pp. 7–21.

Levchenko, V.F. (2010). Origin of life and biosphere is indivisible process (in Russian). In: *Charles Darwin and Modern Biology*, Nestor-Istorya, St. Petersburg, pp. 338–347

Levchenko, V.F. (2011). *Three stages of evolution of life in Earth* (monograph in Russian). Lamberet Academic Publishing, Germany, ISBN 978-3-8433-1707-8

Levchenko, V.F. & Kotolupov, V.A. (2010). Levels of organization of living systems: cooperons. *Zhurn. Evol. Biokhim. Fiziol.*, Vol.46., No.6, pp. 530–538

Levchenko, V.F. & Starobogatov, Ya.I. (1994). Auto-regulated evolution of biosphere. *Dynamics of diversity of organic world in time and space. Proceedings of 40-th session of VPO*, St. Petersburg, pp. 30–32

Lovelock, J.E. (1988). *The Ages of Gaia. A Biography of Our Living Earth*, Oxford University Press, 252 P

Lovelock, J.E. (1991). Gaia: The practical science of planetary medicine. *Gaia book limited*.

Luisi, P. L.; Walde, P. & Oberholzer, T. (1999). Lipid vesicles as possible intermediates in the origin of life. *Current Opinion in Colloid & Interface Science*, Vol.4, pp. 33–39

Ludwig von Bertalanffy (1962). General System Theory – A Critical Review. *General Systems*, Vol.7, pp. 1–20.

Malinovsky, A.A. (1970). The general questions of the systems organization and their value for biology (in Russian). In: *Problems of methodology of system research*, Mysl, Moscow, pp. 146–183

Margulis, L. (1987). *Early Life:* The microbes have priority. In: *Gaia: A way of knowing: political implications of the new biology*, W.I. Thompson, (ed.), Hudson, NewYork, Lindisfarne, pp. 98–109

Margulis, L. & Guerrero, R. (1996). *We are all Symbionts*. In: *Gaia in Action*, P. Bunyard, (ed.), Cromwell Press, Wilts, pp. 167–184

Marochnik, L.S. (1982). Whether position of Solar system in the Galaxy exclusively? *Nature*, No.6, pp. 24–30

Maturana, H. & Varela, F. (1980). *Autopoiesis and Cognition*, Dordrecht, The Netherlands

Maturana, H. & Varela, F. (1987). *The Tree of Knowledge: A new look at the biological roots of human understanding*, New Science Library, Shambhala, Boston.

McKay, D.S.; Gibson, E.K.; Thomas-Keprta, K.L.; Vali, H.; Romanek, C.S.; Clemett, S.J.; Chillier, X.D.F.; Maechling, C.R. & Zare, R.N. (1996) Search for past life on Mars: possible relic biogenic activity in Martian meteorite ALH84001. *Science*, Vol.273, No.5277, pp. 924–930

Mitra, A.N. & Mitra-Delmotte, G. (2010). Magnetism, FeS colloids, and origins of life, In: *The Legacy of Alladi Ramakrishnan in the Mathematical Sciences,* Springer, Dordrecht, The Netherlands, pp. 529–564

Mulkidjanian, A.Y. & Galperin, M.Y. (2009). On the origin of life in the Zinc world: 1. Photosynthesizing, porous edifices built of hydrothermally precipitated zinc sulfide as cradles of life on Earth. *Biology Direct*, Vol.4, No.26. Available from http://www.biology-direct.com/content/4/1/26

Monnard, P.-A.; Apel, C.L.; Kanavarioti, A. & Deamer, D.W. (2002). Influence of ionic solutes on self-assembly and polymerization processes related to early forms of life: implications for a prebiotic aqueous medium. *Astrobiology*, Vol.2, pp. 139–152

Monnard, P-A. & Deamer, D.W. (2002). Membrane Self-Assembly Processes: Steps Toward the First Cellular Life. *The Anatomical Record*, Vol.268, pp. 196–207. Available from http://complex.upf.es/~andreea/2006/Bib/MonnardDeamer.MembraneSelfAssemblyProcesses.pdf

Najdin, D.P.; Pochialainen, V.P.; Kats, Y.I. & Krasilov, V.A. (1986). *Cretaceous. Paleogeography and paleooceanology*, Nauka, Moscow, 262 P

Natochin, Yu.V. (2010). Evolutionary physiology towards the way from "The Origin of Species" to "Origin of Life" (in Russian), In: *Charles Darwin and Modern Biology*, Nestor-Istorya, St. Petersburg, pp. 321–337

Niemann, H. B.; Atreya, S. K.; Bauer, S. J.; Carignan, G. R.; Demick, J. E.; Frost, R. L.; Gautier, D.; Haberman, J. A.; Harpold, D. N.; Hunten, D. M.; Israel, G.; Lunine, J. I.; Kasprzak, W. T.; Owen, T. C.; Paulkovich, M.; Raulin, F.; Raaen, E. & Way, S. H. (2005). The abundances of constituents of Titan's atmosphere from the GCMS instrument on the Huygens probe. *Nature*, Vol.438, Issue 7069, pp. 779–784

Oparin, A.I. (1966). *Origin and Preamble Development of Life* (monograph in Russian), Medicine, Moscow.

Orbeli, L.A. (1979). The Main Problems and Methods of Evolutionary Physiology (in Russian), In: *Evolutionary Physiology*, E.M. Kreps, (ed.), Nauka, Leningrad, USSR, Vol.1, pp. 12–23

Platts, S.N. (2005). The 'PAH World': Discotic polynuclear aromatic compounds as mesophase scaffolding at the origin of life. *NAI 2005 Biennial Meeting Abstracts* Available from http://nai.nasa.gov/nai2005/

Ponnamperuma, C. (1972). *The origin of Life*, E.P.Dutton, New York.

Rasnitsyn, A.P. (1987). Rates of evolution and the evolution theory (hypothesis of the adaptive compromise), In: *Evolution and biocenotic crises*, Nauka, Moscow, pp. 46–64

Rauchfuss, H. (2008). *Chemical Evolution and the Origin of Life*, Springer.

Rautian, A.S. & Zherikhin, V.V. (1997). Models of phylogenesis and lessons of ecological crises of the geological past, *Zhurn. Obsh. Biol.*, Vol.58, No.4, pp. 20–47

Ronov, A.B. (1976). Volcanism, carbonate accumulation, life (in Russian). *Geochemistry*, Vol.8, pp. 1252–1277

Rozanov, A.Y. (1986). *What has happened 600 million years ago* (monograph in Russian), Nauka, Moscow

Rosen, R. (1991). *Life Itself. Comprehensive Inquiry into the Nature, Origin, Fabrication of Life*, Columbia Univ. Press, New York

Rutten, M.G. (1971). *The Origin of Life by Natural Causes*, Elsevier publishing company, Amsterdam, London, New York

Sellgren, K.; Werner, M.W.; Ingalls, J.G.; Smith, J.D.T.; Carleton, T.M. & Joblin, C. (2010). C_{60} in reflection nebulae. *ApJ.*, Vol.722, pp. L54–L57

Semikhatov, M.A. & Chumakov, N.M. (2004). Climate in the Epochs of Major Biospheric Transformations (monograph in Russian), In: *Transactions of the Geological Institute of the Russian Academy of Sciences*, Issue 550, Nauka, Moscow

Sepkoski, J.J. (1996). Patterns of Phanerozoic extinction: a perspective from global data bases, In: *Global Events and Event Stratigraphy*, O.H. Walliser, (ed.), Springer, Berlin, pp. 35–51

Slavinsky, D.A. (2006). *Laws of crisis stages of ecosystems development on an example of dynamics of structurally functional changes* (PhD thesis in Russin), Moscow

Snytnikov, V.N. (2006). Astrocatalysis as initial phase of geobiological processes. Does life create planets? (in Russian), In: *Evolution of Biosphere and Biodiversity. To A. Yu. Rozanov 70-th anniversary*, KMK, Moscow, pp. 49–59

Starobogatov, Ya.I. & Levchenko, V.F. (1993). An Ecocentric Concept of Macroevolution (in Russian). *Zhurn. Obsh. Biol.*, Vol.54, pp. 389–407

Stephan, T.; Jessberger, E.K.; Heiss, C.H. & Rost, D. (2003). TOF-SIMS analysis of polycyclic aromatic hydrocarbons in Allan Hills 84001. *Meteoritics and Planetary Science*, Vol.38, pp. 109–116

Varela, F. J. (1979). *Principles of Biological Autonomy*, North Holland, New York, 306 P

Veis, A.F. (1993) *Organic microfossils of Precambrian - the major component of ancient biota.* In: *Problems of preantropogenic biosphere evolution* (in Russian), Nauka, Moscow, pp. 265–282

Vernadsky, V.I. (1989). *Biosphere and Noosphere* (monograph in Russian), Nauka, Moscow

Viedma, C. (2001). Enantiomeric crystallization from D, L-aspartic and D, L-glutamic acids: implications for biomolecular chirality in the origin of life. *Orig. Life Evol. Biosph.*, Vol.31, pp. 501–509

von Foerster, H. (1974). *Notes for an epistemology of living things,* In: *L'Unite de l'Homme,* E. Morin & M. Piatelli, (eds.), Seuil, Paris

Woolfson, M. (2000). The origin and evolution of the solar system. *Astronomy & Geophysics*, Vol.41, issue 1, pp. 1.12–1.19

Yakovenko, L.V. & Tverdislov, V.A. (2003). Ocean Surface and Factors of Prebiotic Evolution (in Russian). *Biofizika (Biophysics)*, Vol.48, No.6, pp. 1137–1146

Yan, L.; Chary, R.; Armus, L.; Teplitz, H.; Helou, G.; Frayer, D.; Fadda, D.; Surace, J. & Choi, P. (2005). Spitzer Detection of Polycyclic Aromatic Hydrocarbon and Silicate Dust Features in the Mid-Infrared Spectra of $z \sim 2$ Ultraluminous Infrared Galaxies. *ApJ.*, Vol.628, Issue 2, pp. 604–610.

Zavarzin, G.A. (1993). A development of microbial communities in the Earth history (in Russian), In: *Problems of the pre-antropogenic evolution of the biosphere,* Nauka, Moscow, pp. 212–222.

Zherikhin, V.V. (1978). *Development and change of Cretaceous and Cenozoic faunistic complexes: Tracheata, Chelicerata* (in Russian), Nauka, Moscow, 198 P

Zherikhin, V.V. (1984). Ecological crisis - precedent in Mezozoic (in Russian). *Energiya*, No.1, pp. 54–61

Zherikhin, V.V. (1987). Biocenological regulation and evolution (in Russian). *Paleontological Journal*, Vol.1, pp. 3–12

Zolotov, M. & Shock, E. (1999). Abiotic synthesis of polycyclic hydrocarbons on Mars. *J. Geophys. Res.*, Vol.104, No.6, pp. 14,033–14,049

2

The Search for the Origin of Life: Some Remarks About the Emergence of Biological Structures

Mirko Di Bernardo
University of Rome "Tor Vergata"
Italy

1. Introduction

In the chapter "Religious Opinions" of his *Autobiography*, Darwin sustains that, after the law of natural selection has been discovered, the old argument about a "design" of nature as written by Paley falls apart, the same argument which he believed to be decisive in the past. So, in accordance with Darwin, we can no longer argue, for example, that the perfect hinge of a bivalve shell must have been conceived by an intelligent being, like the hinge of a door by man:

A design regulating the variability of living beings and the action of natural selection is not more evident than a design preparing the course of the wind. All that exists in nature is the result of fixed laws. (Darwin, 1958, p. 69).

Darwin's theses, as it is already known, have gradually influenced the entire progress of science. However it is necessary to underline the fact that, at the moment, nobody knows with certainty how life has begun. The question concerning the origins pervades, since the dawn of history, the most elevated exercise of human reason and it represents, for some aspects, a question that appears to interest the entire horizon of reason and, with it, the same dimension that is proper to the category of possibility. In this sense, from the dawn of civilization to the present, every human being and especially every scientist has measured himself within the horizon of that question, laying special emphasis, within his/her own intellectual limits, on offering a satisfying answer to that question itself. The apparent inexhaustibly connected to the question about the origin of life appears to exalt the Dantesque-like contradiction between the impetus towards the necessity for an objective knowledge of the nature of life and the limitation, instead, of the measures established by mankind in order to give a scientific explanation of said question. As a matter of fact, we might never be able to re-build the real historical sequence of the events which allowed the first molecular systems to evolve and reproduce more than 3 billion years ago. Still, if on the one hand the historical course is bound to stay, maybe forever, a mystery, on the other hand it is possible to develop structured theories and experiments to demonstrate, in an almost realistic way, how life may have first crystallized and later spread to the entire globe (Di Bernardo, 2007a, 2007b). In this sense, it is not entirely surprising then that the progress in the understanding of possible paths towards the origins of life extends across several scales from the molecular to the cosmos. The individualization of such paths has resulted in alternative conceptions which have been examined by the researchers. Theses conceptions have led, also through the graft of molecular biology on Darwin's ancient evolutionary ideas to a standard theory; a theory which could be revised and enlarged, as it seems to happen nowadays, to merge with the original complexity theory elaborated first by Kauffman (1993, 1995). Complexity theory in recent years, has come to be appreciated by the scientific community as the most plausible outline capable of giving an answer to the problem of how life emerged from simpler molecular components. The great mystery of biology, according to the American biochemist, lies exactly in the fact that life has emerged and the order we observe has appeared: a theory of emergence should give account of the creation of the astounding order we see from our window as the natural expression of some underlying law. This would tell us if we are "at home in the universe", if here we are "expected" rather than being present against all probability (Kauffman, 1995).

2. Tracing the origins

In the volume *At Home in the Universe: the Search of the Laws of Self-Organization and Complexity*, Kauffman (1995) examines seven hypotheses concerning the attractive question about the origin of life.

The first one dates back to the first half of the twentieth century when scholars focused their attention mostly on the nature of the earth's primitive atmosphere which had given origin to the chemical molecules of life. Much evidence was accumulated to show how the first atmosphere was rich in molecular species such as hydrogen, methane and carbon dioxide, whereas oxygen was almost completely absent (today there are some doubts concerning such evidence). Very soon, the hypothesis that simple organic molecules, which are present in the atmosphere slowly dissolved in recently formed oceans to create a prebiotic broth from which life would have spontaneously emerged was postulated. Nevertheless, Kauffman highlights the main limit of this theory; i.e. the fact that the broth would have been extremely diluted. The quantity of chemicals reactions depends on how quickly the molecular species which reacts are able to meet- and this depends on how high their concentration is. If the concentration of each one is low, the possibility that they collide is really small (Kauffman, 1995).

The second hypothesis is attributed to A. Oparin, a Russian biochemist who proposed a plausible way of dealing with the problem of the diluted broth. When Glycerine is mixed with other molecules it forms some gelatinous structures called coacervates, structures similar to primitive cells. The coacervate, in fact, is able to concentrate in itself organic molecules and exchange them through its external surface. If these microscopic compounds had developed themselves in the primordial broth, they might have concentrated the chemical substances suitable for the formation of metabolisms. Yet, if Oparin laid the foundations to understand how protocells may have formed, he left the question concerning the origin of their contents, i.e. the organic molecules responsible for metabolic activity, completely unsolved. Together with the simple molecules there are several polymers, long molecular chains made almost by the same basic elements. The proteins out of which muscles are made, the enzymes and the structure of the cells, are formed by chains of twenty types of amino acids. The DNA and RNA are composed by chains of four basic elements (the nucleotides): adenine, cytosine, guanine and thymine in the DNA, with uracil instead of thymine in the RNA. Without these molecules, the real material of life, including Oparin's coacervates would have been nothing other than empty shells (Kauffman, 1995).

But where could these basic elements have come from? The third hypothesis attempts to answer this question. In 1952 S. Miller, a young pupil of H. Urely, conducted the first experiment of pre-biotic chemistry: he filled a container with some gas (methane, carbon dioxide and others) which were believed to be in the atmosphere of primitive Earth. He then subjected the container to a "rain" of sparks to simulate lighting as a source of energy and he awaited in the hope of obtaining some proofs of molecular creativity. After a few days he observed that clots of brown matter stuck to the sides and to the bottom of the container. After analyzing them he found out that the material contained a large variety of amino acids. Similar experiments have demonstrated that it is possible, although with extreme difficulty, to form nucleotides of both DNA and RNA, fat molecules and, through them, it is the structural material constituting the cellular membranes. Many other little molecular components of the organisms have been synthesized in the abiogenic way (Kauffman, 1995). However, Robert Shapiro (1986) in the *Origins, a Skeptic's Guide to the Creation of Life on Earth*, underlines the essential problems of this hypothesis. He, in fact, reaffirms that even if scientists would be able to demonstrate how it is possible to synthetize the various ingredients of life, it is not easy at all to fit all this into a single story. One can establish, for instance, that the molecule A can be formed by the molecules B and C with a very low performance in certain conditions; therefore, having demonstrated the feasibility of A, another group begins with an high concentration of molecules and demonstrates that by adding D we can obtain E (again with a small performance and in different conditions); then, still another group demonstrates that E, in high concentrations, can form F in totally different conditions. That being so, Kauffman, quoting Shapiro, writes:

But how, without supervision, did all the building blocks come together at high enough concentrations in one place and at one time to get a metabolism going? Too many scene changes in this theater [...] with no stage manager. (Kauffman, 1995, p. 36).

The fourth hypothesis takes as its theoretical reference-point the discovery of the genetic molecular structure by Watson and Crick in 1953. The discovery of the DNA double helix is, in fact, the final event which has prompted a renewed interest, within the scientific community, in the origin of life. It is hard not to be amazed by how the DNA double-helical structure immediately suggests the way molecules replicate themselves. Each filament specifies the sequence of nucleotides in the complementary filament thanks to the precise pairing of A-T and C-G, respectively. If DNA is a double helix in which each filament is complementary to the other, it means that the DNA double helix might be a molecule able to replicate itself in a spontaneous way. DNA, in brief, becomes the candidate for the first living molecule. It is considered to be the most important molecule in current life, the bearer of the genetic program responsible for the creation of the organism out of a fertilized egg. This magic molecule might have been the first to self-reproduce at the dawn of life. The molecule would have multiplied, finally running into the recipe for making proteins and creating its structure, and for speeding up its reactions by catalyzing them.

However, M. Meselson and F. Stahl (1958) highlighted the limits of the theory by demonstrating that the chromosomal DNA, inside the cell, even if it replicates as suggested by its structure, needs the help of an ensemble of protein enzymes. In other words, the two biochemists found out that DNA alone does not self-replicate. This being the case then, the scholars who were looking for the first living molecule had to look towards RNA or ribonucleic acid which has a leading role in the functioning of the cell. RNA can exist as a single filament or as a double helix and, as it happens in the DNA, the two filaments of the RNA double helix are complementary. In the cell the information necessary to make a protein is copied by DNA in a filament called messenger RNA which is later transferred towards the ribosomes, structures where, with the help of transfer RNA, proteins are produced. Very soon, in the eyes of the scientific community, given the complementarity of the two RNA filaments, it seemed possible that such a molecule could be able to self-replicate without the help of protein enzymes.

The fifth hypothesis thus is called "RNA world" hypothesis and it suggests that life began with a proliferation of RNA molecules (defined by the scholars as "naked genes"). The basic idea of this hypothesis is quite simple: a high concentration of a single filament specified sequence is put into a cylinder (it can be a decanucleotide CCCCCCCCCC) with the subsequent addition of some highly concentrated free nucleotides G; now each G should pair up with one of the nucleotides C of the decanucleotide, - as it is stated by the principle of molecular complementarity postulated by Watson and Crick -, so as to form a group of aligned G adjacent with each other. It is sufficient that the ten nucleotides of G combine with one another through appropriate bonds. Thus we would have the formation of a poly-G decamer. At this point the two filaments (poly-G and poly-C) only need to separate and let the poly C decamer free to align other ten monomers G thus creating another poly G decamer. In the end, the new poly G decamer, in order to obtain a system of molecules which is able to self-replicate, should be able to align any free monomer C which has been added in the cylinder to form a poly C decamer. If what has just been said would actually occur without the adding of any enzyme, we would be in the presence of a naked double-stranded RNA molecule able to self-replicate.

The idea is beautiful and attractive. Anyhow, almost without exceptions the experiment failed. The reasons why it does not work are extremely instructive. Firstly, each of the four nucleotides has a chemical personality of its own which tends to make the experiment fail. Thus a single filament poly-G has the tendency to roll up on its own in order to make two nucleotides G link together. The result is a tangled hank unable of acting as a stamp to self-replicate. Even if the process of copying would not be interrupted by the tangling of the guanine, the naked RNA molecules could be the victims of what we call an error catastrophe: during the copying of a filament into another one, the exchange of some bases- a G in place of a C- would alter the genetic message. In the cells these errors are kept at minimums levels by controlling enzymes assuring the fidelity of the copying process. But without the enzymes which prevent the guanine from tangling, one can only think of the number of copying errors and other mistakes; these errors can result in a RNA message soon lose all meaning. And in a world of RNA in its pure form, where would the enzymes have come from? (Kauffman, 1995).

3. The standard model

The fourth and fifth hypotheses share the idea of mold replication. It is for this reason that, according to many scholars, they meet in what is defined as the standard theory of the origin of life: a hypothesis stating that life should be based on mold replication; i.e. a replication similar to the one we find in the DNA double helix or in the naked double-stranded RNA molecules.

Under some aspects then, in the aperiodic[1] sequence of bases of today it is possible to recognize the Schrodinger's aperiodic solid with its micro code. As imagined by Schrödinger, in fact, the sequence of the bases is arbitrary, therefore it is able to "tell" different things, to encode information (Schrödinger, 1944). The arbitrary sequence can be interpreted as the effective carrier of the genetic code: the triplets of bases are the codons and sixty-one, out of the sixty-four possible triplets of codons, encode the twenty standard amino acids, while the three remaining codons specify stop-signals to the translation. Moreover, the symmetry of the DNA molecule allows the arbitrariness of the bases to be coherent with the mechanism of mold replication. According to this model, the origin of life must have been based on some sort of double chained aperiodic solid. Well, as we have just mentioned, this idea has been proven wrong, nobody has been able to create the experimental conditions which allows a single DNA or RNA chain to align free nucleotides, one at the time, as complementary elements of a single chain; to catalyze the soldering of free nucleotides to form a second chain; to separate the two chains and enter a new replication cycle.

The limits of this model have already been emphasized since the second half of the 1960's by many scholars, among whom Monod undoubtedly stands out. The French biologist, in *Le hasard et la necessite: essai sur la philosophie naturelle de la biologie moderne*, wrote as follows:

One could think that the discovery of the universal mechanisms on which the essential properties of the living beings are based has allowed the resolution of the problem of the origins. As a matter of fact such discoveries, by presenting this question in a new light, a question which today is put in more precise terms, have made the said question more complex than what it seemed before. (Monod, 1970, p. 128).

In the process that must have led to the appearance of the first organisms, the French scientist defined a priori three phases: a) the formation on the earth of the essential chemical constituents of all living beings, precisely nucleotides and amino acids; b) the formation, starting from such substances, of the first macro molecules able to replicate; c) the evolution which has built, around these "replicative structures", a teleonomic apparatus able to lead to the primitive cell. According to the scholar, the first phase is accessible both from a theoretical and experimental point of view. Although we will never know the paths the pre-biotic evolution has really followed, the overall picture is quite clear: the conditions of the atmosphere and those of the earth's crust favoured the accumulation of certain simple compounds of carbon as for instance methane (there were also water and ammonia). Now, according to Monod, from such simple compounds, and in the presence of non-biological catalysts, we obtain quite easily several more complex substances among which we have amino acids and some precursors of nucleotides (azotic bases and sugars). It seems to be demonstrated that, on the earth, at a certain moment, some expanses of water were in the conditions of containing high concentrations of the essential constituents of the two classes of biological macro molecules, nucleic acids and proteins. In this primordial broth different

---

[1] In an attempt to explain the stability of the gene, Schrödinger begins by making some considerations on the quantum states of atoms and the mutagenic effects of ionized radiations. After an estimate of the gene's dimension in terms of number of atoms, the scientist shows that the only sufficiently stable way in which the atoms can be kept together is by forming a molecule (hypothesis already approached by Delbrück). He put forward the famous hypothesis according to which the gene is constituted by an "aperiodic crystal", i.e. a large-sized molecule with a non-repetitive structure, capable of (providing) a sufficient structural stability and a sufficient capability to contain informations (cfr. , Schrödinger, 1944, pp 100, 106-107).

macro molecules were able to form themselves through polymerization of their precursors; i.e. nucleotides and amino acids. In the laboratory we have obtained in fact, in plausible conditions, some polypeptides and polynucleotides with a general structure similar to that of modern macro molecules (Monod, 1970). The second point is about the formation of macro molecules which are able, in the same conditions of the primordial broth, to promote their own replication without the cooperation of a teleonomic apparatus. Such difficulty, according to the French biologist, has been surpassed by demonstrating that a poly nucleotide sequence can actually lead, through a spontaneous pairing, the formation of elements with complementary sequence. Obviously a similar mechanism was slightly efficient and subjected to countless errors, however, from the moment it went into action, according to Monod, the three fundamental processes of the evolution- replication, mutation, selection- started to operate giving a considerable advantage to the macro-molecules which, thanks to their sequential structure, were more suitable to spontaneously replicate .The third phase consists in the gradual appearance of those teleonomic systems which had to build an organism (a primitive cell) from and around the replicative structure:

Here we really once again meet a wall of resistance, since we do not have the smallest idea of what the structure of a primitive cell was. The simplest living system we are aware of, the bacterial cell, an extremely complex and effective small device, had already got to its actual state of perfection more than a thousand million years ago. The development of the metabolic system which, while the primordial broth was gradually being impoverished, had to learn how to mobilize the chemical potential and synthetize the cellular constituents, poses huge problems. The same difficulties apply for the appearance of the membrane equipped with selective permeability and without which the existence of vital cells would not be possible. Yet the most serious problem concerns the origin of the genetic code and its mechanism of translation. More properly, instead of addressing this matter as a problem, we should consider it an enigma. (Monod, 1970, p. 131).

The origin of the code, in Monod's opinion, constitutes that disturbing border of the unknown which the evolution shows to its extreme: the code does not make sense if not translated. The translation mechanism of the modern cell entails at least fifty macro molecular constituents, also encoded in the DNA. It means that the genetic code can only be translated by the same products of translation. This is the modern expression for the *omne vivum ex ovo*. But when and how did this ring close on itself? It is in this moment that Monod transforms the scientific study of life in an interrogation in fact there is still no answer to this question but only hypotheses.

4. Before the genetic code

Hence, given what has been demonstrated up to now, the sixth and seventh hypotheses which we are going to outline consist of two different attempts to exceed the biological aporia highlighted with great shrewdness by the French biologist. Towards the mid-eighties, T. R. Cech, together with his collaborators, found out that the same RNA molecules could act as enzymes and catalyze reactions; very soon, these RNA sequences were defined as ribozymes (Kelly et al., 1982; Atkins et al., 1993, 2011). When the DNA message (the instructions to encode a protein) is copied on a filament of messenger RNA, a certain quantity of information is ignored. The cells do not have only "revising enzymes" but also "editing enzymes". The regions of the sequence containing the genetic instructions (exons) must be separated by the noncoding DNA sequences (introns). Thus, some enzymes remove the introns from the RNA and unify the exons. Subsequently, the sequence of exons is still processed in various ways; it is carried outside the nucleus and translated into a protein in the ribosome. Cech (Atkins et al., 1993) discovered, not without much stupor, that in some cases a protein enzyme is not necessary for the editing, because it is RNA itself which acts as an enzyme, so as to eliminate its own introns. The results amazed the community of molecular biologists. Today we know that a great variety of different ribozymes exists and they are able to catalyze different reactions by acting on themselves or on other RNA sequences. RNA molecules, in the absence of protein enzymes, are rather clumsy when they have to self-reproduce. But maybe a RNA ribozyme can act as an enzyme and catalyze the reproduction of RNA molecules: a similar ribozyme could act on itself to self-replicate. In both cases, a self-reproducing molecule, or a system of molecules would be at hand. Life would be on the right track (Kauffman, 1995). Yet there is a serious problem about the possibility of thinking of the ribozyme with polymerase activity as the original molecule of life. Even if such molecule had developed itself, in fact, could it have protected itself from a deterioration caused by mutations? Could it have evolved? According to Kauffman the answer to this question is probably a negative one. The problem is a form of error catastrophe, described for the first time by the chemist L. Orgel in the context of the genetic code. Let us imagine a ribozyme which is able to function as polymerase and copy every RNA molecule, itself included. This ribozyme, having a source of nucleotides at its disposal, would act as a replicating naked gene. Nevertheless any enzyme only accelerates the correct reaction among the possible alternative secondary reactions which could occur. Errors are inevitable. The self-reproducing ribozyme would necessarily come to reproduce also some mutant variations. But those mutant ribozymes are probably less effective than the normal ribozyme and so they have a high probability of committing errors more frequently. After some cycles, the system could produce an incontrollable spectrum of mutant variations. "Life would have vanished in a runaway error catastrophe" (Kauffman, 1995, p. 42).

Among all the problems related to the sixth hypothesis, the most original is without a doubt the one about the minimal complexity threshold of living creatures: all living organisms, according to Kauffman, seem to have a minimal complexity threshold below which life cannot exist. In nature the simplest free cells are called *pleuromona;* a simplified bacterial species infesting sheep lungs. Unlike viruses which are not living organisms[2], all cells have at least the basic molecular difference of pleuromona, which means they have at least a membrane, a DNA, a code, three hundred assorted genes, a transcription and translation device, a metabolism and a connection of energy fluxes towards and through the inside. Why, therefore, do free cells have an apparent minimal complexity? Kauffman thinks that a positive aspect of the naked gene theory is the simple origin of life, while its negative side is surely the fact that it does not articulate a satisfying answer to the question just posed. In brief, naked RNA, or the naked ribozyme with polymerase activity does not offer large

---

[2] The fundamental characteristic differentiating what is alive from what is not can be found in the self-construction principle. All living organisms, from single cells to multicellulars, can produce their own components through an autonomous process of organization: the internal dynamic of the living being itself is, in the last analysis, responsible for the organization of nature. This would be the reason why viruses, which need to invade the cell in order to reproduce, cannot be considered living organisms to all intents and purposes: they are incapable of self-construction and self-reproduction; as in the case of prions, i.e. the infectious agents of protein nature which are responsible for degenerative brain diseases like "mad cow disease" but they have no nucleic acids and thereby no genetic code.

bases to understand the minimal complexity observed in all living cells. Kauffman considers as a virtue of the theory of the origins the fact that it explains why the matter must reach a certain degree of complexity before life could emerge: "This threshold is not an accident of random variation and selection; I hold that it is inherent to the very nature of life" (Kauffman, 1995, p. 43). G. Wald (1954), in the *The Origin of Life*, wonders how it is possible that a group of molecules can assemble themselves in the exact way to form a living cell and goes on to affirm that, if many attempts are made, what is "inconceivably improbable" becomes "virtually certain". Hence, given so much time, according to this seventh hypothesis, the impossible becomes possible, the possible probable and lastly the probable virtually certain. Therefore, life, for Wald, would have emerged from a pure random combination modeled in stages by time. Such perspective however, is not so distant from Monod's vision who, concluded:

Modern science ignores any immanence. Destiny is written in the moment it is fulfilled and not before. Our destiny was not fulfilled before the appearance of the human species, the only species in the universe able to use a logic system of symbolic communication. Another unique event, which should, exactly for this reason, stops us from any form of anthropocentrism. If this event has really been unique, as the case of the appearance of life, this depends from the fact that, before it showed itself, its possibilities were almost null. The universe was not about to conceive life, nor the biosphere was about to conceive man, our number has come up on a roulette: why, therefore, should we not be aware of the exceptionality of our condition, like the one who has just won a thousand million? (Monod, 1970, pp. 133-134).

However much he wants to reduce all that happens to mere chance, Monod cannot refrain from asking himself a question given the exceptional nature of the phenomenon of self-consciousness. Yet R. Shapiro (1986), answers this question in a very different way: in fact, he calculates that, in the earth's life, $2,5 \times 10^{51}$ casual attempts to create life could have occurred. That is a very large number of attempts. Shapiro goes on with his research trying to calculate the possibility of obtaining, casually, something like *E. coli* (Shapiro, 1986). He then begins with the argument supported by two astronomers, F. Hoyle and N. Wickramasinghe, who, instead of calculating the probability of obtaining a whole bacterium, tried to calculate the possibility of obtaining a functioning enzyme (Hoyle & Wickramasinghe, 1981, 1993, 2000). Thus, if the amino acids are selected and structured in a casual way, what would be, then, the probability of obtaining a real bacterial enzyme with two hundred amino acids? The answer is given by multiplying the probability of each correct amino acids in the sequence, 1 out of 20, for 200 times, so as to produce quite a small probability: 1 out of 20^{200}. Moreover, given the fact that it is not sufficient to create a single enzyme to duplicate a bacterium, but it would rather be necessary to assemble about two hundred functioning enzymes, the probabilities would become 1 out of 10^{40000}, that is to say almost zero. If the total number of attempts made to create life is only 10^{51}, and the possibilities are 1 out of 10^{40000}, then the two scholars come to the conclusion that life would not have been able to develop. In that case, Monod would be right: we would really be a number which came up on a roulette, that is to say, we would be lucky and impossible. However, Kauffman supports a different idea. Hoyle and Wickramasinghe gave up the idea of spontaneous generation of life since the probability of a similar event occurring was comparable to the possibilities that a tornado, by passing over a rubbish dump, would be able to assemble a Boeing 747 with the materials deposited there. The two scientists, however, underestimated the power of self-organization. According to Kauffman, in fact, it is not necessary that a specific group of enzymes is assembled, one after the other, to carry out a specific number of reactions:

There are compelling reasons to believe that whenever a collection of chemicals contains enough different kinds of molecules, a metabolism will crystallize from the broth. If this argument is correct, metabolic networks need not be built one component at a time; they can spring full-grown from a primordial soup. Order for free, I call it. If I am right, the motto of life is not We the improbable, but We the expected. (Kauffman, 1995, p. 45).

These words of the American chemist concerning free order, offer a version of the genial ideas, although intuitive ones, which animated the primitive Darwinian consciousness of complexity. This hypothesis of the origin of life, elaborated by the great scientist, constitutes therefore a radical answer to the disturbing questions posed by Monod in the final part of his volume. Are we really in the exceptional condition of the one who has just won a thousand million? Is life really that miraculous an event which has occurred casually and only once? And finally, since the genetic code can only be translated by the same products of the translation, when and how has such ring closed-in on itself? Most scientists are convinced that the initial life has had simple features and it became complex only later. Whereas, according to Kauffman, life is not bridled by the magic of bases and replications, yet it is based on a deeper logic. It is a natural property of the complex chemical systems: when the number of different molecular species in a chemical broth passes a certain threshold, a net of self-sustaining reactions – an autocatalytic metabolism – suddenly appears. If Kauffman's hypothesis is true it means that life has not appeared in a simple form, but complex and articulated, and it has remained complex and articulated ever since:

The secret of life, the wellspring of reproduction, is not to be found in the beauty of Watson-Crick pairing, but in the achievement of collective catalytic closure. The roots are deeper than the double helix and are based in chemistry itself. So, in another sense, life – complex, whole, emergent – is simple after all, a natural out-growth of the world in which we live (Kauffman, 1995, p. 48).

5. Towards the theory of complexity

An attentive theoretical and experimental work carried out by Kauffman for more than forty years, of which almost twenty were spent at the *Santa Fe Institute*, firmly supports the original theories just mentioned which, under some aspects, are also revolutionary ones. The referring theoretical fire of the great American scientist is represented by what chemistry calls catalysis. Catalysts, as for example enzymes, accelerate chemical reactions which otherwise could only proceed very slowly. In general, many chemical reactions proceed with difficulty: given a great expanse of time, in fact, some molecules of A might combine with B to make C, but, as we have recently pointed out, in the presence of a catalyst (a molecule D) the reaction proceeds way faster. So, as Kauffman says, if D is the catalyst combining A and B to make C, the molecules A, B and C might themselves act as catalysts for other reactions. That being the case, it is possible to define a living organism as that "system of chemicals substances which is able to catalyze its own reproduction". What Kauffman defines as " a collectively autocatalytic system" is something in which molecules speed up the very reactions by which they are formed, that is a system in which, for example, A makes B, B makes C and C makes A again. Now we may consider an entire network of these self-supplied circles; given a source of molecules as food, the network will be able to recreate itself constantly. Such network, according to Kauffman, must be considered like the real metabolic networks residing in living cells, which means that, from a theoretical point of view, it is alive. The American scholar would like to demonstrate that if a group of sufficiently different molecules accumulate in one point, the possibilities of obtaining a self-supporting and self-reproducing metabolism (an autocatalytic system) becomes almost a certainty. According to this thesis the *bios* originates in the property of catalytic closure among different molecular species: if taken alone, in fact, each molecular species is inert, however, once the autocatalytic closure is obtained, the collective system of molecules comes to life.

[…] Every free-living cell, is collectively autocatalytic. No DNA molecules replicate nude in free-living organisms. DNA replicates only as part of a complex, collectively autocatalytic network of reactions and enzymes in cells. No RNA molecules replicate themselves. The cell is a whole, mysterious in its origins perhaps, but not mystical. Except for «food molecules», every molecular species of which a cell is constructed is created by catalysis of reactions, and the catalysis is itself carried out by catalysts created by the cell. To understand the origin of life, I claim, we must understand the conditions that enabled the first emergence of such autocatalytic molecular systems. (Kauffman, 1995, p. 50).

Catalysis alone is not enough to explain life; living systems are in fact open non- equilibrium thermodynamic systems because they exchange matter and energy to reproduce, unlike close thermodynamic systems which are chemical systems in equilibrium since they do not absorb neither matter nor energy from the environment. Usually chemical reactions can be more or less reversible: A transforms into B, but B transforms into A. So, by starting exclusively with molecules A, the concentration of B would increase to the point where the velocity of conversion of A into B would be exactly equal to the velocity of conversion of B into A. This balance is defined as chemical equilibrium: the concentration of A in B does not change with time, but every single molecule of A can transform into B and vice versa as far as thousands times a minute. As we have already mentioned, catalysts (protein enzymes and ribozymes) can accelerate in equal measure the reactions: the equilibrium between A and B is not altered because the enzymes simply increase the velocity to reach the state of balance. Between A and B there is an intermediate state defined as transition state in which some bonds among the atoms of the molecules are tightened and distorted: the molecule in the transition state is quite unhappy and the length of this unhappiness is given by the same energy of the molecule; low energy corresponds to slightly tense molecules, whereas high energy corresponds to tense molecules. This being the case, it is clear how enzymes function by linking and stabilizing the transition phase: they increase the speed to reach the equilibrium of the ratio of concentrations of A and B. For a living system chemical equilibrium equals death. The living beings are dynamic open systems which are constantly far from chemical equilibrium (Di Bernardo, 2010). Thus, such systems obey rules which are very different compared to those of closed systems: here the ratio between A and B will be reversed in comparison with the thermodynamic equilibrium ratio.

Let us consider now a much more complex open system: the living cell. If we are hoping that by understanding the behavior of very simple, open chemical thermodynamic systems we will be able to understand the cell we are really conceited. At present, nobody understands how the complex cellular networks of chemical reactions and their catalysts behave, or which laws govern their behaviors. Ilya Prigogine has called these systems "dissipative" because they continuously dissipate matter and energy to be able to preserve their structure. Unlike simple stationary systems in the thermodynamically open flask, the concentrations of chemical species in a more complex dissipative system may not be able to attain a stationary state, unchanged through time. Concentrations instead, can begin to oscillate up and down in repeated cycles, called limit cycles, which last for long periods of time. Such systems can also generate spatial trends of a certain relevance. (Prigogine, 1967; Prigogine & Stengers, 1988, 1979; Glansdorff & Prigogine, 1971; Nicolis & Prigogine, 1977). For instance, the Belosov-Zabotinskij reaction, recalled several times by Kauffman, develops through some simple organic molecules and it generates two type of spatial structures: in the first one concentric waves stretch by propagating outside, whereas in the second one we have spirals rotating around two central points. The heart is an open system and it can pulse by following a trend similar to the Belosov-Zabotinskij reaction: a sudden death, in fact, caused by a cardiac arrhythmia could correspond to the passage from what is the corner of a concentric rings scheme (heartbeat regularity) to a spiraling scheme inside the myocardium. Even so, Kauffman strongly reaffirms that, although very interesting, such chemical systems are not yet living systems: the cell, in fact, is not only an open chemical system, but a collectively auto- catalytic one. In the cells we have not only the emergence of chemical trends, but cells maintain themselves as entities able to reproduce and head towards a Darwinian evolution. But which laws, which deep principles may have caused the emergence of autocatalytic systems on primordial earth? What we are looking for is the myth of our creation (Kauffman 1995, 2008).

6. Autocatalytic systems and computer simulation

Since the Eighties, Kauffman has dealt with this complex problem by working on more simple models made easily feasible through computer simulation. Thus, through a casual graph, that is a group of points or nodes which are connected in a casual way by a series of lines, the American scholar attempts to build an approximate model which could explain the secret behind autocatalytic systems. Let us consider, for example, 10000 buttons scattered on the floor; we then choose two buttons at random and we connect them through a thread; done with this couple, we choose two other buttons and, as for the previous couple, we connect them through a thread. Going on in fits with this operation, the more we pick up buttons the more the probability to pick up at random two buttons and find out that one of them has been already picked up increases: so after a while the buttons will begin to be connected in big groups. The group which is interconnected in the casual graph is called component; some buttons can appear as not connected to other buttons, whereas others can be connected as a pair, as groups of three or more. The important characteristics of casual graphs show a very regular statistical behavior when the ratio between threads and buttons is modified. In particular, a phase transition occurs when the ratio between threads and buttons exceeds the value of 0.5. At this point, a gigantic group suddenly forms (Kauffman, 1993, 1995). In the 10000 buttons model the gigantic component emerges in the presence of 5000 threads. When such component forms, most nodes are directly or indirectly connected. As the ratio of the threads compared to the buttons keeps increasing above the threshold of 0.5, more and more buttons and small isolated groups will connect with the gigantic component. In this way, the gigantic structure expands itself, but the pace of growth decreases in accordance with the decreasing of the number of components that are not yet isolated. The quite sudden change of dimensions of the bigger group of buttons, when the ratio between threads and buttons exceeds 0.5, according to Kauffman, is a toy version of the phase transition which could have led to the origin of life. If there were an infinite number of buttons, when the ratio between threads and buttons should exceed 0.5 the dimensions of the bigger components would hope discontinuously from the minuscule to the enormous (Langton, 1990). This is a phase transition, like separated water molecules freezing in a single block of ice. Kauffman's thesis is therefore simple: with the increase of the ratio between threads and buttons, suddenly so many buttons become connected that an enormous network is formed in the system. This gigantic component is not a mystery: its formation is the expected natural property of a casual graph (Figure 1). The analog of the origin of life is that when a sufficiently elevated number of reactions are catalyzed in a system of chemical reactions, an enormous network of catalyzed reactions will crystalize immediately. And, as it is, such a network will almost surely be autocatalytic, able to self-sustain and alive (Kauffman, 1995). In the attempt to understand the origin of collectively autocatalytic molecular systems, Kauffman draws some graphs of metabolic reactions in which circles represent chemical substances and squares represent reactions (Figure 2). Besides, he distinguishes the spontaneous reactions which should occur very slowly from the catalyzed ones which, on the contrary, should occur quickly (Figure 3). The goal is to find the conditions in which the same molecules are both the catalysts and the products of the reactions generating the autocatalytic group. This depends on the possibility for each molecule of the system to have a double function: it can act as an ingredient or as a product of a reaction, but it can be the catalyst of another reaction too (Kauffman, 1995). RNA proteins and RNA molecules have this double function: any type of organic molecule, in fact, can be both the substrate and the product of reactions and, at the same time, can act as a catalyst to speed up other reactions. According to Kauffman, if one would know with certainty which group of molecules catalyze certain reactions, it would be possible to foresee

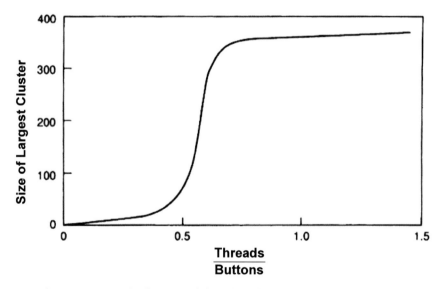

Fig. 1. A phase transition. As the ratio of threads (edges) to buttons (nodes) in a random graph passes 0.5, the size of the connected cluster slowly increases until it reaches a "phase transition" and a giant component crystallizes. For this experiment, the number of threads ranges from 0 to 600, while the number of buttons is fixed at 400 (Kauffman, 1995, p. 57).

with precision which group of molecules could be collectively autocatalytic. Although this knowledge is not available, it is anyway possible to proceed with an approximation by formulating plausible hypotheses and this is because "the spontaneous emergence of self-sustaining webs is so natural and robust that it is even deeper than the specific chemistry that happens to exist on earth; it is rooted in mathematics itself" (Kauffman, 1995, p. 60). Thus, this being the case, it becomes natural to wonder on the probability that a similar self-sustained network of reactions would form naturally. That is to say, is the appearance of collective autocatalytic reactions probable or virtually impossible? Kauffman answers as follows:

The emergence of autocatalytic sets is almost inevitable. […]As the ratio of reactions to chemicals increases, the number of reactions that are catalyzed by the molecules in the system increases. When the number of catalyzed reactions is about equal to the number of chemical dots, a giant catalyzed reaction web forms, and a collectively autocatalytic system snaps into existence. A living metabolism crystallizes. Life emerges as a phase transition. (Kauffman, 1995, pp. 61-62).

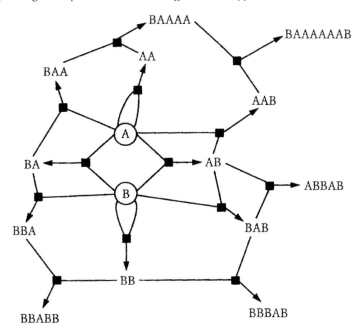

Fig. 2. From buttons and threads to chemicals. In this hypothetical network of chemical reactions, called a reaction graph, smaller molecules (*A* and *B*) are combined to form larger molecules (*AA, AB, etc.*), which are combined to form still larger molecules (*BAB, BBA, BABB, etc.*). Simultaneously, these longer molecules are broken down into simple substrates again. For each reaction, a line leads from the two substrates to a square denoting the reaction; an arrow leads from the reaction square to the product. (Since reactions are reversible, the use of arrows is meant to distinguish substrates from products in only one direction of the chemical flow.) Since the products of some reactions are substrates of further reactions, the result is a web of interlinked reactions (Kauffman, 1995, p. 59).

With the increasing of the diversity and complexity of the molecules in the system also the ratio between reactions and chemical substances in the reaction graph increases: there are, in fact, more reactions through which molecules can be formed than molecules. What happens, therefore, to the ratio between reactions and molecules in the graph when both the complexity and the diversity of such molecules increase? After some short algebraic calculations, it is easy for simple linear polymers to demonstrate that by increasing the length of the molecules the number of molecular species increases exponentially, but the number of reactions through which they transform themselves from one to the other increases even faster. Thus, the chemical system becomes more and more a fertile source of reactions which constantly transform themselves into other molecules. However, in order for the system to generate autocatalytic networks which are able to self-sustain, some of the molecules must act as catalysts by speeding up the reactions. For now, in fact, the system is fertile, but not yet a carrier of life since we do not know with certainty which molecule catalyzes which reaction. Among the different models built by Kauffman the most simple and functional model is the one supposing that each polymer has prearranged possibilities, one out of a million, to be able to function as an enzyme to catalyze a certain reaction. By using this model we decide beforehand which reactions each polymer is able to catalyze. This rule is defined by the American chemist as the "random catalysis rule" and it consists of assigning, at random and in a definite way, to each polymer the reactions it is able to catalyze. In this way, we will be able to distinguish the catalyzed reactions, trace some arrows from the catalysts to the reactions they catalyze and finally wonder if such a model

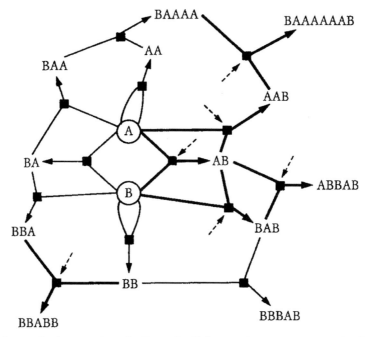

Fig. 3. Molecules catalyzing reactions. In figure 2, all the reactions were assumed to be spontaneous. What happens when we add catalysts to speed some of the reactions? Here the reaction squares indicated by dashed-line arrows are catalyzed, and the heavy, darker lines connect substrates and products whose reactions are catalyzed. The result is a pattern of heavy lines indicating a catalyzed subgraph of the reaction graph (Kauffman, 1995, p. 61).

of chemical system contains a collectively autocatalytic group, that is a network of molecules connected through thick lines also containing the molecules catalyzing the reactions by which the same molecules are formed. In order to make the model more realistic, Kauffman adds a second rule: even if a candidate to the role of ribozyme has a site which fits the right and left end of its substrates, it has only a probability out of a million to have other chemical properties which can make it catalyze the reaction. This suggests the idea that also other chemical characteristics could be necessary, besides the complementarity of the bases, to obtain a catalysis made by the ribozyme. The American biologist calls this "complementary catalysis rule". And here it is the crucial result: it does not matter which of these catalyst rules are used; when a group of model rules reaches a critical diversity, a gigantic component of catalyzed reactions crystalizes, causing the emergence of collectively autocatalytic groups. At this point it is not difficult to understand the reason why this emergence is "virtually inevitable" (Kauffman, 1995). Supposing we establish that any polymer has a possibility out of a million to be able to act as an enzyme for a certain reaction and that we have to use the random catalyst rule, it means that with the increasing of

Fig. 4. An autocatalytic set. A typical example of a small autocatalytic set in which food molecules (*a, b, aa, bb*) are built up into a self-sustaining network of molecules. The reactions are represented by points connecting larger polymers to their breakdown products. Dotted lines indicate catalysis and point from the catalyst to the reaction being catalyzed (Kauffman, 1995, p. 65).

diversity among the molecules of the model system, the ratio between reactions and molecules will increase. When the diversity of the molecules is quite elevated, Kauffman says, the ratio between the reactions and the polymers reaches the level of a million to one. At this level of complexity then, each polymer will be able to catalyze a reaction: when the ratio between catalyzed reactions and chemical reactions is 1, there is an extremely high probability for a gigantic component, a group of collectively autocatalytic molecules, to form. In the inert broth, for example, nothing happens, apart from very slow spontaneous chemical reactions. If both the diversity and the atomic complexity of molecules are increased, more and more often the same elements of the system will catalyze internal reactions (Figure 4). When a certain threshold of diversity is exceeded, an enormous network of catalyzed reactions arises during the phase transition: "the giant component will contain a collectively autocatalytic subset able to form itself by catalyzed reactions from a supply of food molecules" (Kauffman, 1995, p. 64).

7. Conclusion

Thus, the theory about the origin of life, as elaborated by Kauffman[3], has its roots in an irreducible holism which is not the result of reflections of a metaphysical nature, but of mathematical necessity: life has emerged as whole, not a little at a time, and it has remained that way. Therefore, unlike the dominant vision which sees the naked RNA at the origin of life, with its customized evolutionary stories, today we have the hope of explaining why living creatures seem to have a minimal complexity, because nothing simpler than a *pleuromona* can be alive (Kauffman, 1995). However, if life has begun by collective autocatalysis and has incorporated the DNA and the genetic code only later, how can catalytic groups meet with hereditary variations and with a natural selection without yet containing a genome? In other words, is there a way for an autocatalytic group to evolve without all the complications related to the genome?

Richard and Doyne found a natural way for the variation and evolution in such systems to develop. They suggested that a casual and not catalyzed reaction can develop occasionally when an autocatalytic network is active from a metabolic point of view. Such spontaneous fluctuations will tend to generate molecules which are not part of the group; these new molecules, according to the two scholars, can be considered as a sort of "half-light of molecular species" and as a "chemical haze surrounding the autocatalytic group". The original group, by absorbing some of these new molecular species on the inside, would be altered. At this point two cases are possible: A) if one of these new molecules contributed to catalyze its own formation, it would became a member of the network to all intents and purposes and so, a new circuit would be added to the metabolism; B) if the molecular intruder inhibited an already existent reaction, then an old circuit could be eliminated from the group. In both cases a hereditary variation is clearly possible. Moreover, if the result would be a more efficient network, such mutations would be favored and the modified network would take the place of weaker competitors. With reference to such considerations, Kauffman sustains that there are several reasons to believe that autocatalytic groups could evolve even without a genome. Biologists divide cells and organisms by genotype (genetic

---

[3] We must take into consideration that the autocatalysis theory has been recently revisited by Kauffman in 2000, in 2007, and in 2008. For a thorough examination informations of the possible limits of the emergence theory elaborated by the American scholar consult: Carsetti, 2009; Di Bernardo, 2011.

information) and phenotype (enzymes and other proteins, but also organs and morphological characteristics constituting the body). For autocatalytic groups there is no separation between genotype and phenotype. The system acts as its own genome. Nevertheless the capability to incorporate new molecular species, and perhaps eliminate older molecular forms, gives us the promise to generate a network of self-reproducing chemical substances equipped with different characteristics. Darwin teaches us that such systems will evolve by natural selection (Kauffman, 1995). It is inevitable that these proto-cells which are self-reproducing and divided up together with their "daughters", will form a complex ecosystem. It is not metabolic life which has begun as a complex whole but, according to the biochemist, the entire panoply of mutualism and competition we define as ecosystem is blossomed from the beginnings. The story of these ecosystems at all levels is not only the story of evolution, but also of coevolution. These considerations give reasons of the Darwinian intuition concerning the internal constitution of complexity (Darwin, 1902). From what has been said since this moment, in fact, it is clear how, in Kauffman's opinions, autocatalytic systems are not chaotic at all. If life has begun when some molecules spontaneously united to form autocatalytic metabolisms, we will then find a source of molecular order, a fundamental source of internal homeostasis which protects cells from perturbations, and a compromise which let the proto-cellular networks receive light fluctuations without collapsing (Kauffman, 2004a, 2004b). This order that without the genome emerges from the collective dynamics of the network and from the coordinated behavior of associated molecules, as we have abundantly demonstrated through Kauffman's words, is free and is generated by small attractors which, under some conditions, represent the source of order of big dynamic systems. Thus, in this context the concept of catalytic closure in groups of collectively autocatalytic molecules appears as a profound model of the complexity laws, that is those powerful laws which, even if they are not able to supply a detailed prediction of the "three of life" they are able to explain its general form. Life, from this point of view, appears as an emerging phenomenon which develops when the molecular diversity of a pre-biotic chemical system exceeds a given level of complexity. If so much is true, then life does not reside in the individual properties of each single molecule-in details- but it is a collective property of systems of molecules interacting with each other. From this perspective, life has emerged as a whole and has always remained so. From this point of view, life must not be searched in its parts, but in the totality of the emerging properties creating the whole. A group of molecules can have or not the capability of catalyzing its own formation and reproduction starting from simple basic molecules. In the emerging and self-reproducing whole there is no vital force or foreign substance. And anyhow, "the collective system does possess a stunning property not possessed by any of its parts. It is able to reproduce itself and to evolve. The collective system is alive. Its parts are just chemicals" (Kauffman, 1995, p. 24). If these ideas are right, it means that, we are not only "at home in the universe", but we actually have excellent probabilities to share it with some other being which is still unknown (Kauffman, 2001).

8. References

Atkins, J. F., Gesteland, R. F. & Cech, T. R. (ed.). (1993). *The RNA World: The Nature of Modern RNA Suggests a Prebiotic RNA World*, Cold S.H.L.P., New York.

_____(ed.). (2011). *RNA Worlds: From Life's Origins to Diversity in Gene Regulation*, C.S.H.L.P., New York.

Carsetti, A. (ed.). (2009). *Causality, Meaningful Complexity and Embodied Cognition*, Springer, Berlin.

Darwin, C.R. (1958). *The Autobiography of Charles Darwin 1809-1882. With the Original Omissions Restored. Edited and with Appendix and Notes by his Granddaughter Nora Barlow*, Collins, London.

_____(1902). *On the Origin of Species by Means of Natural Selection or The Preservation of Favoured Races in the Struggle for Life*, H. Frowde, Oxford U.P., London.

Di Bernardo, M. (2007a). Simulazione informatica e vita artificiale: è possibile per l'uomo creare la vita reale?. *Dialegesthai. Rivista telematica di filosofia*, http://mondodomani.org/dialegesthai/mdb03.htm.

_____(2007b). *Per una rivisitazione della dottrina monodiana della morfogenesi autonoma alla luce dei nuovi scenari aperti dalla post-genomica*, Aracne, Roma.

_____(2010). Natural Selection and Self-organization in Complex Adaptive Systems. *Rivista di Biologia / Biology Forum*, 103, n. 1, pp. 75-96.

_____(2011). *I sentieri evolutivi della complessità biologica nell'opera di Stuart Alan Kauffman*, Mimesis, Milano.

Glansdorf, P. & Prigogine, I. (1971). *Structure, Stabilité et Fluctuations*, Masson, Paris.

Hoyle, F. & Wickramasinghe, N. C. (1981). *Evolution from Space*, J. M. Dent and Sons, London.

_____(1993). *Our Place in the Cosmos*, J. M. Dent and Sons, London.

_____(2000). *Astronomical Origins of Life: Steps Towards Panspermia*, K.A.P.

Kauffman, S.A. (1993). *The Origins of Order: Self-Organization and Selection in Evolution*, Oxford U.P., New York.

_____(1995). *At Home in the Universe: the Search of the Laws of Self-Organization and Complexity*, Oxford U.P., New York.

_____(2000). *Investigations*, Oxford U.P., New York.

_____(2001). Prolegomenon to a general biology, *Ann. NY Acad. Sci.*, 935, pp. 18-36.

_____(2004a). A Proposal for Using the Ensemble Approach to Understand Genetic Regulatory Networks, *Theoretical Biology*, 230, 4, pp. 581-590.

_____(2004b). The Ensemble Approach to Understand Genetic Regulatory Networks, *Physica*, A 340, pp. 733-740.

_____(2007). Question 1: origin of life and the living state, *Orig. Life Evol. Biosph.*, 37, 4-5, pp. 315-322.

_____(2008). *Reinventing the Sacred. A New View of Science, Reason and Religion*, Basic Books, New York.

Kelly, K. et al. (1982). Self-splicing RNA: Autoexcision and autocyclization of the ribosomal RNA intervening sequence of tetrahymena. *Cell*, 31, n. 1, pp. 147-157.

Langton, G. (1990). Computation to the edge of chaos: Phase transitions and emergent computation. *Physica*, 42D, pp. 12-37.

Monod, J. (1970). *Le hasard et la necessite: essai sur la philosophie naturelle de la biologie moderne*, Ed. du Seuil, Paris.

Nicolis, G. & Prigogine, I. (1977). *Self Organization in Non-Equilibrium Systems*, J. Wiley and Sons, New York.

Prigogine, I. (1967). Structure, Dissipation and Life. *Theoretical Physics and Biology*, Versailles, North-Holland Publ. Co, Amsterdam, 1969.

Prigogine, I. & Stengers, I. (1979). *La nouvelle alliance. Metamorphose de la science*, Gallimard, Paris.

_____ (1988). *Order Out of Chaos: Man's New Dialogue with Nature*, Bantam Books, Toronto.

Schrödinger, E. (1944). *What is life? The physical aspect of the living cell*, Cambridge U. P., Cambridge.

Wald, G. (1954). The Origin of Life. *Sci. Am.*, 191, n. 2, pp. 47-48.

3

The Biosphere: A Thermodynamic Imperative

Karo Michaelian
Instituto de Física, UNAM
Mexico

1. Introduction

The publication of "On the Origin of Species" by Darwin in 1859 set the stage for the emergence of a new paradigm. No longer was biology accredited to the whim of an omnipotent mystical being; for the first time the diversity of life was seen as the result of a plausible physical mechanism which led to complex organisms and species from simpler common ancestors. Darwin's theory assumed a fierce battle for survival of the individual against the hostile biotic and abiotic components of its environment. This fight for survival suggested an implicit metaphysical "will to survive" programmed within each organism, which, together with variability in reproduction, provides sufficient elements for natural selection to drive the evolutionary process. Such a process would lead to ever more apt forms of organisms until reaching what Darwin considered to be a state of "near perfection". The biosphere, under this paradigm, was considered to be an emergent system, the result of the sum of the many individual or collective struggles for survival.

The annunciation of the theory of Gaia by Lovelock (2005) in the 1970's added a completely new dimension to the biosphere. No longer was life subjected to the unrelenting demands of the environment, but living organisms, collectively or individually, could, in fact, alter their physical environment, in such a manner that seemed to be beneficial to their survival. It was found, for example, that organisms had transformed the Earth's atmosphere in such a manner that sunlight could penetrate through to the surface where it was needed by biology. Organisms even regulated the temperature of Earth's surface to keep it comfortable for themselves, and to ensure the presence of the vital liquid water in the biosphere. This biotic regulation of Earth's temperature to within the range for liquid water seemed remarkable considering that the Sun's integrated luminosity had increased by as much as 30% since the very beginnings of life on Earth (Sagan and Chyba, 1997).

Gaia theory was enlightening, it pointed out the many biological controls over the environment; from limiting the maximum concentration of salt in the oceans, and oxygen in the atmosphere (thereby avoiding higher concentrations at which spontaneous combustion occurs) to promoting plate tectonics that has allowed a diversification of ecosystems. From the optics of Gaia, biology was seen to be coevolving together with its abiotic environment. However, an enigmatic question arose, "towards what was the biosphere evolving?" Proponents of the Gaia theory claimed that evolution of the biosphere was proceeding in such a manner so as to provide better conditions for life, and greater stability for all living systems. In the most assertive interpretations of Gaia, the Earth itself was considered as one great living organism practicing homeostatic auto-regulation.

Critics of the Gaia theory countered that the tenets of the theory were in contradiction to those of Darwin's theory of evolution through natural selection. In particular, they argued that evolution was affected by natural selection operating at the individual level, not at the higher coupled biotic-abiotic levels that Gaia theory was claming. How could the biosphere as a whole, or even more enigmatically, the Earth as a whole (in the strongest interpretation of Gaia), be subjected to a kind of natural selection when there was no competition between distinct co-existing biospheres or co-existing Earths? This led to the annunciation of the "problem of the evolution of a system of population of one – the biosphere" (Swenson, 1991).

Proponents of the Gaia theory valiantly contested their critics, explaining Gaian dynamics on the basis of Darwin's theory of evolution through natural selection operating at the organismal level. Lovelock presented his now famous "Daisy World" model to show how Earth's surface temperature could be regulated to optimal values for the daisies by a variant gene that coded for either dark or light colored daisies. Meanwhile, more and more evidence was accumulating (much of it contributed by Lynn Margulis and her collaborators, Margulis; 1986) demonstrating that organisms do, and indeed always have, regulated their environment.

A progressive paleontologist, Gould (2002), saw the problem not in explaining higher level selection from within the Darwinian framework, but in the limitation of the Darwinian theory itself. Arguing his view in "The Structure of Evolutionary Theory", Gould (2002) states (p. 699) "This perspective implies a striking limitation upon the strictly Darwinian style of extrapolative and gradualistic selection that the Modern Synthesis promulgated as an adequate explanation for evolution at all scales of time and effect. ... Once again, we grasp the need for independent macroevolutionary theory…".

Apart from the difficulty of explaining evolution on all hierarchal levels and the co-evolution of a coupled biotic-abiotic system, Darwinian Theory runs into other troubles. For example, there is no doubt that biotic systems are evolving through some kind of natural selection, but the answer to the question of *"what is natural selection actually selecting?"* remains elusive. Even though philosopher Karl Popper(1902-1994) recanted on his original criticism of *survival of the survivors* tautology in the theory of evolution, he still referred to it as a "metaphysical research program" (Popper, 1978), and there still remains a large amount of circularity of argument in Darwin's theory. This circularity cannot be removed from the theory unless a physical foundation for evolution through natural selection can be clearly identified.

Perhaps the most glaring deficiency of Darwinian Theory, however, is its lack of comment on the origin of life itself. Although Darwin expressed in private that life may have started "in a warm little pond" (Darwin, 1887) no experimental or theoretical extrapolation of the Darwinian paradigm to the molecular level has ever produced important insights into life's origin.

This chapter presents an argument for a new paradigm for the biosphere based on the thermodynamic theory of open systems (known as *non-equilibrium* or *irreversible, thermodynamics*) and in particular, on the statistical mechanics of open systems which eventually must provide the foundation for the thermodynamic theory. There are good reasons to believe that life and evolutionary processes occurring within the biosphere can be grounded in thermodynamic theory. First, thermodynamic laws are the most universal of all laws, often referred to as the "laws of laws"; they are derived from fundamental symmetries existing in Nature. There is no doubt that these laws apply to biological processes, since biological processes are composed of chemical, electrical, mechanical, and transport, processes, all of which, indisputably, are under the dominion of these basic symmetries and therefore under the imperative of thermodynamic law. Second, there is empirical evidence for a natural evolution of biological systems to particular stable end states which depend on both the boundary and the initial conditions of the system, similar to the very dynamics driving the evolution of abiotic non-equilibrium processes. Moreover, these end states, known as *stationary states* in non-living systems and as *stasis* in living systems, share similar stability characteristics (Michaelian, 2005).

This chapter argues that biotic and abiotic irreversible processes thermodynamically couple in such a way so as to increase the overall entropy production of Earth in its solar environment. It is suggested that the biosphere arose as a thermodynamic imperative that coupled life and biotic evolution to abiotic irreversible processes in order to remove impediments to greater global entropy production of Earth. It analyzes how biological processes are thermodynamically coupled with abiotic processes within the biosphere. (The term "biosphere" in this chapter is taken to mean that greater entity composing the processes of life, the lithosphere, atmosphere, and hydrosphere.) Examples of biotic-abiotic coupling are; biology catalyzing the hydrological cycle (Michaelian, 2009a, Michaelian, 2011b), and biology catalyzing ocean and wind currents, and the carbon cycle.

Such a thermodynamic view of the biosphere provides an explanation of many intriguing biotic-abiotic associations discovered while accessing the Gaia hypothesis (Lovelock, 2005). It also provides a framework for explaining the observed co-evolution of life with its environment, and for the resolution of the paradox of "the evolution of a system of population one – the biosphere" (Swenson, 1991). Perhaps most importantly, however, it offers a simple physical reason for the origin of life based on entropy production through ultraviolet (UV) light dissipation into heat by RNA and DNA during the Archean (Michaelian, 2009b; 2011a).

Keywords: Biosphere, non-equilibrium thermodynamics, thermodynamic dissipation theory of the origin of life, UVTAR, entropy production, Darwin, Gaia.

2. Co-evolution of the biotic with the abiotic through the thermodynamic imperative

Darwin's theory of evolution through natural selection lacks sufficient elements to explain the origin, persistence, and evolution of the biosphere. In order to understand the biosphere and its evolution, it is first necessary to recognize that biological processes in the biosphere are thermodynamically coupled to abiotic processes, and then understand how biosphere evolution is driven by increases in the entropy production afforded to the Earth through the origin and coupling of its constituent irreversible processes.

No irreversible process, life and the hydrological cycle included, can arise and persist without producing entropy. Entropy production is not incidental to the process, but rather the very reason for its existence. A fundamental characteristic of nature is the search for routes to greater global entropy production, often building on pre-existing routes by coupling new irreversible processes to existing ones. Onsager (1931) has shown how diverse irreversible processes can couple in order to remove impediments to greater global entropy production (Morel and Fleck, 1989). In general, the more complex the dissipative structuring in space and time, i.e. involving many coupled irreversible processes with embedded hierarchal levels and interactions of long spatial and temporal extent, the greater the overall entropy production in the systems interaction with its external environment (Onsager, 1931; Prigogine et al., 1972; Lloyd and Pagels, 1988).

Non-equilibrium thermodynamics shows (Prigogine, 1967) that the entropy production of a system is of a bi-linear form of generalized forces, X, times generalized flows, J, i.e.

$$\frac{dS}{dt} = \sum_i X_i J_i \qquad (1)$$

where the index i runs over all irreversible processes operating in the system. These processes can be, amongst others, flows of mass, heat, or charge, chemical reactions, and photon absorption and dissipation. In ecosystem processes, individuals can be considered as units of entropy production and exchange. The forces X_i are then associated with the populations p_i of species i, and the flows J_i with the flows of entropy (Michaelian, 2005). In this thermodynamic view, selection of a particular irreversible process, or the coupling of irreversible processes, is contingent on increasing the global entropy production of Earth in its solar environment. In general, it is not the individual organism's entropy production, in isolation of all others (if this could even be imagined), which is relevant to selection, but instead the increase in the total entropy production of all coupled processes in the Earth system (the change in the sum of Eq. (1)). However, entropy production is an extensive (additive) quantity and the individual terms $X_i J_i$ are often positive (although not necessarily so if two or more processes occur within the same macroscopic region; in this case a term may be negative as long as a coupled term is positive and of greater magnitude, see (Prigogine, 1967)). Therefore, in this thermodynamic framework, selection based on increases in entropy production can occur at any level, including that of the individual or the biosphere.

In this top-down thermodynamic view, organisms should not be seen as individual entities endowed with a meta-physical "will to survive" competing against others and their environment, but instead as local thermodynamic flows which arise in response to local (on the relevant scale) thermodynamic forces which define their environment. The local thermodynamic potentials providing the forces are created by other irreversible processes dissipating thermodynamic potentials on a still higher level, and so on up until reaching the highest hierarchal level of the Earth in its solar environment. The moment that a local thermodynamic potential wanes, or becomes depleted, the organism (irreversible process spawned) created to dissipate this potential will "suffer", or go extinct. Organisms (irreversible flow processes) on whatever scale, therefore, have higher probability of survival if they are attached to large, robust and stable thermodynamic forces (gradients of potentials). Life, in general, appears to be directly attached to the greatest thermodynamic potential influencing Earth; the solar-space photon potential. As equation (1) above indicates, stronger forces, or stronger flows, imply greater entropy production.

Thermodynamic selection, however, is not "instantaneous"; both biotic organisms and abiotic processes have the ability to adapt to a changing thermodynamic potential. Biotic organisms adapt through their organismal plasticity (e.g. the ability to migrate, or their ability to survive off different thermodynamic potentials - heterotrophy) or through their genetic apparatus and reproduction (plasticity on the species level). In contrast, abiotic processes have an inherent plasticity, for example, a change in size or direction of a hurricane, in response to a change in the size or direction of a temperature gradient. In this thermodynamic view the process is not "striving to survive", it is rather a flow responding to a changing thermodynamic force. The Gaia theory speaks of "life shaping the environment to its own benefit", but the thermodynamic view sees all life coupled to many higher level thermodynamic processes (both biotic and abiotic) which are together evolving to greater levels of global entropy production. It is these higher level potentials that are the creators, deformers, and selectors of those irreversible dissipating processes that we may call abiotic or biotic organisms.

The underlying thermodynamic foundations for evolution are the same for both abiotic and biotic irreversible processes. However, there are differences in mechanism which make them appear very distinct. The difference appears to be related to the nature of the thermodynamic potential that they arise to dissipate. Biotic organisms, in their majority, dissipate directly the photons arriving from the sun. Covalent interactions between the atoms of the organic molecules provide energy gaps compatible with the quantum energies of visible and UV photons. These molecules, when in water, can thus absorb and dissipate efficiently to heat the incident solar photon (hence the coupling of life to the water cycle). If not in water, the molecules remain for a longer time in an excited electronic state where they are vulnerable to destruction through photo-reactions, as well as to radiative decay and other less dissipative means of de-excitation (Middleton et al., 2009). These non-central covalent interactions have directionality properties which allow a great, almost inexhaustible, variety of distinct molecular configurations for pigments that absorb from the far UV to the far infrared (Gatica, 2011).

This diversity of organic molecular configurations allows for a genetic apparatus of biotic individuals that provides information storage and a retrieval system that accumulates information about the other irreversible processes and thermodynamic potentials in the environment. This endows individuals and species with great plasticity in responding (Darwinians would say "adapting") to changes in the thermodynamic potentials of their environment.

Abiotic processes, on the other hand, tend to dissipate secondary heat gradients in which the relevant quantum energy packets are of much lower energy than those of visible or UV photons. Most of the solar photons intercepted by abiotic organisms are in the infrared. Abiotic organisms are usually "glued" together through central and weak hydrogen bonds (e.g. the interaction between water molecules) or through central van der Waals forces. These interactions have no directionality properties, and, therefore, lead to low diversity at the molecular level. This precludes a microscopic molecular genetic code and limits the "organism's" ability to accumulate information about the thermodynamic potentials or other irreversible processes operating in their environment. However, information is nevertheless transmitted at a more macroscopic level, as, for example, in the distribution of ocean surface water temperatures which "remembers" the history of prior recent hurricanes (and other irreversible heat flow processes) and constrains the birth and evolution of new, future hurricanes.

3. Onsager's principle and entropy production in the biosphere

There is an underlying physical basis for expecting nature to evolve towards greater entropy production. Onsager showed that irreversible processes arise and couple to reduce impediments to greater global entropy production of the system (Onsager, 1931; Morel and Fleck, 1989). Onsager's principle should not be confused with the Maximum Entropy Production Principle (MEPP) which suggests that of all possible evolutionary paths, the system will take that particular path which maximizes the entropy production. Onsager's principle simply states that the appearance of a new irreversible process cannot be expected unless the global entropy production of the system increases. Since the Earth system is a complex non-linear system with many thermodynamic potentials generated in a hierarchy of embedded levels, many paths exist in phase space for dissipating the imposed solar photon potential, each path in phase space is separated from others by energy, momentum, or angular momentum barriers (the conserved quantities for classical systems). The actual path taken will depend, because of the non-linearity, very sensitively on the microscopic initial conditions and subsequent perturbations, and there is, therefore, no reason why the chosen path has to be the one of maximum entropy production. While the "maximum entropy production principle" may one day be validated as a probabilistic principle, this principle, at least in its present formulation and at the level of complexity of the biosphere, is not relevant here since we neither have ways of distinguishing the microscopic initial conditions of the Earth system, nor can we recognize and compile the history of subsequent external perturbations that the system experienced. The possibility that we will be able to count and analyze all possible paths available in phase space appears even further remote. Note that there is in principle no absolute maximum to the amount of entropy production of the Earth system. Infinite entropy production would result if the incident photons were converted into infinitely long wavelength photons (constrained, perhaps, by the finite temperature of outer space, the "size" of the Universe, and quantum aspects).

In principle, the thermodynamic framework could be used to predict which individuals, species, ecosystems, clades, or biospheres have higher probability of appearance and survival, however, since almost all irreversible processes occurring on Earth are coupled at many spatial and temporal levels, the Earth system cannot easily be separated into parts. Exceptions are cases in which coupled irreversible processes act on vastly different time scales. In these cases, the entropy production due to the part of short characteristic time scale can be used as a local predictor for the direction of evolution. In practice, however, few processes can be effectively isolated, and it is just as impractical to obtain a prediction for the direction of evolution from this framework as it is from the Darwinian principle of natural selection (of the "fittest").

However, by replacing "selection of the fittest" with "selection of irreversible processes leading to greater global entropy production" the thermodynamic framework avoids the tautology inherent in traditional evolutionary theory. Within the Darwinian framework, selection on a higher level than the individual, e.g. species, populations, communities, ecosystems, etc., is generally attributed to some kind of "emergent" behavior of a complex many-body system. However, it is well known that non-equilibrium "emergent behavior" results from the dissipation of an underlying gradient, a thermodynamic potential. The general struggle for existence leading to evolution through natural selection has not been associated with such a potential and this leads to a certain tautology in Darwinian Theory.

In the following section, an important example is given of how biology is coupled to a particular abiotic irreversible process, the water cycle, and how this coupling augments the entropy production of the Earth system.

4. Biology, solar photons, and the hydrological cycle

All irreversible process, the hydrological cycle included, arise and persist to produce entropy. The steps of the hydrological cycle contributing to the global entropy production are;

1. Sunlight reaching the Earth's surface is absorbed on both biotic and abiotic material and rapidly dissipated into heat, particularly if this occurs on organic material in water. This step contributes to the greatest amount of entropy production in the Earth system, about 63% of the total. Most of the remaining entropy production is due to photon absorption and dissipation on the gasses of the atmosphere (Peixoto et al., 1991, Michaelian, 2011b). Note that if there were no water in the region of photon absorption then the surface of Earth would radiate energy into space at shorter wavelengths or conduct or convect energy to the atmosphere at higher temperature, implying less dissipation and therefore less entropy production. The availability of water in the region also leads to the second step in the water cycle;
2. Heat absorbed by the water, particularly at the surface, converts liquid water into water vapor (the latent heat of vaporization). Water vapor expansion into the greater volume of the atmosphere, and its diffusion into the atmosphere, produce entropy. The cooling of Earth's surface produced by the removal of the latent heat of vaporization implies a more red-shifted black-body surface radiated spectrum.
3. Water vapor is less dense than dry air and so rises to the cloud tops. This diffusion process generates more entropy. As the water vapor rises, its heat is given off to the surrounding air column. This spawns other irreversible processes such as winds, tornadoes and hurricanes, all of which dissipate these secondary heat gradients and therefore generate further entropy.
4. At the cloud tops, at roughly -14°C, the water vapor condenses and releases the heat of condensation into the atmosphere. Much of this energy is then directly radiated into space in a roughly black-body spectrum of -14 °C. Condensation would seem to cancel the entropy production of evaporation (step (2) above), but this condensation occurs at a significantly lower temperature than does the evaporation. The net entropy produced by evaporation-condensation cycle can be roughly calculated as $dS = dQ(1/T_c - 1/T_s)$, where dQ is the heat of vaporization and T_c and T_s are, respectively, the temperature of emission of the energy dQ by condensation at the cloud tops, and the temperature of absorption by liquid water of this energy at the surface of Earth.

Indirectly, the water cycle also leads to entropy production through clouds which are blown over land masses, allowing water to travel inland and provide the essential element for life (particularly trees and plants). This life reduces Earth's albedo due to the organic molecules that absorb and dissipate solar radiation. Old, climax, forests correspond to the regions of lowest albedo on Earth. A greater portion of Earth's surface is thus rendered available for greater entropy production.

5. The origin of life

Darwin foresaw the possibility of a physical theory for the origin of life when he suggested that life could have started in a "warm little pond, with all sorts of ammonia and phosphoric salts, lights, heat, electricity, etc. present, so that a protein compound was chemically formed ready to undergo still more complex changes" (Darwin, 1887). However, nothing in his theory of evolution through natural selection suggests just how life may thus have originated. Many later attempts were made to adapt the Darwinian paradigm of competition to the molecular, or chemical, level, but these attempts have so far been unsuccessful

Most experiments on the origin of life have been carried out in near equilibrium conditions and look for the spontaneous appearance of an auto-catalytic chemical process. However, the mere hope of observing the origin and persistence of an irreversible process such as life without duly considering its thermodynamic function of entropy production is nothing short of expecting a miracle.

If the absorption and dissipation of high energy photons from the sun is the major thermodynamic function performed by life today, then it is plausible that this was also its function in the Archean some 3.8 billion years ago from when there is evidence that life began. Earth's atmosphere at this time was more transparent in the ultraviolet since there was no ozone or oxygen in the atmosphere (Cnossen et al., 2007), and more opaque in the visible. The sun was also perhaps up to 10 times more intense in the far ultraviolet than it is today, principally due to a higher rotation rate. It is then probable that life arose to dissipate photons in the UV and only later developed pigments in the visible when the atmosphere cleared of organic haze and sulfuric acid clouds and became more transparent at these wavelengths. Both RNA and DNA absorb strongly in the UV at 260 nm and dissipate this energy rapidly into heat when they are in water.

These facts have led me to propose a thermodynamic dissipation theory for the origin of life (Michaelian, 2009a, 2011b). In this theory, the reproduction of RNA and DNA is promoted by the entropy production due to dissipation of UV photons by these organic molecules while they floated on the surface of water bodies. Based on the ratio of the concentrations of ^{16}O and ^{18}O isotopes found in sediments of this age, it has been surmised that the ocean surface temperatures were around 80 °C when life began. This is close to the denaturing temperature of intermediate sized DNA fragments. An "ultraviolet and temperature assisted mechanism for replication" (UVTAR) of these first molecules of life can therefore be imagined: Once Earth's surface temperature dropped below the denaturing temperature of RNA or DNA, then during the day, particularly in the afternoon, absorption of UV, visible, and infrared light on the ocean surface would increase the temperature to beyond the denaturing temperature, and the double strands of RNA or DNA would separate into single strands. At night, the two single strands could act as templates for the formation of complementary strands thus providing a new generation of RNA or DNA. This process bears resemblance to a now standard laboratory procedure to amplify (multiply) DNA known as "polymerase chain reation" (PCR).

Biotic evolution in the Archean, as it is now, was driven by increases in the entropy production due to photon dissipation. As the seas cooled further still, it became selectively advantageous for the RNA and DNA molecules to code for aromatic amino acids such as tyrosine and tryptophan that could act as UV photon antennas to augment the local temperature enough to permit denaturation in the late afternoon. The cooling of the ocean surface was thus probably the first promoter of information storage in DNA and RNA.

An experiment is now underway at the Institute of Physics at the UNAM to determine if a PCR-like process can be carried out with a UV light cycle replacing the normal temperature cycling in PCR. If this is successful, the next phase of the experiment would be to attempt amplification without the enzyme polymerase. Although this will certainly slow down the process considerably, other inorganic catalyzers, such as magnesium ions, could have been present and it must be remembered that the first DNA or RNA segments could have been subjected to billions of PCR-like cycles before life as we know it initiated.

6. Conclusions

Arguments have been given to suggest that biology should be considered as a set of irreversible process embedded within, and coupled to, other abiotic irreversible processes, and it is this whole of processes which defines the biosphere. The biosphere and its evolution are the result of a thermodynamic imperative which drives the Earth system towards increases in the global entropy production, in accordance with Onsager's principle. This is achieved principally, but not exclusively, by the organic pigments which absorb and dissipate the incident solar high energy photons into heat in the presence of water at the surface of the Earth. The coupling of the hydrological cycle to biology is thus a natural consequence of this thermodynamic imperative, and it is most probable that this association existed since the time of the first appearance of life on Earth (Michaelian, 2009b, Michaelian, 2011a). No longer should the biosphere be considered as a simple emergent system resulting from individual fights for survival, as Darwinian Theory proposes. Nor should the biosphere be considered as a greater organism, egoistically controlling its environment to increase its own stability, or that of its living components, as Gaian Theory proposes.

This thermodynamic view has the advantage of avoiding tautology in evolutionary theory by identifying a physical and universal foundation for evolution; that of increasing the global entropy production. It allows for selection on all hierarchal biotic levels, and even on coupled biotic-abiotic levels. The solution to the problem of "the evolution of a system of population one" becomes trivial under this framework. There is also an indication, remaining to be investigated further, that this framework could explain stasis and the punctuation of stasis in biological evolution (Gould, 2002) on the basis of thermodynamic stationary states and out-of- equilibrium phase changes, respectively. Finally, this thermodynamic framework provides a plausible explanation for the origin of life on Earth by tying RNA and DNA reproduction, and life's subsequent dispersal into practically all environments, to greater global entropy production.

7. Acknowledgements

The author is grateful for the financial support of DGAPA-UNAM, grant number IN112809.

8. References

Cnossen, I., Sanz-Forcada, J., Favata, F., Witasse, O., Zegers, T., and Arnold, N. F.: The habitat of early life: Solar X-ray and UV radiation at Earth's surface 4–3.5 billion years ago, J. Geophys. Res., 112, E02008, doi:10.1029/2006JE002784, 2007.

Darwin, F., ed. 1887. "The life and letters of Charles Darwin". London: John Murray. Volume 3. p. 18

Gatica Martínez, J., "Contribution of Cyanobacteria to Earth's Entropy Production", Bachelors thesis, UNAM, 2011.

Gould, S. J., "The Structure of Evolutionary Theory", Cambridge, Havard University Press, 2002.

Lloyd, S. and Pagels, H. M.: Complexity as thermodynamic depth, Ann. Phys.. 188, 186–213, 1988.

Lovelock J. E.: Gaia: Medicine for an ailing planet. (2nd ed.) Gaia Books, New York, 2005.

Margulis, L. and Sagan, D., "Microcosmos: Four Billion Years of Microbial Evolution", New York, Summit Books, 1986.

Michaelian, K., Biological catalysis of the hydrological cycle: life's thermodynamic function, Hydrol. Earth Syst. Sci. Discuss., 8, 1093-1123, www.hydrol-earth-syst-sci-discuss.net/8/1093/2011/ doi:10.5194/hessd-8-1093-2011, 2011b.

Michaelian, K.: Thermodynamic dissipation theory for the origin of life, Earth Syst. Dynam., 2, 37–51, 201. doi:10.5194/esd-2-37-2011, 2011a.

Michaelian, K.: Thermodynamic Function of Life. 2009a, (PDF) arXiv. http://arxiv.org/abs/0907.0040.

Michaelian, K.: Thermodynamic Origin of Life. 2009b, (PDF) arXiv. http://arxiv.org/abs/0907.0042.

Michaelian, K.: Thermodynamic stability of ecosystems. Journal of Theoretical Biology 237, 323-335, 2005.

Middleton, C. T., de la Harpe, K., Su, C., Law, Y. K., Crespo-Hernández, C. E., and Kohler, B.: DNA Excited – State dyanmics: from single bases to the double helix, Annu. Rev. Phys. Chem., 60, 217–239, 2009.

Morel, R. E., Fleck, G: Onsager's Principle: A Unifying Bio-theme. J. Theor. Biol., 136, 171–175, 1989.

Onsager, L.: Reciprocal Relations in Irreversible Processes. I., Phys. Rev., 37, 405–426, 1931.

Peixoto J.P., Oort A.H., de Almeida M., Tome A.: Entropy budget of the atmosphere. J. Geophys. Res. 96, 10981–10988, 1991.

Popper, K.: Natural Selection and the Emergence of Mind, Dialectica, 32, no. 3-4, 339-355, 1978.

Prigogine, I., Nicolis, G., and Babloyantz A.: Thermodynamics of evolution (I) Physics Today, 25, 23-28; Thermodynamics of evolution (II) Physics Today, 25, 38–44, 1972.

Prigogine, I.: Thermodynamics of Irreversible Processes, Wiley, New York, 1967.

Sagan, C. and Chyba, C.: The Early Faint Sun Paradox: Organic Shielding of Ultraviolet-Labile Greenhouse Gases, Science, 276, 1217–1221, 1997.

Swenson, R.: End-directed physics and evolutionary ordering: Obviating the problem of the population of one. In "The Cybernetics of Complex Systems: Self-Organization, Evolution, and Social Change", F. Geyer (ed.), 41-60. Salinas, CA: Intersystems Publications, 1991.

Part 2

Ecosystems and Resource Utilization

4

Yukon Taiga – Past, Present and Future

Rodney Arthur Savidge
University of New Brunswick
Canada

1. Introduction

"This land-mass, which I shall hereinafter call Beringia, must have been a good refugium for the biota during the glacial period." Hultén, 1937 (p. 34).

Summary: Primeval Yukon Beringia comprises subalpine taiga over mountainous landscape extending from the Alaska border south-eastward to Mount Nansen and northward to the Arctic Ocean. It is proposed that persistence of populations of *Picea glauca*, *P. mariana* and *Pinus contorta* on subarctic permafrost soils of this never-glaciated region has its explanation in *in situ* ancestries and physio-genetic constitutions, the exceptional morphological diversity and survival fitness tracing back to ancient stock and successive periods of climate change. Based on preliminary chloroplast DNA (cpDNA) evidence, Yukon Beringia's upland conifers were a source of post-glacial reinstatement of subarctic boreal conifers of northwestern North America.

1.1 Objectives

Early explorers observed that northwestern North America and northeastern Siberia lacked evidence, widespread elsewhere in North America and Eurasia, for having been modified by overriding ice sheets (Hayes, 1892; Kryshtofovich, 1935; Bostock, 1936). Hultén (1937) named this region Beringia and, aware of the concept of glacial refugia (Blytt, 1882; Warming, 1888), considered known circumpolar distributions of ~ 2000 arctic and subarctic plant species in relation to perceived postglacial reestablishment east and west of Beringia (Fig. 1). The primary objective here is to highlight the scientific value of populations of spruce trees (*Picea* spp.) in Canada's Yukon Beringia. Although supporting scientific data are wanting, there is nevertheless reason to doubt that interior boreal forest provenances if transplanted into subarctic taiga could display survival fitness comparable to that of Beringia's hardy taiga trees. Thus, major concerns in relation to North American spruce taiga reduce to ancestral lineage and physiology. Are Beringia trees the ancestors of North American taiga conifers? What is the physiological basis for their exceptional subarctic fitness? Answers may well reside in trees populating a never-glaciated portion of 'Primeval' Beringia spanning 62.5 °N to 64.5 °N along the boundary between Yukon Canada and Alaska USA, and extending southeastward to Mount Nansen (Fig. 1).

2. Current state of knowledge

2.1 Geological past of East Beringia

The continental plate containing North America and Eurasia had moved northward from the southern hemisphere to north of the equator ~ 330 million years ago (~ 330 Ma), and during the Mesozoic mountains of East Beringia's Yukon – Tanana Terrane had begun being uplifted (Mortensen, 1992). Subsequent volcanic extrusions, weathering and erosion over millions of years produced the non-glaciated mountainous landscape encountered at present within Yukon Primeval Beringia (Fig. 1). When the Atlantic Ocean began forming and distancing Eurasian and North American continents, ~ 170 Ma, they were approaching their present-day latitudes and shapes. Although Late Cretaceous (~ 100 Ma) brought global warming (Steuber et al., 2005), cold arctic winters followed for 30 million years (Davies et al., 2009).

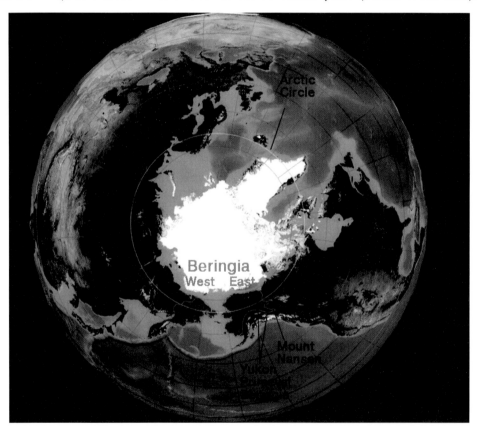

Fig. 1. The outlines of the bounds of West and East Beringia indicate regions where terrestrial refugia could have existed during the Last Glacial Maximum. Approximately 2.9 million years ago (~ 2.9 Ma), Mount Nansen near the southeastern tip of East Beringia stood at the edge of the first and most northwesterly advance of Cordilleran ice sheets toward Yukon Primeval Beringia, a never-glaciated landscape that probably held ancestors of postglacial ecosystem reinstatement eastward across North America.

Relatively warm conditions again prevailed in the Arctic from 70 to 40 Ma, and ~ 45 Ma the high Arctic supported lush temperate to tropical forests containing palms, flowering broadleaved trees and various conifers, including *Picea* spp. (Andrews et al., 1965; Hills et al., 1974; Francis, 1991; LePage, 2001; Jahren, 2007; Taylor et al., 2009). Those ancient forested ecosystems stood far north of the Arctic Circle where present-day tundra is entirely barren of trees. Isotopic data point to the existence during Middle Eocene of a global climate system conducive to warm temperatures and an Arctic Ocean that remained unfrozen even during its extended winter dark period (Jahren et al., 2009). It is probable that changes associated with the Earth's mantle and extraterrestrial cycles affecting axial tilt, precession, and eccentricity of the elliptical orbit of Earth were contributing factors (Milankovitch, 1941; Foulger, 2010).

Stockey and Wiebe (2008) described *Picea*-like fossil needles in Early Cretaceous sediments of Vancouver Island, Canada, evidence that spruce was evolving distantly south of the polar region much earlier than ~45 Ma, an earlier deduction based on Arctic fossil finds (Lepage, 2001). Based on chloroplast DNA (cpDNA) sequences, Ran *et al.* (2006) suggested that the world distribution of spruce reflects two dispersals to Asia via a Bering Strait land bridge during glacial periods; however, additional scientific data are needed to substantiate such a global hypothesis. Nothing is yet known of ecosystems which existed in the domain of Beringia when lush forests were growing in the high Arctic ~ 45 Ma, nor of how temperate trees subsequently acquired the fitness needed to survive contemporary subarctic conditions (Namroud et al., 2008). Those secrets may yet be discovered within Yukon's Primeval Beringia.

2.2 Glacial periods and Primeval Beringia within East Beringia

Early geologists recognized that Yukon glaciations involved several ice sheets (Hayes, 1892; Bostock 1936, 1966). In Table 1, three principal glacial periods of the Yukon are listed. Those three comprise eight ice-sheet advances, each of which moved northwest toward East Beringia beginning ~2.9 Ma (Duk-Rodkin, 1999; Barendregt & Duk-Rodkin, 2004).

Glacial Period	Glacial Maxima (Ma)	Geomagnetic Polarity CHRON	CHRON timespan (Ma)	Unglaciated terrain to the northwest from coordinates (°N, °W):
pre-Reid	2.90-0.78	GAUSS / MATUYAMA	3.58-2.58 2.58-0.78	(61.72, -138.10)
Reid	±0.200	BRUNHES	0.78-0.13	(61.55, -136.92)
McConnell	0.022	BRUNHES	0.13-0.00	(61.32, -136.63)

Table 1. Chronology of Late Cenozoic Cordilleran ice sheet advances toward East Beringia's southeast tip (after Duk-Rodkin, 1999; Barendregt & Duk-Rodkin, 2004).

East Beringia frequently has been considered in relation to landscape modifications produced at the Last Glacial Maximum (LGM) approximately 22,000 years ago (~ 22 ka). However, it is apparent in Table 1 that the bounds of unglaciated East Beringia depend on the timeframe of interest. Occupying a relatively small area of East Beringia on the east side of the Alaska-Yukon boundary is never-glaciated Yukon 'Primeval' Beringia, a mountainous region including Nisling Terrane, Yukon-Tanana Uplands and lands northward along the Alaska boundary to the Firth River and Beaufort Sea (Fig. 1, Table 1). Hayes (1892) was first to note the non-glaciated character of this region. The farthest glacial advances across Nisling Terrane into the Yukon-Tanana Uplands were subsequently found by Bostock (1936) to have terminated in the vicinity of Mount Nansen (62.1064 °N, 137.3031 °W), between the southeastern tip of Primeval Beringia to the northwest and Reid/McConnell terrain to the south and east (Fig. 1; see also Fig. 5). Those advances were referred to by Bostock (1966) as 'Nansen – Klaza' glaciations, and he associated them with the earliest northwestern advance of a Cordilleran ice sheet. That first Cordilleran ice sheet and associated mountain glacial events occurred ~ 2.9 Ma and became known as 'pre-Reid' glaciations (Duk-Rodkin et al., 2004). The pre-Reid glacial period spans two million years (Table 1), and the number and extents of glaciations within its time range remain to be better characterized (Bostock, 1966; Lebarge, 1995; Duk-Rodkin, 1999; Duk-Rodkin et al., 2004).

There is no obvious surficial evidence for Primeval Beringia having ever been altered by overriding glaciers. However, it cannot be concluded that mountain valleys were never affected by major snow accumulations or localized piedmont or valley bottom glaciers. Beringia in general experiences winter and summer extremes. Snag (62.383333 °N, 140.366667 °W) within Yukon lowland Beringia experienced the lowest recorded temperature (-63 °C) in North America (Wahl et al., 1987), and Oymyakon (63.450765 °N, 142.803188 °W) not far west of West Beringia experienced the coldest recorded temperature (-71 °C) in the northern hemisphere (Takahashi et al., 2011). East Beringia is largely semi-arid, shielded from heavy Pacific Ocean precipitation by high mountain ranges in southwestern Yukon. Ongoing uplift of those mountains is probably the principal explanation for progressive reductions in the volume of precipitation hence for concomitant decreases in northwestward advances of Cordilleran ice sheets during successive glacial periods (Armentrout, 1983). In addition, dry arctic winds flowing over East Beringia encourage sublimation, and summers bring long days and temperatures on south-facing slopes reaching to 40 °C. Thus, although glaciers stood in valley bottoms near Mount Nansen as recently as 1970 (R. A. Savidge, unpublished data), it is doubtful that high altitude slopes and mountain tops remained perennially buried beneath snow and ice even during the most severe pre-Reid glacial period.

Angular and only partially rounded cobblestones occur in mineral excavations made in valley bottoms near Mount Nansen (Bostock, 1936; Lebarge, 1995). High altitude cirques exist particularly on northern aspects of some Yukon-Tanana mountains and, evidently, piedmont lobes advanced down slopes and along subalpine valleys (Duk-Rodkin, 1999). The elevations achieved by valley glaciers remain conspicuously marked by tors at > 1400 m a.s.l. (e.g., see Fig. 7C). The pre-Reid ended ~ 0.78 Ma, and an extended interglacial period followed (Duk-Rodkin et al., 2004). Reid glaciations peaked ~ 200 ka; and most recently McConnell glaciations of the Wisconsinan followed (Table 1). Figure 2A indicates the extent of North American Wisconsinan ice sheet coverage 21.4 ka, approximating the LGM for the continent.

During the McConnell glacial period leading to the LGM, both Laurentide and Cordilleran ice sheets of northwestern Canada experienced climate-change episodes when rate of melting exceeded rate of growth, leading to multiple advances and retreats. Coalescence of the two ice sheets occurred in some places (Bednarski, 2008), but at other locations the two

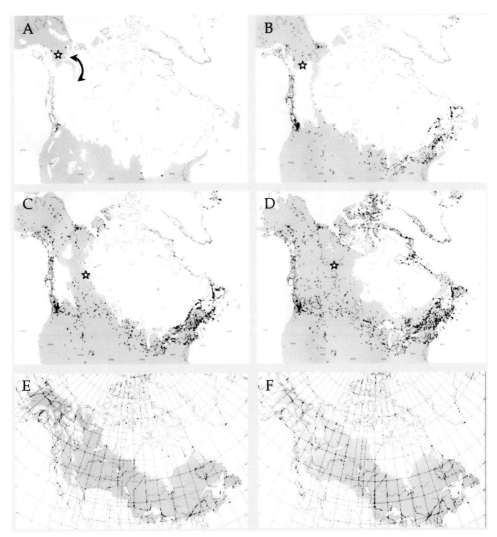

Fig. 2. A-D: Positions of North American ice sheets (white) at successive stages of Wisconsinan deglaciation (after Dyke, 2004). E, F: Native range maps (dark grey) of two boreal forest spruce species. A) 21.4 ka; the star indicates the Klotassin River realm in Yukon Primeval Beringia; the curved double-arrowheaded line indicates a hypothetical ice-free corridor between Cordilleran and Laurentide ice sheets, possibly a refugium for subalpine conifers derived from Beringia. B) 14.1 ka; terrain of the Carmacks region (star) had become available for taiga reestablishment. C) 12.70 ka; terrain of High Level, Alberta (star) had become available for repopulation. D) 8.45 ka; deglaciation was extensive throughout North America, but Yellowknife (star) remained beneath a glacial lake. E) White spruce (after Nienstadt & Zasada, 1990). F) Black spruce (after Viereck & Johnston, 1990). (Maps A, B, C, D © Department of Natural Resources Canada. All rights reserved.)

fronts evidently remained apart (Catto et al., 1996). Consequently, in addition to plausible existence of refugia within Yukon Primeval Beringia, in ice-free corridors and on mountainous nunataks, spots at various locations between Laurentide and Cordilleran ice sheets may have supported 'cryptic' refugia. Before an advancing ice sheet had covered a spot, if the opposing ice-sheet margin melted back to expose barren ground, populations could have shifted position. Thus, sites that today clearly testify to having been glaciated may at one time have sustained cryptic refugia of mobile organisms, and perhaps also of tundra plants and trees (Geml et al., 2006; Loehr et al., 2006; Stewart & Dalén, 2008).

Deglaciation maps based on data compiled by Dyke (2004) illustrate how post-Wisconsinan withdrawal of ice sheets proceeded gradually over thousands of years following the LGM (Figs. 2A-2D). Those data and knowledge of the ranges of conifers within the boreal forest (e.g., Figs. 2E, 2F) provide clues about where and when boreal forest of North America reinstated following the LGM. Barren Canadian landscape to the south of the receding continental ice sheets, in continental USA, undoubtedly began to form new soil for reforestation starting ~ 18 ka. Some subarctic locations remained beneath ice/water for thousands of years thereafter; for example, Yellowknife and the Hudson's Bay region are both part of Canada's present taiga, and both were still beneath ice until ~ 8 ka (Fig. 2D).

At the LGM, an 'ice-free' corridor is thought to have extended between Cordilleran and Laurentide ice sheets from East Beringia to the southern end of the Mackenzie Mountain Range (Fig. 2A). Mackenzie Mountain peaks evidently protruded as an extensive nunatak 'island chain' above the Laurentide ice sheet during the LGM (Szeicz and MacDonald, 2001). The Richardson Mountains of northern Yukon are a plausible alternative connection between Yukon Beringia and the northern terminus of the Mackenzie Mountains (Catto, 1996). Ongoing deglaciation extended that corridor southward along mountainous terrain, to where at ~ 14 ka it was about to merge with another ice-free corridor extending northward along the eastern slope of the Canadian Rocky Mountains (Fig. 2B). Considerable uncertainty attends the assumed LGM nature of the northern ice-free corridor; when and where it actually existed before 13 ka remain open questions (Catto, 1996; Levson & Rutter, 1996; MacDonald & McLeod, 1996; McLeod & MacDonald, 1997; Mandryk et al., 2001). However, the geological evidence indicates that by 12.7 ka a wide north-south ice-free corridor was in place (Fig. 2C). Thus, it can be assumed that generations of conifers had been advancing northward starting ~ 18 ka from seed sources in the south; conceivably, trees had advanced as far north as High Level, Alberta, by 12.7 ka (Fig. 2C). Based on native range maps (Figs. 2E, 2F), with ongoing deglacation a considerable area of boreal forest that presently is co-occupied by white spruce (*Picea glauca* (Moench) Voss) and black spruce (*P. mariana* (Mill.) BSP) had become available for repopulation by 8 ka.

The glaciation history associated with North American taiga and East Beringia remains to be well elucidated, but the preceding considerations raise the possibility that microbes, fungi and plants existing today in North American boreal forest may have very ancient ancestry extending to Yukon's Primeval Beringia. Predecessors of organisms within Primeval Beringia presumably experienced a temperate climate similar to that supporting mixed-wood forests 50 - 40 Ma, and subsequently the ensuing climate change that resulted in familiar frozen arctic environments. Their contribution to taiga remains uncertain, because East Beringia investigations into refugia have focused on accessible lowland Quaternary sites, and little is yet known about either extant organisms or paleobiology of Yukon

Primeval Beringia. Evidence presented below indicates that investigations in the uplands may well clarify many unresolved issues concerning ecosystems of the LGM, such as the plant species which remained available to herbivores (Bradshaw et al., 2003), and the persistent mycorrhizae which enabled plants to survive (Lydolph et al., 2005).

3. Tundra, taiga, and boreal forest

3.1 Tundra, taiga, forest-tundra ecotone and boreal forest nomenclature

'Tundra' refers to treeless arctic/subarctic terrains carpeted with ground vegetation. In common usage, 'taiga' is synonymous with boreal forest (Day, 2006); however, 'boreal forest' includes many distinct ecosystems covering much of northern North America and Eurasia as a circumpolar forested zone (Fig. 3). In other words, boreal forest is not a singular forest. The transitional zone between boreal forest's closed-canopy interior ecosystems and arctic tundra has been called various names (Hustich, 1953) and presently it is referred to as the 'forest-tundra ecotone' (Walker, 2010; Harper et al., 2011). However, 'taiga' has served as a succinct designation (Hustich, 1953; Elliott-Fisk, 2000). For brevity, taiga is hereinafter used to refer specifically to the northern transitional zone of the boreal forest.

3.2 Boreal forest ecosystems

Green shading in the inset of Figure 3 provides an indication of the current approximate position of global boreal forests. LiDAR data of Figure 3 indicate approximate bounds and tree heights of interior boreal forest. Those LiDAR data had 500 m x 500 m pixal resolution (Lefsky, 2010) and, therefore, small trees whether solitary or as small populations are not portrayed.

Boreal forest comprises a plurality of forest ecosystems covering much of northern North America and Eurasia as a circumpolar forested zone (Fig. 3). Thus, although numerous published investigations refer to 'boreal forest,' its heterogeneity makes it doubtful that research findings at any one site have general relevance to the entire circumpolar forest (e.g., see Hustich, 1953; Harper et al., 2011; Viktora et al., 2011). Certainly, every taiga site of pixal area in Figure 3 can be expected to have unique microsite characteristics.

The North American subarctic is a vast taiga domain of which only a very small fraction has been closely investigated. Spruce trees occur distantly north of the Arctic Circle in northwestern Canada and Alaska (Hustich, 1953; MacKay, 1958; Hansell et al., 1971; Cooper, 1986; Cody, 2000; Lloyd et al., 2005). The Firth River valley (mouth 69.5 °N, 139.5 °W) of East Beringia's Ivvavik National Park supports the most northerly spruce populations (Welsh and Rigby, 1971). The record altitude for taiga spruce evidently is held by a small lone specimen on a Yukon mountaintop near Mount Nansen at 1675 m (5495 feet) a.s.l. (Fig. 4D), far above the Yukon's variable spruce timberline of 1220 - 1370 m (4000 - 4500 feet – Bostock, 1936) and well above the "maximum" altitude of 1520 m recorded for interior boreal forest white spruce in the Canadian Rocky Mountains (Nienstadt & Zasada, 1990).

3.3 Evidence for refugia

Two questions of considerable importance in relation to understanding not only the ancestral foundation of boreal forest but also how best to manage it over the long term concern whether upland East Beringian conifers survived the LGM and, if so, whether their progeny propagated outward in all directions, as appears to have been possible (Figs. 1, 2). If East Beringia in fact was a conifer refugium, postglacial reinstatement of taiga may have taken place relatively rapidly by conifer populations already in possession of the physiological hardiness needed for survival in subarctic environments. Alternatively, the possibility exists that East Beringia conditions during the LGM were so stressful that conifers could not survive there.

Fig. 3. Global forests in the first decade of the 21st century, indicated by light detection and ranging (LiDAR) technology from the Ice, Cloud and Land Elevation Satellite Canopy heights are represented by different colours (reproduced courtesy of M.A. Lefsky – see Lefsky, 2010). Inset: North Pole centered view of Earth's boreal forest zone (dark green).

If conifers were entirely eliminated from subarctic taiga during the LGM, boreal forest today must have arisen by northward-only emigration of successive progeny starting from 'mother' trees that survived south of the Wisconsinan ice sheet (Fig. 2A). This northward-only hypothesis has found support in an absence of LGM conifer pollen in cores taken from small lakes in lowland Beringia of the Yukon (Cwynar & Ritchie, 1980; Cwynar 1982, 1988; Ritchie, 1984, 1987; Ritchie & MacDonald, 1986; Pisaric et al., 2001; Brubaker et al., 2005). Further, Hopkins (1982) concluded on the basis of fossil wood ages that white and black spruce and "tree birch" discovered in Alaska and Yukon Beringian lowland sediments evidently were all exterminated ~ 30 ka, trees not reappearing within East Beringia until deglaciation permitted dispersal of new populations northward. Zazula et al. (2006) found evidence for East Beringia spruce trees from 26.0 – 24.5 ka, but continuing existence through the LGM was considered doubtful, perhaps occurring in "rare valley-bottom habitats."

Most investigators of Yukon and Alaska East Beringia have concluded that only steppe vegetation existed following the LGM. MacDonald & McLeod (1996) proposed that boreal forest development was initiated by rapid northward spread of spruce trees starting from western mountains and eastern plains, North America's boreal forest achieving similar occupation to that in existence today ~ 8 ka.

It is not inconceivable that progeny of southern conifers through natural selection gradually acquired the needed fitness to survive extreme subarctic environments while progressively, generation after generation, reinstating denuded postglacial landscapes to the north (e.g., see Delcourt & Delcourt, 1987). On the other hand, comparison of the LGM glacial limit (Fig. 2A) with present-day conifer range maps (Figs. 2E, 2F) reveals a puzzling incongruity about northward-only taiga reinstatement. Excepting regions south of the Great Lakes, current southern limits of black spruce occur well north of latitude 49 °N (Figs. 2E, 2F), whereas at the LGM the ice sheet's southern margin was south of 49 °N (Fig. 2A). If black spruce were non-existent south of the ice sheet in western North America, within the northward-only scenario LGM ancestors of Yukon taiga's spruce trees must have stood south of the Great Lakes or farther east, and their lineage should therefore trace back to ancestors which resided far to the southeast. There is no evidence for such ancestry (see below, section 3.4).

Negative evidence against conifer survival in Yukon throughout the LGM embodies the unstated assumption that, if trees were non-existent in lowlands of East Beringia, they must also have been non-existent everywhere in mountainous Primeval Beringia. However, mountain climatology within Beringia can be markedly distinct from that of lowland valleys. Cold air is denser and moves down mountain sides to pool in valleys and on plateaus; thus, lowlands experience the colder winter temperatures (Takahashi et al., 2011; R. A. Savidge, unpublished data). Relatively strong mountain winds result in variable snow depths, ranging from bare ground to deep drifts and, in general, snow accumulates to greater depths in forested lowlands. Wind-blown mountain slopes accumulate only modest amounts of snow, and it disappears from their south slopes relatively rapidly in springtime.

South-facing slopes and mountain tops extending northwest from Mount Nansen over the western Yukon Plateau presently support white spruce as upright trees, krummholz and ground mats at high altitudes (e.g., see Fig. 8). Moreover, among pre-Reid tors near Mount Nansen, upright white spruce trees of > 5 m height bear seed cones (Figs. 4A, 4B, 4C, 7C, 7D). If trees and other plants survived on southern aspects of nunataks as well as within Yukon Primeval Beringia throughout glacial periods, in effect refugia of Primeval Beringia could have extended southeastward across Yukon-Tanana high altitude terrain as far as Mount Nansen, also northward as far as the Richardson Mountains (Fig. 1). However, no fossil pollen investigations have been done in the uplands of Yukon Primeval Beringia.

White spruce ground mats in upland pre-Reid landscape begin as normal upright trees, and they may bear both male and female cones. On south-facing slopes, such trees not uncommonly reach several meters in height before their trunks die (Figs. 4A, 4B, 4C, 7D). Main-stem dieback evidently occurs when sun-warmed needles lose moisture that cannot be replenished during winter and spring months (R.A. Savidge, unpublished observations). Branches near the ground nevertheless survive beneath winter snow, and in summer those branches extend plagiotropically. The mat specimen shown in Figure 4D, discovered in 2011, had survived for > 20 years on thin soil over permanently frozen rock on a never-glaciated mountaintop at 1675 m a.s.l.

It will be apparent that palynological investigations in lowland East Beringia cannot fully account for presence/absence of conifers during either the LGM or earlier glacial periods. If Beringian lowlands experienced deep accumulations of (non-glacial) snow and ice during any glacial period, perennially buried lowland trees could not be expected to survive continual light exclusion. Trees positioned on south slopes of the uplands could nevertheless have survived and produced seed throughout the LGM, as well as throughout earlier glacial periods, as 'microrefugia' (cf. Holderegger and Thiel-Egenter, 2009). There are no data on travel distances achievable by pollen and seeds as shed from small upright and ground mat trees within mountainous Primeval Beringia, but all sites so far investigated palynologically have been far removed (Williams et al., 2004).

Fig. 4. White spruce mats of Mount Nansen alpine terrain. A) 12 m² mat of plagiotropically growing branches. B) Male and C) female strobili found on the mat. D) A small white spruce mat, age estimated at 20 – 30 years, growing at 1675 m a.s.l.

3.4 DNA evidence for refugia

Genetic investigations into possible refugia supporting post-Wisconsinan reinstatement of subarctic populations of plants, animals and fungi from East Beringia, although increasing in number, remain small, and the findings are also controversial (Senjo et al., 1999; Goetcheus & Birks, 2001; Abbott and Brochmann, 2003; Brochmann et al., 2003; Cook et al., 2005; Edwards et al., 2005; Lydolph et al., 2005; Geml et al., 2006, 2010; Loehr et al. 2006; Politov et al., 2006; Zazula et al., 2006; Eidesen et al., 2007; Shilo et al., 2007, 2008; Waltari et al. 2007; Barnosky, 2008; Stewart and Dalén, 2008; Ickert-Bond et al., 2009; Levsen and Mort, 2009; Tsutsui et al. 2009; Carlsen et al., 2010; Gérardi et al., 2010; Nakonechnaya et al., 2010; Shafer et al., 2010; Stewart et al., 2010; Westergaard et al. 2010).

Chloroplast DNA (cpDNA) investigations provide insight into phylogenetic and biogeographic relatedness among populations of trees (Sigurgeirsson & Szmidt, 1993; Petit et al., 1997). et al., 1997). Based on pollen investigations, refugia of *Salix* and *Betula* were abundantly, *Picea* and *Alnus* spottily, *Populus* questionably and *Pinus* plausibly present in eastern Beringia of Alaska at 21 ka (Williams et al., 2004). In agreement with this, Anderson et al. (2006) concluded on the basis of cpDNA evidence that white spruce in East Beringia of Alaska probably did survive the LGM within favourable microhabitats (of unidentified location).

We investigated the hypothesis that if the mountainous southeastern portion of Yukon Primeval Beringia (site KL in Fig. 5) served as ancestral seed source for progressive reinstatement of postglacial black spruce taiga from west to east across Canada's boreal zone, then diminishing genetic relatedness with increasing geographic distance should be observable (R. A. Savidge, M. Viktora and O.P. Rajora, unpublished). Seven populations were sampled (Fig. 5), using one-year-old needles gathered from 60 well-separated trees within each population. None of the seven investigated sites had been planted with black spruce, nor had any served as a source of seeds for reforestation elsewhere. In each of the 420 samples, cpDNA was probed using three microsatellite markers (loci Pt26081, Pt 63718 and Pt71936, Vendramin et al., 1996) and methods as described (Viktora et al., 2011). Each marker was found to be polymorphic, yielding for all populations 12 alleles for Pt26081, 4 for Pt 63718, and 8 for Pt71936. Three-loci combinations were considered to determine haplotype frequencies at the seven sites (Fig. 5).

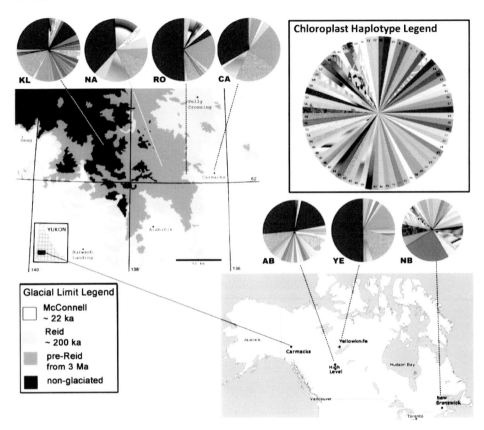

Fig. 5. Chloroplast DNA (cpDNA) haplotype frequencies within seven black spruce populations of Canada's boreal forest, based on three microsatellite markers. KL, Klotassin River; NA, Mount Nansen; RO, Rowlinson Creek; CA, Carmacks; YK, Yellowknife; AB, High Level, Alberta; NB, New Brunswick. Each colour within the chloroplast pie chart legend represents a distinct haplotype, and areas within the pie charts associated with each of the seven indicated sites are relative frequencies of detected haplotypes. The region near Carmacks (see lower right map) is enlarged at upper left to show CA, RO, NA and KL sites sampled across a transect leading from recently glaciated (McConnell) to non-glaciated terrain (see section 2.2 text). Unpublished data of R. A. Savidge, M. Viktora and O.P. Rajora.

Our study was unexpectedly compromised by the fact that most of the 60 sampled Mount Nansen (NA) trees were clonal, yielding eight genets and only eight additional genotypes rather than the expected 60 (Viktora et al., 2011). Each of the eight genets stood on what evidently was never-disturbed tundra and, consequently, each occupied an extensive area. Although data for the NA population indicated it to be distinct from neighbouring populations (Fig. 5), additional sampling of NA is needed before drawing firm conclusions.

In agreement with studies by Juan et al. (2004) and Gérardi et al. (2010), but in disagreement with MacDonald and McLeod (1996), the New Brunswick (NB) population of southeastern Canada was unrelated to black spruce populations in northwestern Canada (Fig. 5). Our cpDNA data unexpectedly revealed the AB population near High Level, Alberta, to be paternally related to the KL population, more than to Rowlinson Creek (RO) and Carmacks (CA) populations, despite RO and CA sites being geographically between KL and AB. Within the northward-only hypothesis for subarctic postglacial spruce taiga reinstatement, assuming that AB was the first populated of the sampled northwestern populations, CA, RO and NA locations were in geographic line for repopulation before KL, and the greater affinity between KL and AB seems to argue against northward-only reinstatement. Thus, data of Figure 5 constitute the first evidence that Yukon Primeval Beringia was a source for post-glacial taiga re-establishment of northwestern Canada's boreal black spruce forest.

As indicated in Figure 5, YE was more related to RO than to other sites, despite the fact that YE remained beneath a glacial lake for several thousand years after AB had been deglaciated (compare Figs. 2C and 2D). Thus, existence of the RO - YE relationship is an indication that RO progeny had earlier established in the Richardson or Mackenzie Mountains and thereafter survived the LGM as refugia with priority access to barren YE ground. At the LGM, widespread Cordilleran, Laurentide and local glaciers affected Richardson and/or Mackenzie Mountains at somewhat higher altitudes than those reached by lobes of the Mount Nansen region's piedmont glaciers, but the fact that spruce mats can grow at 1675 m a.s.l. (Fig. 4D) is evidence that subalpine/alpine south slopes of the Richardson and Mackenzie Mountains could also have contained nunatak refugia genetically related to the hardy conifers currently standing within Primeval Beringia.

Ongoing phylogenetic investigations in the mountains of Primeval Beringia are needed to answer the crucial still controversial question: Did conifers such as those represented by the KL population persist on Yukon Primeval Beringia soil during the LGM? With informed selection of sites, it can be predicted that the answer will eventually be found to be in the affirmative. If this proves to be the case, there will be incentive to probe even deeper into the past, eventually perhaps linking Primeval Beringia to high Arctic Eocene fossil conifers (Gugerli et al., 2005).

3.5 Phenotypic variation

Spruce trees are almost the only conifer species in the Mount Nansen region, where mature white spruce phenotypes are remarkably and conspicuously diversified (Figs. 6A, 6B), and similar variation also exists within black spruce. If crown form is in fact inherited rather than controlled by environment (Kärki & Tigerstedt, 1985), spruce manifests broad genetic diversity.

In addition to spruce trees, encountered rarely in upland pre-Reid landscape are scattered century-old (by annual ring counts – R. A. Savidge, unpublished data) pine trees similar to *Pinus contorta* Dougl. (Fig. 6C). Those high altitude pines stand on pre-Reid landscape and are geographically far removed from relatively young pine stands of McConnell-glaciated terrain to the east. Cones of upland pre-Reid pine trees are notably shorter than those of lowland lodgepole pine trees (Fig. 6D); however, both upland and lowland trees display three-needled as well as two-needled fascicles, a trait exclusive to Yukon lodgepole pine.

Although lowland pollen data indicated that lodgepole pine entered the Yukon from the south only recently (MacDonald & Cwynar, 1985; Cwynar & MacDonald, 1987), it remains possible that the actual ancestral source was Yukon Primeval Beringia.

Fig. 6. Conifers of pre-Reid landscape near Mount Nansen. A) Varied white spruce crown forms at 1200 m a.s.l.; B) White spruce downswept branch form at 1330 m a.s.l.; C) Upland pine at 1430 m a.s.l.; white spruce in background; D) Pine seed cones and needles.

4. Taiga population fitness and soils

4.1 Taiga overview

Figures 7A and 7B demonstrate that the transitional zone between taiga and tundra can be readily if coarsely distinguished. Taiga skirts the northern boundary of the interior boreal forests and has a circumpolar length exceeding 14,000 km (Figs. 1, 3). Taiga boundaries are presently moving in response to climate change (Harper et al., 2011), and the total areas occupied by both taiga and tundra are therefore uncertain.

Considered in longitudinal section, boreal forest everywhere displays four distinct conifer subzones which, in effect, constitute community-level phenotypes. Proceeding from north southward, Hustich (1953) described those four zones as A) limit of species, B) treeline, C) biological limit of forest, and D) economic limit of forest. Emphasizing ecophysiology, the four zones are seen as: 1) The northern zone where tundra transitions southward into widely spaced small individual conifers; 2) A zone comprising variably sized patches of forest within tundra, otherwise seen as islands or populations of conifers having 100% live crowns but nevertheless lacking closed-canopy state at maturity; 3) A broad zone of interior boreal forest ecosystems where broadleaved tree species commonly co-occur with mature closed-crown conifers, and where lower spreading branches of conifers die and abscise in response to light exclusion by the closed canopy; and 4) A southern transitional zone, where dominantly coniferous old growth closed-canopy ecosystems merge with temperate-zone mixed-wood or steppe ecosystems. Similarly, in alpine regions, a timberline to tundra transitional taiga zone is found high on mountain slopes, and at lower altitudes are more or less widely spaced full-crowned conifers (Figs. 7A, 7B). Subalpine slopes often grade at lower altitude into closed-canopy stands having a broadleaved tree component; however, taiga valleys rarely if ever display ecosystems akin to those of boreal forest southern latitude transitional regions. Valley bottoms in mountain taiga tend to be devoid of conifers, evidently a response to accumulation of heavier cold air and freezing of boggy soils. Willows nevertheless survive in such locations.

Depending on how taiga and tundra are distinguished – for example, based on arbitrary values for 'tree' height (commonly 3 m), between-tree spacing distance for a 'stand', and area for a 'forest' – taiga's northern boundary can be delineated on high resolution satellite images (Fig. 7A). However, ground proofing of hyperspectral images (nominally 0.6 m resolution, Fig. 7B) revealed that > 50% of actually present taiga conifers having heights ≤ 5 m were not detectable on such images, nor could broadleaved trees standing as solitary individuals in tundra be resolved. Clumps of closely spaced small trees were discernible on images, but within clumps many trees of similar size and form to those shown in Figure 7C could not be resolved. Solitary upland trees such as that shown in Figure 7D were generally overlooked (R. A. Savidge, unpublished data). It remains no minor concern, particularly in relation to understanding effects of climate change on taiga dynamics, that immense areas of tundra and taiga remain to be resolved at small-tree resolution using either airborne or satellite imagery. In other words, mapping the distributions of small conifers, whether young or old within taiga and tundra, continues to require on-the-ground surveys.

Fig. 7. Yukon taiga on the southeastern edge of Primeval Beringia. A) High resolution (0.6 m) QuickBird satellite overview, and B) close-up of white spruce (*Picea glauca*) treeline at approximately 1330 m a.s.l. The white arrow indicates a clone of *Picea mariana* within a white spruce population (cf. Viktora et al., 2011). C) White spruce at > 1400 m a.s.l. near a tor; D, Alpine white spruce at > 1500 m a.s.l. (satellite data provided by DigitalGlobe, Inc.).

In relation to defining the taiga treeline within North America, conifers because of their evergreen visibility have received the greater attention. However, willows (*Salix* spp.) and birches (*Betula* spp.) occur as shrubs and, arguably (depending on one's definition of 'tree'), also as trees at latitudes far north of conifer treelines, and also at taiga altitudes well above alpine timberline (Polunin, 1948; Porsild 1955, 1964; MacKay, 1958; Beschel & Webb, 1963; Kuc, 1974). Where conifers for whatever reason cannot establish independently in tundra, willows and birches aid in subsequent conifer establishment (Nilsson, 2005). Both willows and birches contribute organic matter, hence aiding topsoil development. Willows beyond conifer treeline may exceed several metres in height and as such, they give the impression of being the hardiest of all northern trees. Willows are engendered with high levels of salicylic acid, a compound conferring disease resistance and serving in internal heat generation (Raskin, 1992). Thus, in relation to physiological tolerance of subarctic and far-northern environments, it would seem to follow that where willows cannot survive, conifers cannot be expected to.

Tree species diversity within Yukon Primeval Beringia is greater than that within upland pre-Reid populations (Fig. 8). Primeval Beringia timberline at ~ 1300 m a.s.l. on southern aspects occurs as a mixture of spruce, poplars (*Populus tremuloides* and *P. balsamifera* ssp. *trichocarpa*), birches and willows (Fig. 8A). As is the case for pre-Reid terrain, white spruce dominates (Figs. 8B, 8C), but small-diameter black spruce trees occur as monotypic populations on muskeg flats. On well-drained valley soils, mixtures of *P. glauca* var. *albertiana*, *P. glauca* var. *porsildii*, poplars, willows and alders occur. Most spruce species hybridise where ranges overlap (Farjon, 1990), but a hybridisation barrier evidently exists between the *albertiana* and *porsildii* varieties of white spruce, as the two co-exist side by side in Yukon Primeval Beringia as distinct phenotypes.

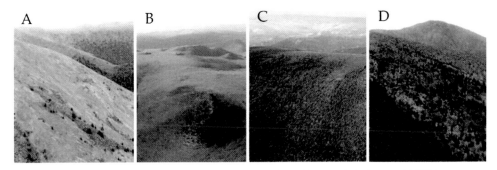

Fig. 8. Examples of September landscapes within Yukon Primeval Beringia. A) White spruce on patchy carpets of birch-willow vegetation on tallus slopes; B) Conifer populations as discrete aggregates scattered over the uplands; C) Lineated 'streams' of conifers aligned by slope solifluction; D) Mixed woods at high altitudes in the Klotassin River valley.

North-facing slopes of both Primeval Beringia and pre-Reid landscapes at 1200 m a.s.l. and higher altitudes are generally devoid of all but short-branched, weakly foliated, widely spaced, single-stemmed, small diameter columnar-form white spruce trees having heights commonly less than 5 m. Tree growth is influenced indirectly by, for example, the extent of activity possible within a cool root system (Wareing 1985, Junttila and Nilsen 1993), but it can also be a function of additional, microsite-dependent extrinsic factors (Germino et al. 2002). For example, growth may be limited if soil water potential is either too high or low, nutrient imbalance or deficiency arises, root-system integrity is altered by solifluction or frost heaving, allelopathic inhibition occurs, or soil microflora become imbalanced (Ivarson,

1965). Considerable microsite variation evidently is normal. For example, within- and between-tree annual increments of multi-centurion subalpine white spruce growing on a south-facing slope near Mount Nansen were poorly correlated chronologically (Yuan, 2012).

Since 1990, *Populus tremuloides* and *P. balsamifera* ssp. *trichocarpa* have invaded pre-Reid uplands from lower Reid and McConnell glaciated altitudes, and they now are found as tree-sized individuals in subalpine tundra above spruce timberline. Valleys in Yukon Primeval Beringia and relatively low altitude (<1200 m a.s.l.) terrain of Reid/McConnell glaciated landscapes also have abundant alders (*Alnus* spp.). However, in contrast to *Populus* movement, no alder of any species was found during either 1968-1971 or 2007 – 2010 investigations on pre-Reid uplands (R. A. Savidge, unpublished observations).

4.2 Population fitness

The populations of diverse cryptobiotic organisms which survive within regions such as Yukon Primeval Beringia manifest adaptation to processes of natural selection which evidently operated over very long time frames. Ecophysiological considerations related to the fitness of a tree species to survive in the subarctic include tree tolerance of the north's seasonally changing light extremes (spectrum, intensity, photoperiod), desiccation and winter abrasion by wind-blown ice particles, low water potentials particularly in relation to evapo-transpirational demands and albedo/sunscald-related responses during periods when soil or trunk xylem water is frozen, deleterious effects on the root system (e.g., low root-zone temperature, low pH, anaerobiosis, limiting nutrient(s) microflora and/or fauna, allelopathy), herbivory, and insect and disease challenges (Larcher, 1995).

Various explanations have been advanced to explain positions of treelines, but consensus for any has yet to be achieved. Air temperature hypotheses have long been a favoured avenue to explain climatic tolerance (Weiser, 1970; Coursolle et al., 1998; Arft et al., 1999; Körner, 2007; Brandt 2009). However, low winter temperature *per se* is an improbable factor, because a variety of tree species of good size occur at sites such as Snag, Yukon, and Oymyakon, Siberia, locations of record temperature minima in the northern hemisphere. There nevertheless is preponderant agreement that treeline position somehow correlates with growing-season air temperature. Dwarf shrubs, including krummholz conifers, supersede upright trees wherever mean air temperature of the warmest month is less than about 10 °C. Expressed another way, treeline usually occurs below that altitude or latitude where mean temperature of the growing season does not exceed 7 °C (Körner, 2007). As noted earlier, however, a focus on tree height in relation to the treeline and fitness overlooks advantages conferred to morphotypes that survive near the ground.

Larcher (1995) noted for arctic/subarctic plants that photosynthesis and respiration both continue at temperatures a few degrees below the freezing point. At lower temperatures, photosynthetic activity ceases whereas respiration continues. Compounding this, northern plants have relatively high respiration rates compared to plants of southern climes (Larcher, 1995). In addition, little if any photosynthesis occurs throughout long dark taiga winters, and for tree survival it is therefore essential that sufficient storage reserves are present by the end of each summer's growing season (Lloyd et al., 2002; Röser et al., 2002; Zarter et al., 2006). Following their long winter, subarctic conifers are further drained of vital stored chemical energy due to their intrinsic tendency to commence root and cambium growth in late spring before buds have broken and new needles matured (Savidge & Förster, 1998). Standing dead trees (i.e., snags) are common in taiga and, in the absence of disease or obvious physical damage, it appears that the taiga-tundra treeline manifests that point where loss of metabolic reserves exceeds a tree's ability to support its respiratory needs. In other words, retention of 100% live crown by taiga trees probably is not merely incidental to their wide spacing, rather an essential survival attribute in order for the annual photosynthesis to respiration ratio of the whole tree to be maintained near or above unity.

Based on extensive researches done in more southern realms and limited investigations in the north, root-system survival during winter must surely be a critical factor for every subarctic tree. Root systems are more susceptible than above-ground organs to freezing-induced mortality (Tyron and Chapin, 1983; Wareing, 1985; Delucia, 1986; Husted and Lavender, 1989; Lopushinsky and Max, 1990; Camm and Harper, 1991; Junttila and Nilsen, 1993; Pregitzer et al., 2000). Not only snow cover but also the nature and depth of soil influence its ability to insulate the root system from extremely low air temperatures commonly experienced in the taiga zone during winter (see Fig. 10). The limits of root system tolerance of low temperatures by taiga trees remain poorly investigated but, as addressed further below, white and black spruce and lodgepole pine can survive and grow on permafrost despite always cold root systems.

Fitness to survive in the north undoubtedly involves many interacting environment and genetic factors, but physiological understanding remains incomplete. Considerable progress was made during the period 1950 - 2000 toward understanding how plants tolerate low temperatures and other stresses (Scholander et al., 1953; Billings & Mooney, 1968; Raskin, 1992; Larcher, 1995; Arft et al., 1999; Kimball et al., 2007). However, only a minor fraction of the investigations involved controlled experimentation with *in situ* taiga/tundra plants (Arft et al., 1999). Moreover, there exists a major gap between the impressive ability of molecular biology technologies to generate huge amounts of DNA data and science's quite limited ability to interpret those data such that they are relevant to ecophysiology. Phenomena of seed production, seed dormancy, seedling establishment, vegetative reproduction, photosynthesis, respiration, drought resistance, plant dormancy, storage reserve accumulation and utilization, growth, partitioning between shoot, root and reproductive systems, soil nutrient roles, tissue cold hardiness, flowering and pollination all remain fascinatingly curious and quite incompletely understood physio-genetic phenomena of taiga and tundra ecosystems.

Taiga is heterogeneous, but by definition its central unifying theme is that trees do survive and complete their lifecycles within it. Nature, in the absence of human intervention, has maintained healthy sustainable taiga forests by means of two mechanisms, intrinsic physiological fitness and genetic diversity. Physiological fitness is an individual attribute maintained through ongoing natural selection and propagated within both sexually reproductive and clonal populations. Combinations of physiological fitness and genetic diversity ensure that trees (and all forest organisms) are able to survive the ranges of interacting environmental factors they are exposed to throughout lifecycles in taiga. Again, however, it is still entirely unclear which intrinsic factors contribute to physiological fitness within taiga. Intrinsic tolerance and adaptability unique to each species and each individual plant are, in general, genetic/epigenetic attributes (Wareing 1985; Durzan 1993).

4.3 Taiga soils

Boreal forest presently holds more carbon per unit area than any other terrestrial area (Shiklomanov and Rodda, 2003; Carlson et al., 2009). Although carbon is abundantly present in tissues of trees and ground vegetation, most boreal forest carbon occurs as subsurface dead organic matter in cold peatlands, and in taiga and tundra soils.

Tundra and taiga vegetation exist as communities in precariously balanced struggles of survival, and their fitness to tolerate cold soils is a distinguishing attribute of all subarctic plants. Figure 8A provides an example of the obligatory relationship between taiga soil formation and ensuing establishment of conifers. In order for the sequence of tundra – taiga – forest succession to occur on barren rock, accumulation of organic matter in support of soil formation is essential. The process undoubtedly requires many decades, probably centuries, beginning with bedrock weathering, disintegration, and coating of rock fragments with bacteria, lichens and mosses. Hardy grasses, sedges, *Salix*, *Betula*, *Dryas* and other plant species occur in the high Arctic where no conifers are to be found (Polunin 1948, Beschel and Webb 1963, Porsild 1964, Kuc 1974), and there can be little doubt that establishment of such plants serves as preliminaries to the overall process of soil formation.

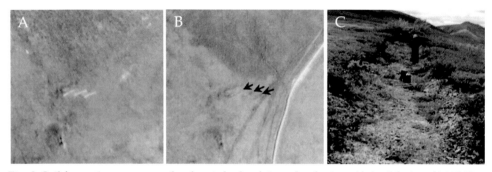

Fig. 9. Soil formation on exposed volcanic bedrock in upland taiga. A) Aerial view (A11069-92, National Air Photo Library - Natural Resources Canada) of three bulldozer trenches made at 62.0733 °N, 137.1666 °W prior to August 24, 1947; B) Same location on August 10, 2003 (A28501-106, National Air Photo Library - Natural Resources Canada), with positions of the three trenches arrowed. C) Ground inspection of one of the three trenches in 2008, showing patchy soil and incomplete establishment of ground vegetation.

Figure 9 provides direct evidence of the lengthy duration required for organic soil formation to occur on nutrient-rich volcanic rock fragments following disturbance of subalpine taiga ground vegetation. Three parallel shallow trenches were bulldozed some time before August 1947 at an altitude of 1400 m a.s.l., in a pre-Reid upland never-glaciated region near Mount Nansen (Fig. 9A). The excavations removed a pre-existing thin carpet of tundra vegetation and exposed frozen, weathered coarsely fragmented bedrock beneath. Analysis of air photographs taken in 2003 indicated that tundra healing was occurring (Fig. 9B). Field investigations around the perimeters of the trenches during 2008 revealed a remarkable diversity of densely packed broadleaved woody shrubs, vascular and non-vascular ground plants, fungi, and lichens. However, within the trenches only isolated patches of thin (< 1 cm) organic matter were found, and only a small fraction of the many tundra plant species

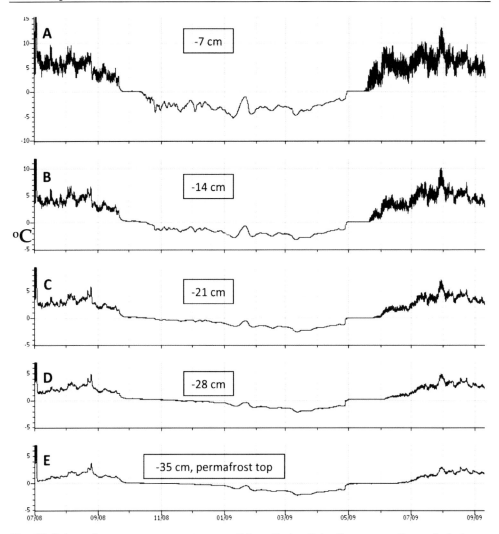

Fig. 10. Sub-surface temperatures at sequential vertical soil depths on a gentle south-facing slope of subalpine taiga (1320 m a.s.l.) underlain by permafrost and supporting a stand of mature white spruce trees near Mount Nansen, Yukon. Data loggers were placed at the gametophytic stem bases of live mosses (-7 cm), at the upper surface of the humic horizon (-14 cm), and progressively deeper into the humic horizon to where it was solidly frozen (- 35 cm). The root zone consisted mainly of compacted brown humus at various stages of decay. Macro spruce tree roots nearby the excavation made to insert the data loggers existed only near the ground surface, probably because permafrost until recent decades extended to the ground surface (R. A. Savidge, unpublished data). Fine roots were most abundant in humic matter between 0 – 20 cm depths, but some were also present immediately above the – 35 cm permafrost table (Fig. 10). Light data (not shown) served to verify that the data loggers had remained undisturbed throughout the monitoring period.

surrounding the trenches had established, covering < 50% of bedrock area within the trenches (Fig. 9C). Clearly, soil formation and plant establishment on barren rock fragments in a subarctic environment is a lengthy process even when a plethora of plant species and other soil-forming agents are already present nearby.

Closed-canopy interior boreal forest trees experience cold soil mainly during winter, whereas permafrost soils of tundra and taiga are perennially cold (Fig. 10). To gather the data of Figure 10, Pendant Hobo light/temperature data loggers (Onset Corp.) spaced 7 cm apart were inserted vertically into ground beneath a stand of mature subalpine white spruce of pre-Reid taiga landscape near Mount Nansen, logging ground temperatures every 20 minutes from early July 2008 to early September 2009. On the basis that soil water freezes near -3 °C, data of Figure 10 demonstrate that when the humus layer in taiga is sufficiently deep, water in liquid state is always available in the active layer over permafrost, not only in support of transpiration and photosynthesis during the growing season (Sugimoto et al., 2002), but even during winter periods. Taiga tree root systems evidently tolerate growing season soil temperatures that, farther south, would jeopardize tree survival (Stattin & Lindström, 2000).

Permafrost interfaces with an unfrozen active layer, but taiga tree roots in cold humic matter have little or no direct access to mineral soil nutrients. A similar deficiency undoubtedly also applies to plants growing on debris-covered glaciers (Fickert et al., 2007). Thus, fungal and bacterial associations are of great importance in determining nutrient uptake within taiga (Ivarson, 1965; Read et al., 2004). As noted above, scattered white spruce trees of unusual cylindrical phenotype grow on cold north slopes in pre-Reid and Primeval Beringia taiga. The trees grow slowly and produce only short branches. This may be a cold soil or limiting nutrient response; the precise physiological explanation remains wanting.

5. The future subarctic forest

If the present warming trend continues, much of what for centuries has been tundra will, in not very distant future, convert into taiga followed by mixed conifer – broadleaved forests. Similar climate change has happened before (Cwynar 1982, 1988; Burn, 1994). However, exploration and other human activities constitute unknown environmental impacts which introduce uncertainty into any prediction of what the future may hold, particularly in relation to the ability of taiga to retain the germplasm needed to survive through the next glacial period, an event that will undoubtedly occur sooner or later (Crucifix 2011).

In association with climate change and permafrost thawing, southern species are becoming more common and invasive in the north (Hickling et al., 2006). Boreal forest holds abundant fresh water in support of plant growth (Shiklomanov and Rodda, 2003) and, as refugial ecosystems within Yukon Primeval Beringia become increasingly disrupted by man, it can be predicted that those ecosystems will also be invaded by new species. However, landscape warming is likely to be the major catalyst for displacement of hardy indigenous populations from Beringia. Because there is no historical basis for understanding the impact on taiga populations of anthropogenic modifications in association with climate change, the future integrity of the taiga wilderness cannot be assumed to be sustainable. Until more has been learned, the wisest option at present would be to manage locations such as Yukon Primeval Beringia as wilderness preservation sites. A related question concerns whether hardy northern germplasm can be conserved over the long term, for example in seed banks or as cryo-preserved micropropagules.

Testing progeny of a population in regions different from the one that the population is indigenous to is known as provenance testing (Hagman, 1993). Over the last two centuries, economic forestry practices in closed-crown boreal forest ecosystems have disrupted natural population structures, shuffling provenances and exerting still-uncertain effects on both physiological fitness and genetic diversity in relation to trees' abilities to tolerate environmental change. Northern taiga by virtue of its remoteness, harsh climate and small incentive for economic forestry exploitation has so far been largely exempted from those manipulations, and that inadvertent preservation may be the primary hope for reinstatement of boreal forest ecosystems following a future glacial period.

Provenance trials in relation to taiga remain a novelty (Walker, 2010). A noteworthy trial involved planting white spruce seeds and seedlings of a central Alaskan provenance at several locations north of treeline in northern Alaska, in a region where only tundra vegetation occurs. None of the white spruce seeds produced seedlings, and only two of 100 seedlings planted in calendar year 1968 were still alive when observed in 2001 (Wilmking and Ibendorf, 2004). Very little is yet understood about the physiology of taiga tree establishment (Walker, 2010).

6. Conclusions

Taiga is a highly complex transitional zone covering an immense circumpolar expanse and distinguished by exceptionally hardy widely spaced conifers that occupy both lowland and mountainous regions between the northern hemisphere's boreal forest and tundra. Although a major reservoir of sequestered carbon, fresh water and biodiverse ecosystems comprising species physiologically adapted to tolerate northern conditions, taiga remains an altogether mysterious realm holding secrets about past climate change and how its organisms acquired fitness to tolerate those changes. Within the taiga zone is Beringia landscape shielded from the Wisconsinan ice sheet advance, and deeper within East Beringia is never-glaciated Yukon Primeval Beringia. The latter is a relatively small area of mountainous taiga that may well hold the key to linking polar forests that existed in the high Arctic 45 million years ago with those of the present day, and to aiding in future management of the biosphere by enabling increased understanding of both the ancestry and the physio-genetic basis for survival of trees, and life in general, within the subarctic. Although its never glaciated character is potentially of immense scientific value, Yukon Primeval Beringia presently is an entirely virgin region in relation to scientific investigations aimed at deciphering the ancestry of North American boreal forests. Because the physiological basis for population fitness of both taiga and interior boreal forest trees remains unknown, it is difficult to estimate how important conservation of Yukon Primeval Beringia ecosystems could become when Earth approaches the next glacial period. Presently, taiga permafrost is thawing in response to rapid warming, and should this trend continue, it is expected that major modification of the taiga zone will occur over coming decades.

7. Acknowledgements

Data concerning Mount Nansen and Primeval Beringia were gathered during International Polar Year 2007-2008 under Government of Canada Program for International Polar Year support of PPS Arctic Canada, part of the PPS Arctic international research team (ppsarctic.nina.no). Yukon field assistance to the author was provided by G.J. Muff, H. Foerster, P. Foerster, X. Yuan, M. Hancox and A. E. Savidge. Yukon accommodation to facilitate performance of field research was provided by G.J. & B. Muff and B. & G. Wheeler. C. Braden, R. Braden and A. J. Savidge contributed time and expertise to Yellowknife field sampling. Black spruce populations from High Level, Alberta and New Brunswick were sampled by O.P. Rajora and M. Viktora, respectively. Chloroplast haplotype data (Figure 5) were produced by M. Viktora supported by Natural Sciences and Engineering Research Council of Canada Discovery grant, New Brunswick Innovation Foundation Research Assistant Initiative funds and Canada Research Chair (CRC) Program funds to O. P. Rajora. Air North supported travel. Digital Globe provided QuickBird satellite imagery. A. Duk-Rodkin kindly shared her vast knowledge about northwestern Canada's glacial periods.

8. References

Abbott, R. J. & Brochmann, C. (2003). History and evolution of the arctic flora: in the footsteps of Eric Hultén. *Molecular Ecology* 12: 299-313.

Anderson, L.L., Hu, F. S., Nelson, D. M., Petit, R. J. & Paige, K.N. (2006). Ice-age endurance: DNA evidence of a white spruce refugium in Alaska. *Proceedings of the National Academy of Sciences of the USA* 103: 12447-12450.

Armentrout, J.M. (1983). Glacial lithofacies of the Neogene Yakataga Formation, Robinson Mountains, southern Alaska Coast Range, Alaska. In: *Glacial-Marine Sedimentation.* B. F. Molnia, (ed.), pp 629-665, Plenum Press, New York.

Andrews, H.N., Phillips, T.L, & Radforth, N.W. (1965). Paleobotanical studies in Arctic Canada. 1. *Archaeopteris* from Ellesmere Island. *Canadian Journal of Botany* 43:545-556.

Arft, A.M., Walker, M.D., Gurevitch, J., Alatalo, J.M., Bret-Harte, M.S., Dale, M., Diemer, M., Gugerli, F., Henry, G.H.R., Jones, M.H., Hollister, R.D. Jónsdóttir, I.S., Laine, K. Lévesque, E., Marion, G.M., Molau, U., Mølgaard, P. Nordenhäll, U., Raszhivin, V., Robinson, C.H., Starr, G., Stenström, A., Stenström, M., Totland, Ø. Turner, P.L., Walker, L.J., Webber, P.J., Welker, J.M. & Wookey, P.A. (1999). Response of tundra plants to experimental warming: meta-analysis of the International Tundra Experiment. *Ecological Monographs* 69:491-511.

Barendregt, R. W. & Duk-Rodkin, A. (2004). Chronology and extent of Late Cenozoic ice sheets in North America: A magnetostratigraphic assessment. In *Quaternary Glaciations – Extent and Chronology, Part II.* J. Ehlers and P.L. Gibbard, (eds.), pp 1-7, Elsevier.

Barnosky, A.D. (2008). Climatic change, refugia, and biodiversity: where do we go from here? An editorial comment. *Climatic Change* 86:29-32.

Bednarski, J. M. (2008). Landform assemblages produced by the Laurrentide Ice Sheet in northeastern British Columbia and adjacent Northwest Territories – constraints on glacial lakes and patterns of ice retreat. *Canadian Journal of Earth Sciences* 45:593-610.

Beschel, R.E. & Webb, D. (1963). Growth ring studies on arctic willows, In: *Prelimary Report 1961-1962, Axel Heiberg Island Research Report*s, F. Müller, (ed.), pp 189-198, McGill U., Montreal.

Billings, W.D. & Mooney, H.A. (1968). The ecology of Arctic and alpine plants. *Biological Reviews* 43: 481 – 529.

Blytt, A. (1882). Die Theorie der wechselnden kontinentalen und insularen Klimate. *Botanische Jahrbücher* 2: 1–50.

Bostock, H. S. (1966). Notes on glaciation in central Yukon Territory. *Geological Survey of Canada Paper* 65-56, 18 p.

Bostock, H.S. (1936). *Carmacks District, Yukon.* Geological Survey Canada, Memoir 189, 67p.

Bradshaw, R.H.W., Hannon, G.E. & Lister, A.M. (2003) A long-term perspective on ungulate-vegetation interactions. *Forest Ecology and Management* 181: 267-280.

Brandt, J.P. (2009). The extent of the North American boreal zone. *Environmental Reviews* 17: 101-161.

Brochmann, C., Gabrielsen, T.M., Nordal, I., Landvik, J. Y. & Elven, R. (2003). Glacial survival or tabula rasa? The history of North Atlantic biota revisited. *Taxon* 52: 417-450.

Brubaker, L. B., Anderson, P. M., Edwards, M. E. and Lozhkin, A. V. (2005). Beringia as a glacial refugium for boreal trees and shrubs: new perspectives from mapped pollen data. *Journal of Biogeography* 32: 833-614 848.

Burn, C.R. (1994). Permafrost. tectonics, and past and future regional climate change. Yukon and adjacent Northwest Territories. *Canadian Journal of Earth Science* 31: 182-191.

Camm, E.L. & Harper, G.J. (1991). Temporal variations in cold sensitivity of root growth in cold-stored white spruce seedlings. *Tree Physiology* 9:425-431.

Carlsen, T., Reidar, E. & Brochmann, C. (2010). The evolutionary history of Beringian *Smelowskia* (Brassicaceae) inferred from combined microsatellite and DNA sequence data. *Taxon* 59:427-438.

Carlson, M. Wells, J. and Roberts, D. (2009). *The Carbon the World Forgot: Conserving the Capacity of Canada's Boreal Forest Region to Mitigate and Adapt to Climate Change.* Boreal Songbird Initiative and Canadian Boreal Initiative, Seattle, WA and Ottawa, 33 pp.

Catto, N. R. (1996). Richardson Mountains, Yukon-Northwest Territories: the northern portal of the postulated 'ice-free corrrdior'. *Quaternary International* 32: 3-19.

Catto, N.R., Liverman, D.G.E., Bobrowsky, P.T. & Rutter, N. (1996). Laurentide, Cordilleran, and montane glaciations in the western Peace River – Grande Praisie region, Alberta and British Columbia, Canada. *Quaternary International* 32: 21-32.

Cody, W. J. (2000). *Flora of the Yukon Territory, 2nd ed*. NRC Research Press, Ottawa.

Cook, J. A., Hoberg, E.P., Koehler, A., Henttonen, H., Wickström, L., Haukisalmi, V., Galbreath, K., Chernyavski, F., Dokuchaev, N., Lahzuhtkin, A., MacDonald, S.O., Hope, A., Waltari, E., Runck, A., Veitch, A., Popko, R., Jenkins, E., Kutz, S. & Eckerlin, R. (2005). Beringia: Intercontinental exchange and diversification of high latitude mammals and their parasites during the Pliocene and Quaternary. *Mammal Study* 30: 533-544.

Cooper, D. J. (1986). White spruce above and beyond treeline in the Arrrigetch Peaks region, Brooks Range, Alaska. *Arctic* 39: 247-252.

Coursolle, C., Bigras, F.J. & Margolis, H.A. (1998). Frost tolerance and hardening capacity during the germination and early developmental stages of four white spruce (*Picea glauca*) provenances. *Canadian Journal of Botany* 76: 122-129.

Crucifix, M. (2011). How can a glacial inception be predicted? *Holocene* 21:831-842.

Cwynar, L.C. (1982). A Late-Quaternary vegetation history from Hanging Lake, northern Yukon. *Ecolological Monographs* 52: 1-24. 622

Cwynar, L.C. (1988). Late Quaternary vegetation history of Kettlehole Pond, southwestern Yukon. *Canadian Journal of Earth Science* 18: 1270- 1279.

Cwynar, L. C. & MacDonald, G. M. (1987). Geographical variation of lodgepole pine in relation to population history. *American Naturalist* 129:463-469.

Cwynar, L.C. & Ritchie, J.C. (1980). Arctic steppe-tundra: a Yukon perspective. *Science* 208: 1375–1377.

Davies, A., Kemp, A. S. & Pike, J. (2009). Late Cretaceous seasonal ocean variability from the Arctic. *Nature* 460: 254-258.

Day, T. (2006). *Taiga*. Chelsea House, New York, ISBN-10:0-8160-5329-4.

Delcourt, P. A. & Delcourt, H.R. (1987). *Long-term Forest Dynamics of the Temperate Zone. A Case Study of Late-Quaternary Forest in Eastern North America*. Ecological Studies 63, Springer-Verlag, ISBN 3-540-96495-9. 439 p.

Delucia, E.H. (1986). Effect of low root temperature on net photosynthesis, stomatal conductance and carbohydrate concentration in Engelmann spruce (*Picea engelmannii* Parry ex Engelm.) seedlings. *Tree Physiology* 2:143-154.

Duk-Rodkin, A. (1999). *Glacial limits map of Yukon Territory*. Geological Survey of Canada, Open File 3694, Indian and Northern Affairs Canada Geoscience Map 1999-2, scale 1:1000000.

Duk-Rodkin, A., Barendregt, R.W., Froese, D.G., Weber, F., Enkin, R., Smith, I.R., Zazula, G.D., Waters, P. & Klassen, R. (2004). Timing and extent of Plio-Pleistocene glaciations in north-western Canada and east-central Alaska. In *Quaternary Glaciations – Extent and Chronology, Part II*. J. Ehlers and P.L. Gibbard, (eds.), pp 313-335, Elsevier.

Durzan, D. J. (1993). Molecular bases for adaptation of coniferous trees to cold climates. In: *Forest Development in Cold Climates*. Alden, J., Mastrantonio, J. L., Odum, S. (eds.), pp 15-42, Plenum Press, New York and London, ISBN 0-306-44480-1.

Dyke, A. S. (2004). An outline of North American deglaciation with emphasis on central and northern Canada. In *Quaternary Glaciations – Extent and Chronology, Part II*, eds. J. Ehlers and P.L. Gibbard, Elsevier B.V., pp 373 – 424.

Edwards, M.E., Brubaker, L.B., Lozhkin, A.V. & Anderson, P.M. (2005). Structurally novel biomes: a response to past warming in Beringia. *Ecology* 86: 1696-1703.

Eidesen, P.B., Carlsen, T., Molau, U. & Brochmann, C. (2007). Repeatedly out of Beringia: *Cassiope tetragona* embraces the arctic. *Journal of Biogeography* 34: 1559-1574.

Elliott-Fisk, D. L. (2000). The taiga and boreal forest. In *North American Terrestrial Vegetation*, 2nd ed. M. G. Barbour and W. D. Billings (eds), pp 41-74, Cambridge University Press, Cambridge, U.K.

Farjon, A. (1990). Pinaceae: drawings and descriptions of the genera *Abies, Cedrus, Pseudolarix, Keteleeria, Nothotsuga, Tsuga, Cathaya, Pseudotsuga, Larix* and *Picea*. Koeltz Scientific Books, 633 Königstein.

Fickert, T., Friend, D., Grüninger, F., Molnia, B. & Richter, M. (2007). Did debris-covered glaciers serve as Pleistocene refugia for plants? A new hypothesis derived from observation of recent plant growth on glacier surfaces. *Antarctic & Alpine Research* 39:245-257.

Foulger, G.R. (2010). *Plates vs Plumes: A Geological Controversy.* Wiley-Blackwell, 364 pp., ISBN 978-1405161480.

Francis, J.E. (1991). The dynamics of polar fossil forests: Tertiary fossil forests of Axel Heiberg Island, Canadian Arctic. In: *Fossil Forests of Tertiary age in the Canadian Arctic Archipelago,* Christie, R.L. and McMillan, N.J. (eds.), pp 29-38, Geological Survey of Canada, Bulletin 403.

Geml, J., Laursen, G.A., O'Neill, K., Nusbaum, H.C. & Taylor, D.L. (2006) Beringian origins and cryptic speciation events in the fly agaric (*Amanita muscaria*). *Molecular Ecology* 15: 225-239.

Geml, J., Kauff, F., Laursen, G.A. & Taylor, D. L. (2010). Genetic studies point to Beringia as a biodiversity hotspot for high-latitude fungi. *Alaska Park Science* 8: 37-41.

Gérardi, S., Jaramillo-Correa, J.P., Beaulieu, J. & Bousquet, J. (2010). From glacial refugia to modern populations: new assemblages of organelle genomes generated by differential cytoplasmic gene flow in transcontinental black spruce. *Molecular Ecology* 19: 5265-5280.

Germino, M.J., Smith, W.K. & Resor, A.C. (2002). Conifer seedling distribution and survival in an alpine-treeline ecotone. *Plant Ecology* 162:157-168.

Goetcheus, V. G. & Birks, H. H. (2001). Full-glacial upland tundra vegetation preserved under tephra in the Beringia National Park, Seward Peninsula, Alaska. *Quaternary Science Reviews* 20: 135-147.

Gugerli, F., Parducci, L. & Petit, R. J. (2005). Ancient plant DNA: review and prospects. *New Phytologist* 166:409-418.

Hagman, M. (1993). Potential species and provenances for forest development in cold climates. pp 251-263. In Alden, J., Mastrantonio, J. L., Odum, S. 1993. *Forest Development in Cold Climates.* Plenum Press, New York and London, ISBN 0-306-44480-1.

Hansell, R.I.C, Chant, D.A. & Weintraub, J. (1971). Changes in the northern limit of spruce at Dubawnt Lake, N.W.T. *Arctic* 24:233-34.

Harper, K.A., Danby, R.K., De Fields, D. L., Lewis, K.P., Trant, A.J., Starzomski, B.M., Savidge, R. & Hermanutz, L. (2011). Tree spatial pattern within the forest-tundra ecotone: a comparison of sites across Canada. *Canadian Journal of Forest Research* 41: 479–489.

Hayes, C. W. (1892). An expedition through the Yukon district. *National Geographic Magazine* 4: 117-162, plates 18-20.

Hickling, R, Roy, D.B., Hill, J.K, Fox, R. and Thomas, C.D. (2006). The distributions of wide-ranging taxonomic groups are expanding polewards. *Global Change Biology* 12: 450-455.

Hills, L. V., Klovan, J.E. & Sweet, A.R. (1974). *Juglans eocinerea* n. sp., Beaufort Formation (Tertiary), southwestern Banks Island, Arctic Canada. *Canadian Journal of Botany* 52: 65-90.

Holderegger, R. & Thiel-Egenter, C. (2009). A discussion of different types of glacial refugia used in mountain biogeography and phylogeography. *Journal of Biogeography* 36:476-480.

Hopkins, D. M. (1982). Aspects of the paleogeography of Beringia during the Late Pleistocene, pp 3- 28 In *Paleoecology of Beringia*, ed. D. M. Hopkins, J. V. Matthews, C. E. Schweger and S. B. Young, Academic Press, 489 p.

Hultén, E. (1937). *Outline of the History of Arctic and Boreal Biota During the Quaternary Period.* Lehre J. Cramer, Stockholm, New York, 168 pp.

Husted, L. and Lavender, D.P. (1989). Effect of soil temperature upon the root growth and mycorrhizal formation of white spruce (*Picea glauca* (Moench) Voss) seedlings grown in controlled environments. *Annales des Sciences Forestières* 46:750s-753s.

Hustich, I. (1953). The boreal limits of conifers. *Arctic* 62: 139-162.

Ickert-Bond, S.M., Murray, D.F. & DeChaine, E. (2009). Contrasting patterns of plant distribution in Beringia. *Alaska Park Science* 8:26-32.

Ivarson, K. C. (1965). The microbiology of some permafrost soils in the Mackenzie Valley, N.W.T. *Arctic* 18: 256-260.

Jahren, A.H. (2007). The Arctic forest of the Middle Eocene. *Annual Reviews of Earth and Planetary Science* 35:509–540.

Jahren, A.H., Byrne, M.C., Graham, H.V., Sternberg, L.S.L. & Summons, R.E. (2009). The environmental water of the middle Eocene Arctic: Evidence from δD, $\delta^{18}O$ and $\delta^{13}C$ within specific compounds. *Palaeogeography, Palaeoclimatology, Palaeoecology* 271: 96–103.

Juan, P., Jaramillo, C., Jean, B. & Bousquet, J. (2004). Variation in mitochondrial DNA reveals multiple distant glacial refugia in black spruce (*Picea mariana*), a transcontinental North American conifer. *Molecular Ecology* 13: 2735-2747.

Junttila, O. and Nilsen, J. (1993). Growth and development of northern forest trees as affected by temperature and light. In: *Forest Development in Cold Climates*. Alden, J., Mastrantonio, J. L., Odum, S. (eds.), pp 45-57, Plenum Press, New York and London, ISBN 0-306-44480-1.

Kärki, L. & Tigerstedt, P.M.A. (1985). Definition and exploitation of forest tree ideotypes in Finland. In: *Attributes of Trees as Crop Plants*, Cannell, M.G.R. & Jackson, J.E. (eds.), pp 102-109, ITE, NERC, Abbots Ripton, England, ISBN 0-904282-83-X.

Kimball, J.S., Zhao, M., McGuire, A.D. and Heinsch, F.A. (2007). Recent climate-driven increases in vegetation productivity for the western Arctic: Evidence of an acceleration of the northern terrestrial carbon cycle. *Earth Interactions* 11 (4): 1-30.

Körner, C. (2007). Climatic treelines: conventions, global patterns, causes. *Erdkunde* 61:316-324.

Kryshtofovich, A. N. (1935). A final link between the Tertiary floras of Asia and Europe. *New Phytologist* 34: 339-344.

Kuc, M. (1974). Noteworthy plant records from southern Banks Island, N.W. T. *Arctic* 27:146-150.

Larcher, W. (1995). *Physiological Plant Ecology*, 3rd ed. Springer-Verlag, Berlin Heidelberg New York, ISBN 3-540-58116-2 3, 506 p.

Lebarge, W.P. (1995). *Sedimentology of Placer Gravels near Mt. Nansen, Central Yukon Territory.* Bulletin 4, Exploration and Geological Services Division, Northern Affairs Program, Yukon Region, Whitehorse, 155 p.

Lefsky, M. A. (2010). A global forest canopy height map from the Moderate Resolution Imaging Spectroradiometer and the Geoscience Laser Altimeter System. *Geophysical Research Letters* 37, L15401: 1-5.

LePage, B. A. (2001). New species of *Picea* A. Dietrich (Pinaceae) from the Middle Eocene of Axel Heiberg Island, Arctic Canada. *Botanical Journal of the Linnean Society* 135: 137-167.

Levsen, N. D. & Mort, M. E. (2009). Inter-simple sequence repeat (ISSR) and morphological variation in the western North American range of *Chrysosplenium tetrandrum* (Saxifragaceae). *Botany* 87:780-790.

Levson, V.M. & Rutter, N. W. (1996). Evidence of Cordilleran Late Wisconsinan glaciers in the 'ice-free corrridor'. *Quaternary International* 32: 33-51.

Lloyd, A.H., Wilson, A.E., Fastie, C.L. & Landis, R.M. (2005). Population dynamics of black and white spruce in the southern Brooks Range, Alaska. *Canadian Journal of Forest Research* 35:2073-2081.

Lloyd, J., Shibistova, O., Zolotoukhine, D., Kolle, O., Arneth, A., Wirth, C., Styles, J.M., Tchebakova, N.M. & Schulze, E.D. (2002). Seasonal and annual variations in the photosynthetic productivity and carbon balance of a central Siberian pine forest. *Tellus* 54B: 590-610.

Loehr, J., Worley, K., Grapputo, A., Carey, J., Veitch, A. & Coltman, D.W. (2006). Evidence for cryptic glacial refugia from North American mountain sheep mitochondrial DNA. *Journal of Evolutionary Biology* 19: 419-430.

Lopushinsky, W. & Max, T.A. (1990). Effect of soil temperature on root and shoot growth and on budburst timing in conifer seedling transplants. *New Forests* 4:107-124.

Lydolph, M.C., Jacobsen, J., Arctander, P., Gilbert, M.T.P., Gilichinsky, D.A., Hansen, A. J., Willerslev, E. & Lange, L. (2005). Beringian paleoecology inferred from permafrost-preserved fungal DNA. *Applied and Environmental Microbiology* 71: 1012-1017.

Mackay, J. R. (1958). The valley of the lower Anderson River, N.W.T. *Geographical Bulletin* 11:37-56.

MacDonald, G.M. & Cwynar, L. C. (1985). A fossil pollen based reconstruction of the late Quaternary history of lodgepole pine (*Pinus contorta* spp. *latifolia*) in the western interior of Canada. *Canadian Journal of Forest Research* 15: 1039-1044.

MacDonald, G.M. & McLeod, T.K. (1996). The Holocene closing of the 'ice-free' corridor: a biogeographical perspective. *Quaternary International* 32: 87-95.

Mandryk, C.A.S., Josenhans, H., Fedje, D.W. & Mathewes, R.W. (2001). Late Quaternary paleoenvironments of northwestern North America: implications for inland versus coastal migration routes. *Quaternary Science* Reviews 20:301-314.

McLeod, T. K. & MacDonald, G. M. (1997). Postglacial range expansion and population growth of *Picea mariana, Picea glauca* and *Pinus banksiana* in the western interior of Canada. *Journal of Biogeography* 24: 865-881.

Milankovitch, M. (1941). *Kanon der Erdbestrahlungen und seine Anwendung auf das Eiszeitenproblem*. (New English Translation (1998): *Canon of Insolation and the Ice Age Problem, with An Introduction and Biographical Essay by Nikola Pantic*, Alven Global, 636 pp.. ISBN 86-17-06619-9.)

Mortensen, J.K. (1992). Pre-Mid-Mesozoic tectonic evolution of the Yukon-Tanana Terrane, Yukon and Alaska. *Tectonics* 11:836-853.

Nakonechnaya, O.V., Kholina, A.B., Koren, O.G., Janeček, V., Kohutka, A., Gebauer, R. & Zhuravlev, Y.N. (2010). Characterization of gene pools of three *Pinus pumila* (Pall.) Regel populations at the range margins. *Russian Journal of Genetics* 46: 1417-1425.

Namroud, M.-C., Beaulieu, J., Juge, N., Laroche, J. & Bousquet, J. (2008). Scanning the genome for gene single nucleotide polymorphisms involved in adaptive population differentiation in white spruce. *Molecular Ecology* 17: 3599-3613.

Nienstaedt, H. & Zasada, J.C. (1990). *Picea glauca* (Moench) Voss, white spruce. In: *Silvics of North America 1. Conifers,* Burns, R.M. and Honkala, B.H. (eds.), 54 p, Agriculture Handbook 654, USDA Forest Service, Washington, D.C.

Nilsson, M.C. (2005). Understory vegetation as a forest ecosystem driver, evidence from the northern Swedish boreal forest. *Frontiers in Ecology and the Environment* 3: 421-428.

Pisaric, M.F.J., MacDonald, G.M., Velichko, A.A. & Cwynar, L.C. (2001). The lateglacial and postglacial vegetation history of the northwestern limits of Beringia, based on pollen, stomata and tree stump evidence. *Quaternary Science Reviews* 20: 235-245.

Politov, D.V., Belokon, M.M & Belokon, Yu. S. (2006). Dynamics of allozyme heterozygosity in Siberian dwarf pine *Pinus pumila* (Pall.) Regel populations of the Russian far east: comparison of embryos and maternal plants. *Russian Journal of Genetics* 42: 1127-1136.

Polunin, N.V. (1948). Botany of the Canadian Eastern Arctic, Part III; Vegetation and Ecology, *National Museum of Canada, Ottawa, Bulletin 104, Biological Series 32,* 304 p.

Petit, R.J., Pineau, E., Demesure, B., Bacilieri, R., Ducousso, A. & Kremer, A. (1997). Chloroplast DNA footprints of postglacial recolonization by oaks. *Proceedings of the National Academy of Sciences USA* 94: 9996-10001.

Porsild, A. E. (1955). The vascular plants of the western Canadian Arctic Archipelago. *National Museum of Canada, Ottawa, Bulletin 135, Biological Series 45,* 226 p.

Porsild, A. E. (1964). *Illustrated Flora of the Canadian Arctic Archipelago.* National Museum of Canada, Ottawa, Bulletin 135, Biological Series 45, 226 p.

Pregitzer, K.S., King, J.S., Burton, A.J. & Brown, S.E. (2000). Responses of tree fine roots to temperature. *New Phytologist* 147:105-115.

Ran, J.-H., Wei, X.-X. & Wang, X.-Q. (2006). Molecular phylogeny and biogeography of *Picea* (Pinaceae): Implications for phylogeographical studies using cytoplasmic haplotypes. *Molecular Phylogeny and Evolution* 41: 405-419.

Raskin, I. (1992). Role of salicylic acid in plants. *Annual Review of Plant Physiology* 43:439-463.

Read, D.J., Leake, J.R. & Perez-Moreno, J. (2004). Mycorrhizal fungi as drivers of ecosystem processes in heathland and boreal forest biomes. *Canadian Journal of Botany* 82: 1243-1263.

Ritchie, J. C. (1987). *Postglacial Vegetation of Canada.* Cambridge U. Press, Cambridge, ISBN 0 521 20868 2.

Ritchie, J. C. (1984). *Past and Present Vegetation of the Far Northwest of Canada.* University of Toronto Press, Toronto.

Ritchie, L.C. & MacDonald, G.M. (1986). The patterns of post-glacial spread of white spruce. *Journal of Biogeography* 13: 527-540.

Röser, C., Montagnani, L., Schulze, E.D., Mollicone, D., Kolle, O., Meroni, M., Papale, D., Marchesini, L.B., Federici, S. & Valentini, R. (2002). Net CO_2 exchange rates in three different successional stages of the "Dark Taiga" of central Siberia. *Tellus* 54B:642-654.

Savidge, R. A. & Förster, H. (1998). Seasonal activity of uridine 5'-diphosphoglucose : coniferyl alcohol glucosyltransferase in relation to cambial growth and dormancy in conifers. *Canadian Journal of Botany* 76:, 486 – 493.

Scholander, P.F., Flagg, W., Hock, R.J. & Irving, L. (1953). Studies on the physiology of frozen plants and animals in the arctic. *Journal of Cellular and Comparative Physiology* 42 (S1): 1-56.

Senjo, M., Kimura, K., Watano, Y., Ueda, K. & Shimizu, T. (1999). Extensive mitochondrial introgression from *Pinus pumila* to *P. parviflora* var. *pentaphylla* (Pinaceae). *Journal of Plant Research* 112: 97-105.

Shafer, A. B. A., Côté, S. D. & Coltman, D. W. (2010). Hot spots of genetic diversity descended from multiple Pleistocene refugia in an alpine ungulate. *Evolution* 65: 125-138.

Shiklomanov, I.A., & Rodda, J.C. (2003). *World Water Resources at the Beginning of the Twenty-first Century*. Cambridge, UK: Cambridge University Press.

Shilo, N.A., Lozhkin, A.V., Anderson, P.M., Brown, T.A., Pakhomov, A. Yu. & Solomatkina, T.B. (2007). Glacial refugium of *Pinus pumila* (Pall.) Regel in northeastern Siberia. *Doklady Earth Sciences* 412:122-124.

Shilo, N.A., Lozhkin, A.V., Anderson, P.M., Vazhenina, L.N., Glushkova, O. Yu. & Matrosova, T.V. (2008). First data on the expansion of *Larix gmelinii* (Rupr.) Rupr. into arctic regions of Beringia during the Early Holocene. *Doklady Earth Sciences* 423:1265-1267.

Sigurgeirsson, A. & Szmidt, A. E. (1993). Phylogenetic and biogeographic implications of chloroplast DNA variation in *Picea*. *Nordic Journal of Botany* 13: 233-246.

Stattin, E. & Lindström, A. (2000). Influence of soil temperature on root freezing tolerance of Scots pine (*Pinus sylvestris* L.) seedlings. In: *The Supporting Roots of Trees and Woody Plants: Form, Function and Physiology*, A. Stokes (ed.), pp 259-268, Kluwer, Dordrecht.

Steuber, T., Rauch, M., Masse, J.-P., Graaf, J. & Malkoč, M. (2005). Low-latitude seasonality of Cretaceous temperatures in warm and cold episodes. *Nature* 437: 1341-1344.

Stewart, J. R. & Dalén, L. (2008). Is the glacial refugium concept relevant for northern species? A comment on Pruett and Winker 2005. *Climatic Change* 86:19-22.

Stewart, J.R., Lister, A.M., Barnes, I. & Dalén, L. (2010). Refugia revisited: individualistic responses of species in space and time. *Proceedings of the Royal Society B* 277: 661-671.

Stockey, R. A. & Wiebe, N.J.P. (2008). Lower Cretaceous conifers from Apple Bay, Vancouver Island: *Picea*-like leaves, *Midoriphyllum piceoides* gen. et sp. nov. (Pinaceae). *Canadian Journal of Botany* 86: 649-657.

Sugimoto, A., Yanagisawa, N., Naito, D., Fujita, N. & Maximov, T.C. (2002). Importance of permafrost as a source of water for plants in east Siberian taiga. *Ecological Research* 17: 493-503.

Szeicz, J. M. & MacDonald, G.M. (2001). Montane climate and vegetation dynamics in easternmost Beringia during the Late Quaternary. *Quaternary Science Reviews* 20:247-257.

Takahashi, S., Sugiura, K., Kameda, T., Enomoto, H., Kononov, Y. Anunicheva, M.D. & Kapustin, G. (2011). Response of glaciers in the Suntar–Khayata range, eastern Siberia, to climate change. *Annals of Glaciology* 52: 182-192.

Taylor, T.N., Taylor, E.L, & Krings, M. (2009). *Paleobotany: the Biology and Evolution of Fossil Plants*, 2nd ed., Elsevier, ISBN: 978-0-12-373972-8.

Tsutsui, K., Suwa, A., Sawada, K., Kato, T., Ohsawa, T.A. & Watano, Y. (2009). Incongruence among mitochondrial, chloroplast and nuclear gene trees in *Pinus* subgenus *Strobus* (Pinaceae). *Journal of Plant Research* 122:509-521.

Tyron, P.R. & Chapin, F.S. (1983). Temperature control over root growth and root biomass in taiga forest trees. *Canadian Journal of Forest Research* 13:827-833.

Vendramin, G.G., Lelli, L., Rossi, P & Morgante, M. (1996). A set of primers for the amplification of 20 chloroplast microsatellites in Pinaceae. *Molecular Ecology* 5: 595-598.

Viereck, L.A. & Johnston, W.F. (1990) *Picea mariana* (Mill.) B.S.P, black spruce. In: *Silvics of North America 1. Conifers,* Burns, R.M. and Honkala, B.H. (eds.), Agriculture Handbook 654: 227-237, USDA Forest Service, Washington, D.C.

Viktora, M., Savidge, R. A. & Rajora, O. P. (2011). Clonal and nonclonal genetic structure of subarctic black spruce (*Picea mariana*) populations in Yukon territory. *Botany* 89: 133–140.

Wahl, E.E., Fraser, D.B., Harvey, R.C. & Maxwell, J.B. (1987). *Climate of Yukon.* Climatological Studies, 40. Environment Canada, Atmospheric Environment Service, 313p.

Walker, X. (2010). The reproduction, establishment, and growth of white spruce in the forest tundra ecotone of the Inuvik-Tuktoyaktuk region. M.Sc. thesis, University of British Columbia, Vancouver, B.C.

Waltari, E., Hoberg, E. P., Lessa, E. P. & Cook, J.A. (2007). Eastward ho: phylogeographical perspectives on colonization of hosts and parasites across the Beringian nexus. *Journal of Biogeography* 34: 561-574.

Wareing, P. F. (1985). Tree growth at cool temperatures and prospects for improvement by breeding. In: *Attributes of Trees as Crop Plants*, M.G.R. Cannell and J. E. Jackson (eds.), pp 80-88, ITE, NERC Press, Abbots Ripton, England. ISBN 0-904282-83-X.

Warming, E. (1888). Om Grønlands vegetation. *Meddelelser om Grønland* 12: 1–245.

Weiser, C.J. (1970). Cold resistance and injury in woody plants. *Science* 169:1269-1278.

Welsh, S.L. & Rigby, J.K. (1971). Botanical and physiographic reconnaissance of northern Yukon. *Brigham Young University, Science Bulletin, Biological Series* 14(2), 64 p.

Westergaard, K. B., Jørgensen, M.H., Gabrielsen, T.M., Alsos, I.G. & Brochmann, C. (2010). The extreme Beringian/Atlantic disjunction in *Saxifraga rivularis* (Saxifragaceae) has formed at least twice. *Journal of Biogeography* 37:1262-1276.

Williams, J.W., Shuman, B.N., Webb, T., Bartlein, P.J. & Leduc, P.L. (2004). Late Quaternary vegetation dynamics in North America: scaling from taxa to biomes. *Ecological Monographs* 74: 309-334.

Wilmking, M. & Ibendorf, J. (2004). An early tree-line experiment by a wilderness advocate: Bob Marshall's legacy in the Brooks Range, Alaska. *Arctic* 57: 106-113.

Yuan, X. (2012). Growth patterns of timberline *Picea glauca* on a site in the Yukon Plateau: Correlations within tree and between trees. M.Sc.F. thesis, University of New Brunswick, Fredericton, Canada.

Zarter, C.R., Demmig-Adams, B., Ebbert, V., Adamska, I. & Adams, W.W. III. (2006). Photosynthetic capacity and light harvesting efficiency during the winter-to-spring transition in subalpine conifers . *New Phytologist* 172: 283–292.

Zazula, G.D., Telka, A.M., Harington, C.R., Schweger, C.E. & Mathewes, R.W. (2006). New spruce (*Picea* spp.) macrofossils from Yukon territory: implications for Late Pleistocene refugia in Eastern Beringia. *Arctic* 59: 391-400.

5

Trace Metals and Radionuclides in Austrian Forest Ecosystems

Stefan Smidt[1], Robert Jandl[1], Heidi Bauer[2], Alfred Fürst[1], Franz Mutsch[1], Harald Zechmeister[3] and Claudia Seidel[4]
[1]*Federal Research and Training Centre for Forests, Natural Hazards and Landscape*
[2]*Vienna University of Technology*
[3]*University of Vienna*
[4]*University of Natural Resources and Life Sciences, Vienna*
Austria

1. Introduction

An abundance of trace metals in the environment is a characteristic property of only a few terrestrial ecosystems. Serpentinite contains high concentrations of trace metals - especially Ni and Cr - that are released by weathering. Other natural sources of trace metals are volcanic eruptions and the deposition of marine aerosols. In the soil some trace metals are strongly adsorbed to the mineral matrix or fixed in stable complexes with soil organic matter, other ones are quite mobile. Trace metals show a high affinity to humic acids, iron oxides, clay minerals and carbonates in the soil.

Trace metals have gained in importance due to human activities that have fundamentally changed their biogeochemical cycling. They are emitted into the environment by numerous industrial processes and are dispersed by the application of sewage sludge, biogenic waste, pesticides and fertilizers.

Most trace metals play an ambivalent role in the environment: Depending on the concentration they are both essential micronutrients and toxic components for plants and soil organisms (VDI, 1984; Hock & Elstner, 1995; Weigl & Helal, 2001; Larcher, 2003; Alloway & Reimer, 1999; Ormrod, 1984; Elling et al., 2007). Therefore the emission and deposition of trace metals and their cycling in the environment are important issues in environmental research.

Naturally as well as artificially produced radionuclides can be found in all compartments of the environment. The interest in the content of radionuclides in *Picea abies* needle increased after the Chernobyl accident in 1986, because parts of the European forests were highly contaminated especially by Cs-137.

The role of trace metals in ecosystems depends on the degree and type of dispersion as gaseous compound or aerosols in the atmosphere and on their solubility and mobility in wet depositions. Some metals accumulate in plants and in soils. High local emission rates, e.g. by foundries, lead to acute damages of plants and to a long-lasting contamination of soils

(Kazda & Glatzel, 1984) and are consequently found in game. In pristine areas, especially in the Alps, deposition rates of trace metals (e.g. of Pb) may be surprisingly high due to long-range transport of air masses (Herman et al., 2000; Smidt & Herman, 2003).

The goal of this chapter is to give an overview about the activities to monitor the input and distribution of trace metals and radionuclides in Austrian forest ecosystems and to describe temporal trends of the contamination of Austrian forests. For this purpose we use deposition measurements, bioindicators, and soil analyses.

1.1 Limit values, classification values and critical loads

A set of legal and effect-related limit values was established defining the trace metal concentrations that have adverse effects in terrestrial ecosystems:

1.1.1 Vegetation

Legal limit values for deposition rates and concentrations in the ambient air have been established in Austria, Switzerland and Germany (Table 1).

Trace metal	Austria Forest culture	Switzerland Vegetation *)	Germany General *)	EU Human beings, environment
	g ha^{-1} yr^{-1}	g ha^{-1} yr^{-1}	g ha^{-1} yr^{-1}	ng m^{-3}
Pb	2,500	365	365	
Cd	50	7.3	7.3	5
Cu	2,500	-		
Zn	10,000	1,460		
Tl	-	7.3	7.3	
Ni	-	-	54.8	20
As	-	-	14.6	6
Hg	-	-	3.65	

*) Original limit value (annual mean value) transformed from µg m^{-2} d^{-1} into kg ha^{-1} a^{-1}
Austria: Forstgesetz, Zweite Verordnung gegen forstschädliche Luftverunreinigungen, BGBl. 199/1984;
Switzerland: Schweizerische Luftreinhalteverordnung, 1985;
Germany: Technische Anleitung Luft (2002);
European Union: EU Daughter Directive 2004/107/EG (goals for 2012)

Table 1. Legal limit values for atmospheric deposition of trace metals established in Austria, Switzerland, Germany and for concentrations in the air by the European Union.

The limits vary in terms of the subject of protection. Limits for Hg, Pb and Cd concentrations in needles of *Picea abies* are based on the data of the Austrian Bio-Indicator Grid and long-term investigations. No established limits exist for the concentration of radionuclides in non-edible forest vegetation.

1.1.2 Soils

For the classification of soils typical ranges in the trace metal content and Critical Levels were established (Table 2).

	Very low (1) mg kg^{-1}	Low (1) mg kg^{-1}	Medium (1) mg kg^{-1}	High (1) mg kg^{-1}	Very high (1) mg kg^{-1}	Critical Loads (2) g ha^{-1} a^{-1}
As	< 5	5 - 10	10 – 15	15 - 20	> 20	<10 -> 100 (A, H)
Cd	< 0.10	0.10 - 0.25	0.25 - 0.50	0.50 - 1.00	> 1.00	<1 -> 5 (E, H)
Cr	< 30	30 - 50	50 – 60	60 - 100	> 100	<20 -> 120 (F, EF)
Co	< 7	7 - 12	12 – 20	20 - 50	> 50	
Cu	< 15	15 - 25	25 - 50	50 - 100	> 100	< 10 -> 60 (F, EF)
Hg	< 0.15	0.15 - 0.25	0.25 - 0.50	0.50 - 1.0	> 1.0	< 0.05 - > 0.30 (E, H)
Mo	< 0.3	0.3 - 0.6	0.6 - 2.0	2.0 - 5.0	> 5.0	
Ni	< 20	20 - 30	30 - 40	40 - 60	> 60	< 20 - > 120 (A, H)
Pb	< 10	10 - 20	20 - 50	50 - 100	> 100	< 5 - > 30 (H)
Zn	< 60	60 - 90	90 - 150	150 - 300	> 300	< 100 -> 600 (E, EF)

(1): Blum et al. (1997), (2): Hettelingh et al. (2006)
A: agriculture, E: environment, EF: ecosystem function, F: forest, H: human health

Table 2. Evaluation of trace metal concentrations in the soil and Critical Loads.

Some classifications distinguish between the trace metal content in the mineral soil and the organic surface layer (De Vries & Bakker, 1998; Dosskey & Adriano, 1991). In most cases the trace metal contents of soils refer to the acid extractable fraction, usually aqua regia (Blum et al., 1997; Vanmechelen et al., 1997). With this method both the "weatherable stocks" and the trace metals from atmospheric deposition are detected. According to the Austrian Standard "L 1075 – Principles for the evaluation of the content of selected elements in soil" trace metal contents belong to the acid extractable fraction.

1.2 The detection of heavy metals and radionuclides in the environment

The trace metal input into ecosystems can be determined by chemical analyses of aerosols, rain water, by means of bioindication and by plant and soil analyses, respectively. Bioindication is a low cost option for plants, fungi and animals accumulating trace metals (Markert et al., 2003). The anthropogenic impact on the soil can be detected by comparison of concentration levels of lower and upper horizons. If anthropogenic impacts of trace metals exist, higher concentrations in the "O-layers" and upper mineral soil layers (0-5 or 0-10 cm) than in deeper soil layers (deeper 30 or 50 cm) are observed. The relation between the concentrations in the upper and deeper layers is the enrichment factor (EF). An EF > 1 indicates accumulation of a specific trace element, an EF < 1 means a depletion. In general the EF should be at least > 1.5 for an indication of anthropogenic inputs (Walthert et al., 2004; Mutsch, 1998).

Radioactive elements are detectable by gamma-ray spectrometry in various compartments of the environment. It was described in the framework of the investigation programme "Artificial and natural radionuclides in spruce needles in Upper Austria from 1983 to 2008 – an application for radioecological monitoring" (Seidel, 2010). *Picea abies* needle samples of the Austrian Bio-Indicator Grid were analysed to investigate the geographical and temporal distribution of radionuclides in spruce needles during the last 25 years. While Cs-137 plant uptake has been widely studied in the last few decades, natural radionuclides have received less attention, despite ubiquitously occurring in the environment. Hence *Picea abies* needle samples were also analysed for the following natural radionuclides: K-40, Pb-210, Ra-226, Ra-228, U-238.

1.3 Effects of selected trace metals

Deficiency of essential trace metals, e.g. Cu, Fe, Mn, Mo, Ni, Zn, causes reduced growth of shoots and roots as well as visible symptoms (e.g. chlorosis). Excess concentrations in the soils and plant material, respectively, lead to numerous disturbances: Many trace metals inhibit enzymes due to the substitution of essential metals and/or reaction with SH-groups. Furthermore, they injure plasma membranes, promote the formation of reactive oxygen species and inhibit the stomatal function. General eco-toxicological risks are disturbances of microbial communities and reduced microbial biomass in the soil, decreasing litter decomposition, reduction of plant growth, damages of the roots and mycorrhiza, reduced species-diversity in the soil followed by inhibition of nutrients and water uptake, and accumulation of metals in the food chain (DeKok & Stulen, 1992; Hock & Elstner, 1995; Andre et al., 2006; Table 3). Plants provide several mechanisms to prevent heavy metal injuries (Larcher, 2003).

	Industrial use / production (examples)	Plant use	Uptake, role (1)	Symptoms (2)
	Micronutrients			
Co	Alloys	Vitamine B_{12}, enzymes	Moderate transport from the roots to other plant parts	Inhibition of enzymes, discolouration
Cr	Alloys, surface protection of metals	Growth stimulation, formation of chlorophyll	Low translocation. Especially Cr(VI) is toxic	Discolouration of leaves, root injuries
Cu	Electrical industry, alloys, pigments	Enzymes (e.g. of the nitrogen cycle)	Active and passive uptake and moderate translocation, organic chelates	Small and discoloured leaves, reduced root growth
Fe	Most important industrial metal	Enzymes (e.g. electron transport chain)	Negative effects of excess Fe in the soil/plants are of low relevance	Chlorosis
Zn	Household and industrial instruments, alloys	Cell metabolism	Uptake as Zn^{2+} or organic chelates. Very mobile in acid soils, easily allocated to different plant parts	Reduced root and shoot growth, chlorosis
Ni	Alloys, catalysts	Part of enzymes, membranes and nucleic acids	Essential for plants with N-fixing symbionts	Discolouration, brown roots, reduced root growth

	Non-nutrients	Uptake, role (1)	Symptoms (2)
Pb	Alloys, batteries, pigments	Very low mobility, uptake probably passive. Deposition and foliar uptake may contribute significantly to leaf concentrations.	Discolouration, reduced leaf and root growth
Cd	Byproduct, pigment, semiconductors, batteries	Highly toxic, uptake passively or metabolically, inhibits enzymes. Very mobile within plants. In the soil it is easily adsorbed by clay minerals and organic matter.	Discolouration of leaves and roots, reduced growth
Hg	Batteries, NaOH-production, Combustion, mining, sintering of ores	Inhibits enzymes. Retained in soils as organo-complex. Uptake elemental (Hg^0), promoted by methylation and the formation of other methylated organo-compounds. In acid soils harmful for microbes and roots. Low Hg uptake from soil, roots are significant adsorption sites and a barrier for Hg transport to foliage; transported in the atmosphere as Hg^0 and Methyl-Hg.	Discolouration of leaves
Radioactive isotopes		Cs-137: For humans hazardous, similar to K, slow downward migration in soil, accumulation in soil organic matter. K-40: Similar to Cs. Pb-210: Mainly deposited by natural fallout, transfer via roots is neglible. Ra-226, Ra-228: Similar to Ca, mobile, accumulation in the clay fraction, strong affinity to Fe- and Mn-hydroxides. U-238: Slow migration in soils.	Mutations in trees, light-brown leaves and needles (exclusively found near Chernobyl)

Table 3. Use of micronutrients and non-micronutrients, uptake and symptoms caused by trace metal uptake by plants (Hock & Elstner, 1985). (1) Markert et al. (2003); (2) Weigel & Helal 2001.

1.4 Metal cycles in forest ecosystems

Ecosystems are effective sinks for trace elements and radioactive isotopes such as Cs-137. The vegetation as well as the soil is able to accumulate and transfer trace elements within the ecosystems. The "rough surfaces" of forests are effective filters for atmospheric contaminations, dissolved substances and particles. Deposition, litterfall and stemflow affect soils by the transfer of trace elements in the surface soil layer and accumulate there.

Wash-off and leaching due to precipitation, litterfall, percolation in the soil and uptake by plants determine the internal cycle of trace elements. The input into the biogenic cycle takes place by wet, dry and occult deposition, output by harvesting, runoff into the groundwater and to a minor part by volatilisation (e.g. carbonyles and/or alkyles of Hg, Ni, Pb). Plants are able to take up trace metals as well as radionuclides actively via the roots and passively via the leaves.

The internal cycle is determined by the pH-dependent mobility of the respective element or radionuclide in the soil. The mobility in plants depends on the type of their chemical bonding, e.g. as ion or ligand or bound to an organic molecule like citric acid, oxalic acid, amino acids or peptides. Some of the metals can be translocated through the xylem and phloem into other parts of the plant (Ziegler, 1988; Riederer, 1991). Because of high phosphate concentrations in the phloem the mobility of various elements (e.g. Ca, Sr, Pb, Po) is limited, whereas alkali metals, chlorine and phosphorus are rapidly translocated in plants via xylem and phloem (Gerzabek, 1992; Thiessen et al., 1999).

Most of the Chernobyl-derived Cs-137 is still fixed in the upper soil layers which is an indication for the low migration rates and discharge via drainage water. Forest soils act as a long-term sink for Cs-137 and as the main source for radioactive contamination of forest vegetation. The soil contamination by Cs-137 can only be reduced by physical decay, by the export of biomass that previously had adsorbed Cs-137, and by soil erosion (Seidel, 2010). The export in the biomass can be harvesting of fungi and fruits, the consumption by game, and timber harvest (Strebl et al., 2000; IAEA, 2002).

1.5 Objectives

Regarding the fate of selected metals and radioactive isotopes, the following questions arise:

- What is the current input into forest ecosystems?
- Can a risk for forest ecosystems be derived from the input data?
- Which trace metals are of special relevance and how is the feasibility for the use of spruce needles and mosses for environmental radioactive monitoring?
- Can trace metals from anthropogenic or geogenic origin in soils be distinguished?
- What are the patterns of pollution in scale and time within the last twenty years?
- How is the temporal distribution of Cs-137 in *Picea abies* needles in Upper Austria from 1983 – 2008 and what is the radioactive contamination of *Picea abies* needles due to natural radionuclides?
- How is the vertical distribution of artificial and natural radionuclides in soils?

2. Monitoring networks

Trace metals in Austrian forests are monitored within the scope of different programmes with different spatial and temporal resolution (Figure 1, Table 4).

3. Data sets, methods and results

Data sets of atmospheric depositions, forest soils and plant material enable the assessment of the input and the risk for forest ecosystems at a local and regional scale in Austria.

3.1 Deposition

Airborne trace metals and other components are removed from the atmosphere by wet, dry and occult deposition and enter the ecosystems. Wet deposition is the washout of components by rain and snow and is calculated by multiplying volume-weighted annual mean concentrations with the precipitation heights. Dry deposition comprises the settling of gases and particulate matter. The removal of airborne components via fog and hoarfrost (ice formation on cold surfaces) is the occult deposition. It is calculated by the deposition velocities and the average concentrations in the gas phase and fog phase, respectively.

Trace Metals and Radionuclides in Austrian Forest Ecosystems

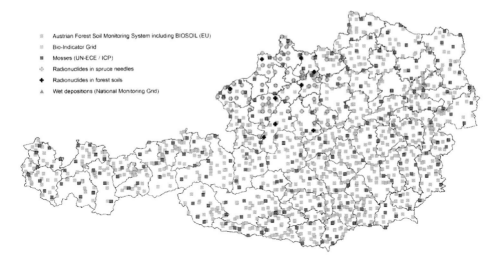

Fig. 1. Distribution of plots where investigations were conducted.

	Deposition	Spruce needles (Hg)	Spruce needles (radioactive elements)	Mosses	Forest soils	Forest soils – (radioactive elements)
Project title	National Monitoring Grid	Austrian Bio-Indicator Grid	Radionuclides in spruce needles [1]	UNECE / ICP [2]	Austrian Forest Soil Monitoring System; BioSoil (EU) [3]	Radionulides in forest soils [1]
From – to	annual since 1984	1986, 1996, 2006	1983-2008	1995, 2000, 2005, 2010	1987/89 - 2006/07	2008
Number of plots/samples	8	763 plots	71	220	511 (1987/89); 139 (2006/07)	9
Pb, Cd, Cu, Zn	+			+	+	
Co, Cr, Ni, Fe	+			+		
Hg		+		+	+	
Cs-13 [4], K-40 [5], Pb-210 [5], Ra-226 [5], Ra-228 [5], U-238 [5]						+

[1] Programme "Artificial and natural radionuclides in spruce needles in Upper Austri – an application for radioecological monitoring"
[2] Monitoring of atmospheric metal pollution by mosses in Austria within the UNECE ICP monitoring programme
[3] Austrian Forest Soil Monitoring System
[4] Artificial radionulide; [5] natural radionuclides

Table 4. Trace metals and radionuclides measured in Austria.

According to meteorological conditions, geographical situation, type of vegetation, type of trace metal etc. different input pathways may be relatively more important for the deposition loads at a given site. In Austria wet deposition predominates at many sites. In an extensive field campaign covering a mountain and a valley site along an altitudinal gradient in western Austria, the total deposition of 10 trace metals was assessed for four seasons and at both sites. Wet deposition was the most important input pathway (Table 5; Bauer et al., 2008).

Element, type of deposition	Christlumkopf (1786 m a.s.l.)	Mühleggerköpfl (920 m a.s.l.)
	g ha^{-1} yr^{-1}	g ha^{-1} yr^{-1}
Pb, wet deposition	14.07	9.32
Pb, dry deposition	0.35	0.12
Pb, occult deposition *)	4.20 (7.70 *)	0.04 (0.05 *)
Cd, wet deposition	1.42	0.94
Cd, dry deposition	0.01	0.003
Cd, occult deposition	3.58 (6.80 *)	0.10 (0.12 *)

Table 5. Wet, dry and occult deposition rates (g ha^{-1} yr^{-1}) of the analyzed heavy metals at the site Christlumkopf (1786 m a.s.l.) and Mühleggerköpfl (920 m a.s.l.), October 1997 - September 1998, also taking the "edge-effect" (*) into account (edge effect: increased deposition rates at forest margins).

A long-term study (1998-2008) carried out at the ICP-Forest site Zöbelboden (Upper Austria), showed that the relative importance of the various deposition pathways studied were quite variable for the most abundant metals (Al, Pb and Zn). Over all the relative importance of wet deposition was much higher for Al, equal for Pb and much lower for Zn compared to dry deposition (Offenthaler et al., 2008).

3.1.1 Wet deposition of Pb and Cd - Trends in Austria

Due to their hazardous effects on humans and ecosystems the deposition of Pb and Cd has been monitored for more than 20 years. In 1985 total Pb emissions in Austria amounted to 327 Mg and even to 219 Mg in 1990. Since then Pb emissions have decreased continuously to 15 Mg in 2008, which corresponds to a reduction of 93 % within 18 years. Cd emissions, however, were reduced by 51 % in the period 1985-1990, from 3.1 Mg to 1.58 Mg and since 1990 only by 27 % to 1.15 Mg in 2008 (Anderl et al., 2010). One reason for the strong decrease of Pb in rain and snow samples at rural and urban sites in Austria was the ban of leaded gasoline. This measure led to a decrease of Pb concentrations. Kalina et al. (1999) observed reductions in rural sites between 60 and 80 % within a period of 10 years (1984 - 1995) and around 90 % within 10 years in urban areas. Overall, an annual decline between 4.5 and 16 % was found (Table 6).

The reduction of the Pb concentration in precipitation is also reflected in the decrease of Pb loads by means of wet deposition. In Austria the wet deposition of Pb has been reduced by 80 % to 90 % at rural and 95 % at urban sites since 1990. At the European continental background station Mt. Sonnblick a decrease of 70 % since 1990 was observed (Figure 2).

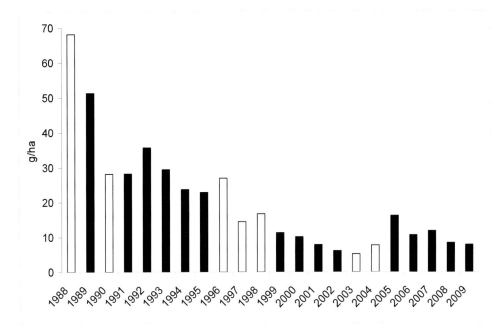

Fig. 2. Annual trend of wet deposition of Pb at the background station Mt. Sonnblick (3105 m a.s.l.); light bars represent incomplete data sets (missing data for 1-3 months were extrapolated).

According to the trend of Cd emissions, Cd loads by means of wet deposition did not decrease to the same extent as Pb loads. The reductions of Cd input by wet deposition since 1994 have been between 4 % and 10 % at rural sites and 20 % – 36 % at urban sites.

		Pb	Cd
Site (altitude)	Measuring period	Mean / annual changes (g ha^{-1} yr^{-1})	
Bisamberg / Vienna (310 m)	1990 - 2006	11.7 / – 2.14 ***	2.5 / -0.18*
Lainz / Vienna (230 m)	1987 - 2006	43.4 / -6.12 ***	2.5 / -0.006 n.s.
Lobau / Vienna (155 m)	1987 - 2006	17.6 / -3.57 ***	1.9 / 0.02 n.s.
Nasswald / Lower Austria (600 m)	1988 - 2006	39.2 / -6.26 ***	3.9 / 0.01 n.s.
Haunsberg / Salzburg (510 m)	1984 - 2006	21.8 / -2.20 ***	
Sonnblick / Salzburg (3105 m)	1988 - 2006	23.4 / -2.39 ***	
Werfenweng / Salzburg (940 m)	1984 - 2006	11.2 / -1.26 ***	

Table 6. Average inputs of Pb and Cd in Austria by means of wet deposition (means and mean annual changes) and statistical significances of the trends. Further data are reported in www.waldwissen.net/wald/klima/immissionen/

3.2 Bioindication

3.2.1 Hg in the needles of *Picea abies*

The atmosphere is in most cases the only source of Hg contamination in vegetation. Hg is transported in the atmosphere as Hg^0 and as Methyl-Hg. Its uptake from the soil is limited, with roots acting as a significant adsorption site and as a barrier for Hg transport to foliage (Grigal, 2002). These factors are the reason why Hg as a leading element for environmental pollution was selected. By means of a quick and easy analytical method (LECO-AMA 254) and the availability of archived samples it was possible for the first time to evaluate the situation throughout Austria. The analytical method is sensitive enough to determine naturally occurring Hg concentrations (Rea & Keeler, 1998; Rea et al., 2002). The analysed samples were collected within the framework of the Austrian Bio-Indicator Grid in the years 1986, 1996 and 2006 (Fürst et al., 2003).

A significant Hg impact can be found in the areas with iron- and steel production in the region around Linz (Upper Austria) and Leoben/Donawitz (Styria). Surprisingly the contamination is detectable in 100 km distance from Linz downstream the Danube (see maps 1986 and 1996; Figure 3). According to industry records VOEST-Alpine (Upper Austria), 720 kg Hg were emitted by the sintering plant in Linz in the year 2000 (Federal Environment Agency, 2004). Presently, the sintering of Hg-containing Austrian iron ore is a major Hg source in Austria (Fürst, 2007).

Of historical interest is the case of the elevated Hg levels in Brückl (Carinthia). Here amalgam electrodes were used in the chloralkali process of the chlorine plant in the late 1990ies. This source of Hg pollution was eliminated by a technological change. However there are still elevated Hg levels found in the surroundings. The second Austrian electrolysis plant in Hallein (Salzburg) was shut down at the end of the nineties and is shown in the maps from 1986 and 1996. In the Inn Valley (Tyrol) elevated Hg levels were detected in the areas around Schwaz (former silver mining) and Brixlegg (metal recycling by the Montanwerke).

In the vicinity of the capital cities Vienna and Graz (Styria) areas with Hg pollution were detected as well, however for the identification of the individual sources the density of the Bio-Indicator Grid is here not sufficient.

The Hg levels detected in Austria ranged from 0.006 to 0.174 mg kg^{-1} in 1986, from 0.005 to 0.245 mg kg^{-1} in 1996 and from 0.005 to 0.066 mg kg^{-1} in the needles in 2006. The percentage of the plots in the lowest class increase from 27.0 % in 1986 and 49.2 % in 1996 to 73.7 % in 2006.

These concentrations correspond to the levels detected in the needles in other European countries. Around 90 % of the results obtained in different European studies (n = 63) were in the concentration range between 0.005-0.100 mg Hg kg^{-1} (Grigal, 2002).

3.2.2 Radionuclides in the needles of *Picea abies*

In the period from 1986 to 2008, 782 *Picea abies* needle samples were analysed by gamma-ray spectrometry, being provided by the Austrian Bio-Indicator Grid to investigate the spatial and temporal distribution of radionuclides in needles with the main focus on the radioactive contamination before and after the Chernobyl fallout in 1986. The samples were analysed for Cs-137, K-40, Pb-210, Ra-226, Ra-228 and U-238.

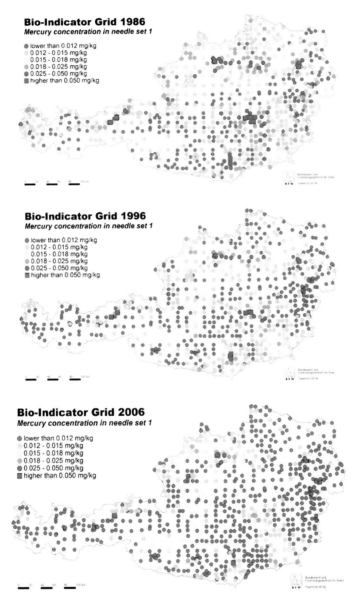

Fig. 3. Hg concentrations in the needle sets 1 of the sampling years 1986 (a), 1996 (b) and 2006 (c).

Due to the half-life time of 30.2 years and the high amount of deposition, Cs-137 can be well detected. The Cs-137 activity concentrations range from the detection limit (D.L.; approx.

2 Bq kg⁻¹) to 5150 Bq kg⁻¹ dry matter (D.M.) during the investigation period between 1986 and 2008 (1 Bq = 1 Becquerel = 1 decay per second). The maximum value was detected for needle samples of needle set 2 (sprout season to autumn of the subsequent year) in the year of the Chernobyl fallout. In the following years the Cs-137 activity concentrations in the younger sprouts of needle set 1 (sprout season to autumn) were higher than those in the older sprouts of needle set 2 (Figure 4).

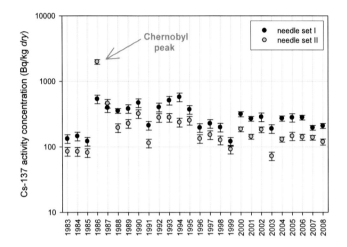

Fig. 4. Cs-137 activity concentrations of *Picea abies* samples at a site in the Bohemian Massif (Upper Austria, 1983-2008).

This observed age-dependency is in general agreement with former studies. The Cs-137 activity concentrations in *Picea abies* needles decreased at all measuring sites in the years after Chernobyl-derived contamination by various ecological half-live times. At several sites the Cs-137 activity concentrations decreased fast after the Chernobyl-derived contamination. Other sites are characterized by a low decrease of Cs-137 activity concentrations during the last 22 years which is exclusively due to the decay of Cs-137. Hence it can be assumed that these sampling sites are situated in closed forest ecosystems.

The process of Cs-137 recycling due to the needle fall to soil was detected at all measuring sites. The mean duration of Cs-137 cycles varies between 4-5 years depending on site and concentration. At some sites high Cs-137 activities were observed in the needle samples in the years before the Chernobyl accident occurred. These high contaminations could be attributed to the global fallout of nuclear weapon tests in the 1950ies and 1960ies.

With the knowledge of the current background radiation as well as temporal variations of Cs-137 activity concentrations in *Picea abies* needles over the last years, low percentages of Chernobyl-derived contamination (at some sites less than 1 %) are necessary to detect additional Cs-137 deposits in the youngest needles (Seidel, 2010).

K-40 activity concentrations are in the range of the D.L. (approx. 15 Bq kg⁻¹ D.M.) and 294 Bq kg⁻¹ D.M. as it could be observed that K-40 activity concentrations are strongly influenced by the geological background and natural variations over the years. The trend for constantly increased activity concentrations in the younger sprouts, which was observed for series of Cs-137, was not found for K-40.

Pb-210 activity concentrations range from D.L. (approx. 5 Bq kg^{-1} D.M. to 45 Bq kg^{-1} D.M.). Pb-210 is generated by the decay of Ra-226, which is found in most soils and rocks and which produces short-lived gaseous Rn-222 as a daughter. Most of Rn-222 decays with several intermediate radioactive progenies to Pb-210 within the soil, producing supported Pb-210, which is essentially in equilibrium with the parent Ra-226. Some of the Rn-222 diffuses upwards into the atmosphere and decays to Pb-210. This Pb-210 is deposited by fallout, and since it is not in equilibrium with the parent Ra-226, it is commonly termed unsupported or excess Pb-210 (Pb-210*ex*). Pb-210*ex* was determined in spruce needles to quantify the atmospheric input. Up to 100 % of the Pb-210 content in *Picea abies* needles were of atmospheric origin. Significant higher Pb-210*ex* activity concentrations were mostly detected in the older sprouts, which was interpreted as a result of longer atmospheric exposition.

Ra-226 activity concentrations range from the D.L. (1-4 Bq kg^{-1} D.M.) to 106 Bq kg^{-1} D.M. For needle set 1 significant higher Ra-226 activity concentrations were detected than for older sprouts of needle set 2. Due to the chemical similarity of Ra with the essential nutrient Ca, it is possible that the plant takes up Ra-226 from soil instead of Ca and transports it along with the transpiration stream to the newly developed sprouts.

All Ra-228 as well as U-238 activity concentrations in spruce needles are below D.L. (approx. 4-5 Bq kg^{-1} D.M.).

The total radioactivity in spruce needles is strongly influenced by the amount of the radionuclide Cs-137. A ranking list for the radionuclide content in spruce needles could be compiled: Cs-137 > K-40 > Pb-210 > Ra-226. It can be concluded that 52 % of the measured radioactivity was of artificial origin and 48 % of natural origin in the spruce needle samples from 1986-2008 (Seidel, 2010).

3.2.3 Heavy metals in mosses

The suitability of bryophytes for the quantification of trace metal deposition is based on their accumulation abilities derived from a series of morphological and physiological properties (e.g. cationic exchange properties). Since its introduction, the method has been tested and improved in many respects (e.g. sampling strategies, preparation of the samples, growth of indicator species, analyses and statistical tests), leading to a large number of papers supporting its applicability. Detailed literature reviews of methods and applications have been given e.g. by Onianwa (2001) and Zechmeister et al. (2003).

Beside trace metals listed in this article moss has been taken for the indication of a wide range of substances that has been analysed in many studies (Zechmeister et al., 2004; Zechmeister et al., 2005; Zechmeister et al., 2006), e.g. PAHs, dioxins, and trace metals originating mainly from traffic (e.g. Pt, Pd, Sb, Sn, Mo).

Mosses were used for bioindication of atmospheric trace metals in 1991, 1995, 2000, 2005, 2010 at 220 sites throughout Austria, where approximately 160 plots out of them are located in the Austrian Alps. Samples were taken within a Europe-wide project coordinated by the

UNECE ICP-Vegetation (e.g. Harmens et al., 2010) and according to respective guidelines (ICP Vegetation, 2005). Concentrations of Al, As, Cd, Co, Cr, Cu, Fe, Hg, Ni, Pb, Sb, V, and Zn were determined in five moss species: *Hylocomium splendens, Pleurozium schreberi, Hypnum cupressiforme, Abietinella abientina* and *Scleropodium purum*.

Before analysis moss was dried and digested by the use of nitric and perchloric acid in a Kjeldatherm. Analysis was performed by inductively coupled plasma atomic emission spectrometry and by an atomic absorption spectrometer with a graphite furnace (for details see Zechmeister et al., 2008).

Mean values from all sampling sites decreased for most of the analysed metals in the overall investigation period (1991-2010). Nevertheless, some elements like Cu or Mo showed even increases and some elements did not show a clear trend for the overall period (e.g. Hg). Despite the general trend, local patterns varied strongly, according to local pollution sources like highways or major industries. Compared to other regions, the Northern Tyrolean Limestone Alps show the highest rates of metal deposition in Austria. Eastern parts of Austria are also influenced by major deposition rates which derive partly from long range transport from eastern countries as well as from agricultural sources.

Additionally, mosses were sampled across several transects in the Austrian Alps in 1991. By that time significant trends could be shown, among them an increase in the concentrations of Pb, Cd and Zn with increasing altitude (see the trend of Pb in Figure 5). Examples for deposition patterns in time and space are given in Figures 6 and 7.

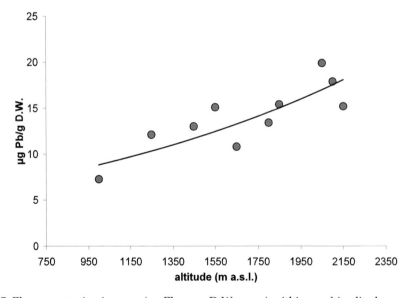

Fig. 5. Pb-concentration in moss (µg Pb per g D.W. moss) within an altitudinal transect (numbers refer to m a.s.l.) at the Gasteiner Tal (Salzburg) and the respective regression line (modified from Zechmeister, 1995).

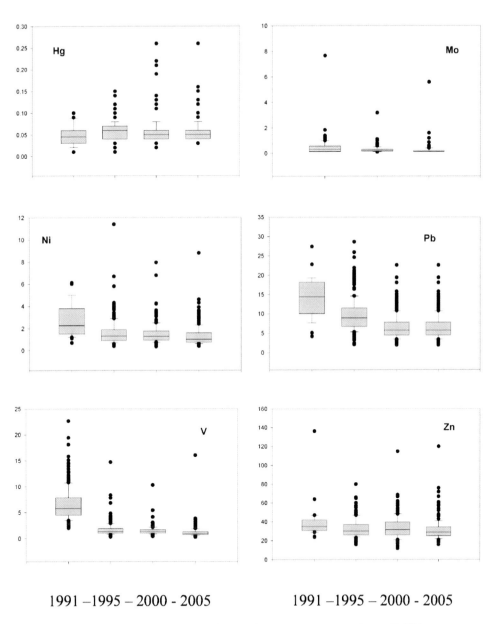

1991 – 1995 – 2000 - 2005   1991 – 1995 – 2000 - 2005

Fig. 6. Temporal trends of Hg, Mo, Ni, Pb, V and Zn concentrations ($\mu g\ g^{-1}$ D.W.) in mosses between 1991 and 2005 (n = 220).

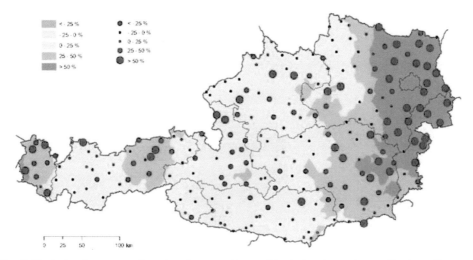

Fig. 7. Variation in pattern showing the overall sum (deviation from the median) of all trace metals (Al, Pb, V, S, Zn, Fe, Cu, Cr, Ni, Co, Mo, As, Hg, Cd) analysed by the moss method at 220 sites in Austria in 2005 (derived from Zechmeister et al., 2009). Dots show the measured values at individual experimental sites, the areas the result of ordinary kriging.

3.3 Trace metals in forest soils

Trace metal data were obtained from the Austrian Forest Soil Monitoring System which is part of the Pan-European forest soil condition inventory. The first survey (organic surface layer; synon. 'litter layer', 0-10, 10-20, 20-30 and 30-50 cm) was conducted in 1987/89 at 511 plots describing the trace metal situation in Austrian forest soils (Mutsch 1992, 1998). A second investigation was conducted in 2006/2007 within the framework of the EU Project "BioSoil" with a reduced number of plots (139) of the former Austrian Forest Soil Monitoring System. At this time samples were taken from the litter layer and from the depth layers: 0-5, 5-10, 10-20, 20-40 and 40-80 cm. Data of both investigations were compared in order to detect temporal changes. Due to partly different sampling procedures of both investigations the following approach was necessary: the depth layer 0-10 cm of the first investigation was compared to the mean value of the depth layers 0-5 and 5-10 cm of the second one. In a similar way the depth layer 20-40 cm of the second investigation was compared to the mean value of the depth layers 20-30 and 30-50 cm of the first investigation. Thus it was possible to find out possible temporal changes within a period of about 20 years.

Depending on the concentrations in different soil depths the origin of the atmospheric trace metal input can be distinguished by the enrichment factor of the surface layer and upper mineral soil layer compared to the deepest investigated mineral soil layer (ratio of Pb- and Cd-content of the surface layer and of 0-10 cm compared to 30-50 cm). Within the frame of the first investigation (Austrian Forest Soil Monitoring System – 1987/89) the following mean enrichment factors for the surface layer were found: Zn (1.8 in carbonate influenced soils), Pb (3.5 in acidic soils) and Cd (5.6 in acidic soils). Moreover, the altitude of the plots showed an obvious elevation-dependent increase of Pb- and Cd-concentrations, especially in the upper mineral soil and thus an important influence of atmospheric deposition. This relatively high accumulation occurs in regions where high flow of transboundary air masses and high precipitation amounts are characteristic ("Nordstaulagen" of the Alps; Mutsch, 1992; Herman et al., 2000).

For all investigated heavy metals and for nearly all depth layers a declining trend of concentrations was detectable within the time interval 1987/89 and 2006/07. The most pronounced decrease was observed in the litter layer. In carbonate soils the decrease seems to be more distinct (Table 7).

Carbonate soils	Litter layer		0-10 cm		10-20 cm		20-40 cm	
	1987/89	2006/07	1987/89	2006/07	1987/89	2006/07	1987/89	2006/07
Cu	15.1	11.5	16.3	15.1	15.4	14.1	15.4	13.8
Zn	155	111	139	131	124	121	86	87
Pb	114	70	97	85	79	68	56	49
Cd	1.27	0.99	1.64	1.29	1.31	1.08	1.02	0.88
Hg	0.32	0.24	0.27	0.25	0.23	0.22	0.22	0.21

Silicate soils	Litter layer		0-10 cm		10-20 cm		20-40 cm	
	1987/89	2006/07	1987/89	2006/07	1987/89	2006/07	1987/89	2006/07
Cu	16.1	13.6	18.5	17.9	21.0	19.9	23.5	23.3
Zn	91	75	73	73	75	77	81	82
Pb	98	61	49	48	35	34	24	25
Cd	0.70	0.54	0.38	0.20	0.29	0.15	0.28	0.13
Hg	0.35	0.28	0.16	0.16	0.13	0.12	0.12	0.11

Carbonate soils	Litter layer		0-10 cm		10-20 cm		20-40 cm	
	1987/89	2006/07	1987/89	2006/07	1987/89	2006/07	1987/89	2006/07
Cu	15,1	11,5	16,3	15,1	15,4	14,1	15,4	13,8
Zn	155	111	139	131	124	121	86	87
Pb	114	70	97	85	79	68	56	49
Cd	1,27	0,99	1,64	1,29	1,31	1,08	1,02	0,88
Hg	0,32	0,24	0,27	0,25	0,23	0,22	0,22	0,21

Silicate soils	Litter layer		0-10 cm		10-20 cm		20-40 cm	
	1987/89	2006/07	1987/89	2006/07	1987/89	2006/07	1987/89	2006/07
Cu	16,1	13,6	18,5	17,9	21,0	19,9	23,5	23,3
Zn	91	75	73	73	75	77	81	82
Pb	98	61	49	48	35	34	24	25
Cd	0,70	0,54	0,38	0,20	0,29	0,15	0,28	0,13
Hg	0,35	0,28	0,16	0,16	0,13	0,12	0,12	0,11

Table 7. Mean values (mg kg^{-1}) of selected trace metals 1987/89 and 2006/07 of the BioSoil-plots.

The results proved the emission-reductions of Cu, Zn, Pb and Cd and consequently of deposition rates. The question was whether these reductions were also reflected by soils. For this purpose the differences of the element contents of approximately 130 plots (slightly different number of plots dependent on depth layer and element) were compared after the interval of about 20 years. Because the selected metals show a different behaviour of carbonate and silicate the evaluation is done separately for these two groups. The Figures 8a-e show the mean temporal changes between 1987/89 and 2006/07 for Cu, Zn, Pb, Cd and Hg for the litter layer and for the mineral soil layers 0-10 cm, 10-20 cm and 20-40 cm.

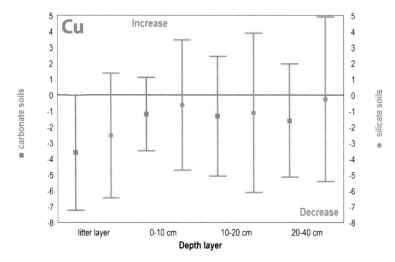

Fig. 8a. Differences (mg kg^{-1}) of the Cu-content 2006/07 – 1987/89 and standard deviations.

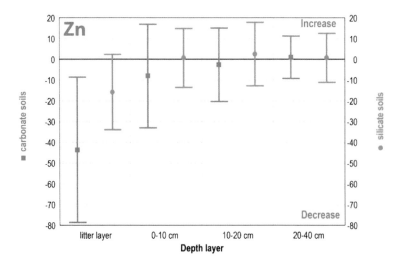

Fig. 8b. Differences (mg kg^{-1}) of the Zn-content 2006/07 – 1987/89 and standard deviations.

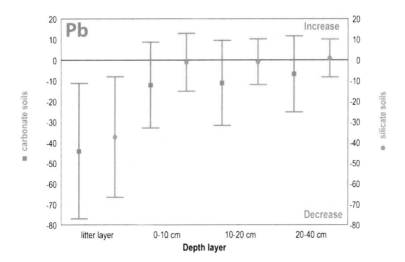

Fig. 8c. Differences (mg kg⁻¹) of the Pb-content 2006/07 – 1987/89 and standard deviations.

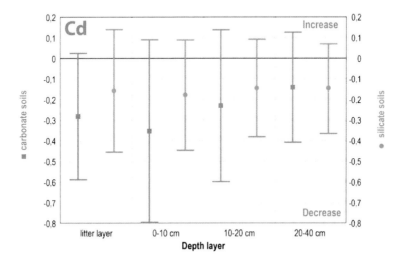

Fig. 8d. Differences (mg kg⁻¹) of the Cd-content 2006/07 – 1987/89 and standard deviations.

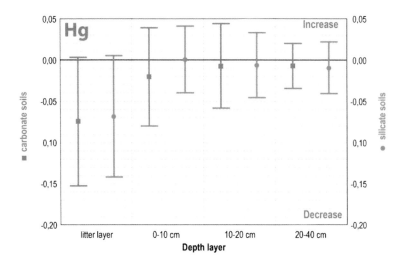

Fig. 8e. Differences (mg kg⁻¹) of the Hg-content 2006/07 – 1987/89 and standard deviations.

Cu: A decrease of Cu-contents is derived in both carbonate and silicate soils for all depth layers. The decrease in carbonate soils is more distinct than in silicate soils in terms of absolute values and relative changes. It can be hypothesised that in the acidic silicate soils the background contents of heavy metals are commonly higher and therefore the relative change by reduction of emitted Cu smaller (Figure 8a).

Zn: The decrease of Zn within 20 years in the litter layer of the carbonate soils is more distinct than in silicate soils. In the mineral soil layers no obvious changes could be detected: possible temporal changes are overlapped by local variation (Figure 8b).

Pb: A very clear decrease of Pb can be proved for the litter layers. It is the most distinct decrease found in this study. This tendency is weaker in carbonate mineral soils, whereas no change is detectable in acidic mineral soils. It seems that even the deposited but very immobile Pb disappears with time from the soil whereby a dynamic balance between input and output exists (Figure 8c).

Cd: Whereas for all the other discussed heavy metals the most obvious decrease can be found in the litter layers, for Cd also the decrease in the mineral soil layers is very clear. Cd is a very mobile element. Due to the current reduced Cd immissions the high Cd-levels measured in the past have not been maintained (Figure 8d).

Hg: Regarding Hg contents, the most distinct decrease is observed in the litter layers. For the mineral soils the tendency of changes is not as evident (Figure 8e).

Figure 9 gives an overview of the Hg contents in Austrian forest soils in different layers. In most cases the Hg contents are higher in the upper layers which indicates atmospheric import. Higher concentrations can be found nearby industrial regions.

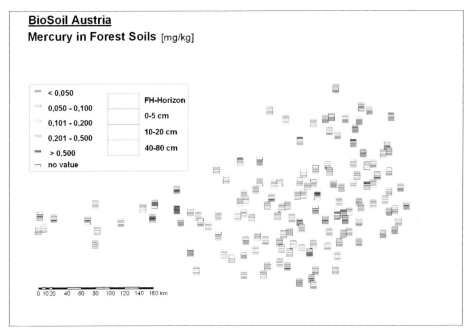

Fig. 9. Concentrations of Hg in forest soils (mg kg^{-1}). FH-horizon: soil litter layer.

Radionuclides: A typical depth distribution of radionuclides from a forest site in the Bohemian Massif (Upper Austria) is given in Figure 10.

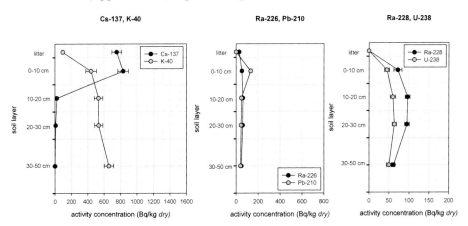

Fig. 10. Depth distribution of radionuclides in soil at a sample site in the Bohemian Massif, Upper Austria (2008).

The vertical depth distribution of Cs-137 showed that between 50 % and 90 % of the Chernobyl Cs-137 can be found in the uppermost soil layers with a peak in the organic rich layer of 0–10 cm depth. In a few soils, Cs-137 mass activity concentrations, which can be attributed to the global fallout of the nuclear weapon tests in the 1950ies and 1960ies, were detected down to a depth of 50 cm. The vertical depth distribution of K-40 showed that the amount of K-40 contamination is strongly influenced by local geological situations. Generally the K-40 mass activities increase with depth, which gives a further indication for the geogenic origin of K-40. Lowest K-40 mass activities were detected for the litter layers, whereas these low amounts are probably caused by a very rapid K-40 uptake via fine roots. Thus, the measured low K-40 mass activities in litter layers can be an indication for a very efficient cycling-mechanism. The vertical depth distribution of Pb-210 is similar to Cs-137. The maximum values were detected in the litter layer as well in the uppermost soil layers. According to the determined Pb-210ex values, it was concluded that the Pb-210 detected in the uppermost soil layers is mostly of atmospheric origin and in deeper soil layers of geogenic origin. The natural radionuclides Ra-226, Ra-228 and U-238 were almost exclusively detected in the mineral soil horizons.

The Chernobyl-derived contamination is still present and well detectable in Austrian forest ecosystems. The vertical depth distribution of forest soils in Upper Austria showed that up to 90 % of the Chernobyl Cs-137 can be found in the litter and in the uppermost soil. In general, the amount of Cs-137 is decreasing exponentially with depth. In only a few soils, Cs-137 mass activities – which can be attributed to the global fallout of the nuclear weapon tests in the 1950ies and 1960ies - were detected down to a depth of 50 cm.

4. Conclusions

General statements

The evaluation of the relevance of trace metals in Austrian forest ecosystems can be achieved by comparing the deposition input and the pools in the soil with applicable threshold values. In addition, plants themselves are bioindicators. We used the needle chemistry of both *Picea abies* and mosses which are known for their ability to accumulate trace metals.

Deposition

The input of Pb and Cd by means of wet deposition remained far below legal limits established in Austria and Switzerland and decreased during the last 20 years. The reduction of Pb loads was significant. Nevertheless, considering the permanent input of Pb into soils especially at "hot spots" nearby metal industry may lead to negative effects on soil microbes after a long term input over decades.

Bioindication with *Picea abies* needles

The Hg levels in the needle samples of the Austrian Bio-Indicator Grid are mostly low and below the toxic range. Only in the proximity of specific pollution sources elevated levels could be detected. The results show a significant downward trend from 1986 to 2006. A decrease during an almost 30 year period (1963-1991) from approx. 3,5 ppm down to 0,1 ppm was also found in the content of Pb in spruce needles of six plots in Austria (Herman, 1998). The inclusion of legal limiting values for Hg in Norway spruce needles would be useful.

Bioindication with mosses

Compared to other bioindicators (e.g. fungi, needles), mosses have the main advantage in the accumulation and assessment of actual wet, occult and dry deposition only. The uptake by mosses is a solely passive process and not interfered by any physiological activity of the moss. Another big advantage (e.g. compared to lichens) is the possibility to detect the annual increments of the monitoring species. The analysis of these increments provides information on the deposition within this period. There is no interference with deposits from previous periods because mosses gain the measured trace elements by atmospheric deposition only. Furthermore, aerogene soil particles deposited on the moss can be distinguished by a set of metals (e.g. Al, Fe).

A big task of current research will be the calibration of metal concentrations in mosses with data originating from time consuming and expensive established methods. Such investigations should enable to implement the bioindication method in legislation and to replace the established methods.

Trace metals in forest soils

Temporal decrease of Cu, Zn, Pb, Cd and Hg were proven between the first (Austrian Forest Soil Monitoring System – 1987/89) and the second soil inventory (BioSoil – 2006/07). The decrease is strong in the surface layer and applies to the mineral soil. Because emissions accumulate in (forest) soils these results are an indicator for the reduction of anthropogenic trace metal emissions in the last two decades and show that forest soils are a useful indicator for atmogenic pollution.

The presented results also demonstrate that temporal changes of selected parameters are detectable in soils. The obvious decrease of heavy metal concentration in the surface layer and no increase in deeper soil layers indicate that no accumulation takes place. Continuing decrease of heavy metal deposition will probably lead to distinct decreases in deeper layers in the future. The changes are explainable by cause-effect relationships (reduction of emissions) and support every effort in this direction. The success of environmental policy can be scrutinized by forest soil monitoring and forest soil investigations.

Radionuclides in forest soils

In Austrian forest ecosystems the Chernobyl-derived Cs-137 is still plant available due to the slow migration of the radionuclide, its retention in soil organic matter and the continuous Cs-137 cycling in forest ecosystems. Thus forest soils act as a long term sink for radio-caesium and as the main source for radioactive contamination of forest vegetation. Spruce needles are an appropriate indicator for radiation monitoring and integrated in environmental surveillance programmes, which are constructed for radiation protection of the public and the environment.

Abbreviations: Al: aluminium, As: arsenic, Ca: calcium, Cd: cadmium, Co: cobalt, Cr: chrome, Cs: caesium, Cu: copper, Fe: iron, Hg: amalgamate, K: potassium, Mo: molybdenum, Ni: nickel, Pb: lead, Po: polonium, Pt: platinum, Ra: radium, S: sulphur, Sn: tin, Sr: strontium, Tl: thallium, U: uranium, V: vanadium, Zn: zinc

5. References

Alloway B.J.; Reimer T. (1999). Schwermetalle in Böden: Analytik, Konzentration, Wechselwirkungen. Springer Berlin, ISBN 3-540-62086-9.

Andre O., Vollenweider P., Guenthardt-Goerg M.S. (2006). Foliage response of heavy metal contamination in Sycamore Maple (Acer pseudoplatanus L.). *For. Snow Lands. Res.* 80 (3), 275-288.

Anderl M., Köther T., Muik B., Pazdernik K., Stranner G., Poupa S., Wieser M. (2010). Austria's Informative Inventory Report (IIR) 2010, Submission under the UNECE Convention on Long-Range Transboundary Air Pollution Report. REP-0245.

Austrian Standards Institute / ON (2004). ON L 1075 – Principles for the evaluation of the content of selected elements in soils.

Bauer H., Smidt S., Stopper S., Herman F., Puxbaum H. (2008). Schwermetalleinträge in den Nordtiroler Kalkalpen. *CBl. F.d. ges. Forstwes.* 125 (2), 103-120.

Blum W.E.H., Klaghofer E., Kochl A., Ruckenbauer P. (1997). Bodenschutz in Österreich. Bundesministerium für Land- und Forstwirtschaft.

De Kok L., Stulen I. (eds., 1998). Responses of plant metabolism to air pollution and global change. Backhuys Publishers Leiden.

De Vries W., Bakker D.J. (1998). Manual for calculating critical loads of heavy metals for terrestrial ecosystems. DLO Winand Staring Centre, Wageningen, The Netherlands, SC Report 166, 1-144.

Dosskey M.G., Adriano D.C. (1991). Trace metal impact on plants: Mediation by soil and mycorrhizae. In: The deposition and fate of trace metals in our environment. Symposium Philadelphia, PA, October 8, 1991. Proceedings 105-115.

Elling W., Heber U., Polle A., Beese F. (2007). Schädigung von Waldökosystemen. Auswirkungen anthropogener Umweltveränderungen und Schutzmaßnahmen. Elsevier Amsterdam, New York, Tokio.

Federal Environment Agency (2004). Medienübergreifende Umweltkontrolle in ausgewählten Gebieten. *Monographien* M-168 Umweltbundesamt GmbH.

Fleck J.A., Grigal D.F., Nater E.A. (1999). Mercury uptake by trees: An observational experiment. *Water, Air, and Soil Pollution* 115, 513-523.

Fürst A., Smidt S., Herman F. (2003). Monitoring the impact of sulphur with the Austrian Bio-Indicator Grid. *Environmental Pollution* 125, 13-19.

Fürst A. (2007). Quecksilber in Fichtennadeln als Immissionsmarker. *Forstschutz aktuell* 41, 18-20. BFW-Wien, ISSN 1815-5103

Gerzabek M.H. (1992). Caesium-137 in soil texture fractions and its impact on Cesium-137 soil-to-plant transfer. *Commun. Soil Sci. Plant Anal.* 23, 321-330.

Grigal D.F. (2002). Inputs and outputs of mercury from terrestrial watersheds: A review. *Environ. Rev.* 10, 1–39.

Harmens H., Norris D.A., Steinnes E., Kubin E., Piispanen J., Alber R., Aleksiayenak Y., Blum O., Coşkun M., Dam M., De Temmerman L., Fernández A., Frolova M., Frontasevya M., González-Miqueo L., Grodzinska K., Jeran Z., Korzekwa, S., Krmar, M., Kvietkus, K., Leblond S, Liiv S, Magnússon, S., Maňkovská B., Pesch R., Rühling A., Santamaría J., Schröder W., Spiric Z., Suchara I., Thöni L, Urumov V., Yurukova L., Zechmeister H.G. (2010). Mosses as biomonitors of atmospheric heavy metal deposition: spatial and temporal trends in Europe. *Environmental Pollution* 158, 3144-3156.

Herman F. (1998). Investigation of the lead content of spruce needles in remote and rural areas over a thirty year period. *Environ. Sci. & Pollut. Res.*, Special Issue 1, 70-74.

Herman F., Smidt S., Huber S., Englisch M., Knoflacher M. (2000). Evaluation of pollution-related stress factors for forest ecosystems in Central Europe. *Environmental Science and Pollution Research* 8 (4), 231-242.

Hettelingh J.P., Denier van der Gond H., Groeneberg B.J., Ilyin I., Reinds G.J., Slootweg J., Travnikov O., Visschedijk A., de Vries W. (2006). Heavy metal emissions, depositions, Critical Loads and exceedances in Europe. VROM-DGM, Directie Klimaatverandering en Industrie, IPC 650, P.O. Box 20951, +32.30.274.3048.

Hock B., Elstner E.F. (1995). Pflanzentoxikologie. BI Wissenschaftsverlag, Bibliographisches Institut Mannheim - Wien - Zürich.

IAEA (2002). Modelling the migration and accumulation of radionuclides in forest ecosystems. Report of the Forest Working Group of the Biosphere Modelling and Assessment (BIOMASS) Theme 3. IAEA-BIOMASS-1, International Atomic Energy Agency, Vienna.

ICP Vegetation (2005). Heavy metals in European mosses: 2005/2006 survey. Monitoring manual. ICP Vegetation Programme Coordination Centre, CEH Bangor, UK. http://icpvegetation.ceh.ac.uk.

Kalina M., Puxbaum H., Tsakovski S., Simeonov V. (1999). Time trends in the concentrations of lead in wet precipitation from rural and urban sites in Austria. Chemosphere 38, 2509-2515.

Kazda M., Glatzel G. (1984). Schwermetallanreicherung und Schwermetallverfügbarkeit im Einsickerungsbereich von Stammablaufwasser in Buchenwäldern (Fagus sylvatica) des Wienerwaldes. *Zeitschrift für Pflanzenernährung und Bodenkunde* 147 (6), 743-752.

Krommer V., Zechmeister H.G., Roder I., Hanus-Illnar A. (2007). Monitoring atmospheric pollutants in the Biosphere Reserve Wienerwald by a combined approach of biomonitoring methods and technical measurements. *Chemosphere* 67, 1956-1966.

Larcher W. (2003). Physiological Plant Ecology. 4th Edition. Springer Berlin.

Markert B.A., Breure A.M., Zechmeister H.G. (2003). Bioindicators and Biomonitors. Principles, concepts and applications. Elsevier Amsterdam.

Mutsch F. (1992). Schwermetalle. In: Österreichische Waldbodenzustandsinventur, Waldbodenbericht. *Mitteilungen der Forstlichen Bundesversuchsanstalt* 168/II, 145-188.

Mutsch F. (1998). Indication of long-range transport of heavy metals based on the Austrian Forest Soil Monitoring System. *Environ. Sci. Pollut. Res.*, Special Issue 1, 81-87.

Offenthaler I., Dirnböck T., Grabner M.-T., Kobler J., Mirtl M., Riemer S. (2008). Long-trerm deposition of trace metals at the integrated monitoring site Zöbelboden. Umweltbundesamt, Report REP-0246.

Onianwa P.C. (2001). Monitoring atmospheric metal pollution: a review of the use of mosses as indicators. *Environmental Monitoring and Assessment* 71, 13–50.

Ormrod D.P. (1984). Impact of trace element pollution on plants. In: *Air Pollution and Plant Life* (M. Treshow, ed.), 291-319. John Wiley & Sons Chichester, New York, Brisbane.

Rea A.W., Keeler G.J. (1998). Microwave digestion and anaysis of foliage for total mercury by cold vapour atomic fluorescence spectroscopy. *Biogeochemistry* 40, 115-123.

Rea A.W., Lindberg S.E., Scherbatskoy T., Keeler G.J. (2002). Mercury accumulation in foliage over time in two northern mixed-hardwood forests. *Water, Air, and Soil Pollution* 133, 49-67.

Riederer M. (1991). Die Kutikula als Barriere zwischen terrestrischen Pflanzen und der Atmosphäre. *Naturwissenschaften* 78, 201-208.
Seidel C. (2010). Artificial and natural radionuclides in spruce needles in Upper Austria from 1983 to 2008 – an application for radioecological monitoring. Dissertation Universität für Bodenkultur Wien, 153 S.
Smidt S., Herman F. (2003). Evaluation of air pollution-related risks for Austrian mountain forests. *Environmental Pollution* 130, 99-112.
Strebl F., Bossew P., Kienzl K., Hiesel E. (2000). Radionuklide in Waldökosystemen. *Monographien* Band 59, Umweltbundesamt Wien, 73pp.
Thiessen K.M., Thorne M.C., Maul P.R., Pröhl G., Wheater H.S. (1999). Modelling radionuclides distribution and transport in the environment. *Environmental Pollution* 100, 151-177.
Vanmechelen L., Groenemans R., Van Ranst E. (1997). Forest Soil Condition in Europe. EC-UN/ECE, Brussels, Geneva.
VDI (Verein Deutscher Ingenieure; 1984). Schwermetalle in der Umwelt. UFOPLAN-Nr. 104 03 186. VDI Kommission Reinhaltung der Luft.
Walthert L., Zimmermann S., Blaser P., Luster J., Lüscher P. (2004). Waldböden der Schweiz Band 1. Grundlagen und Region Jura (150-163). Birmensdorf, Eidgenössische Forschungsanstalt WSL. Bern, Hep Verlag. 768 pp.
Weigel H.J., Helal H.M. (2001). Schwermetalle. In: Terrestrische Ökosysteme. Wirkungen auf Pflanzen, Diagnose und Überwachung, Wirkungen auf Tiere, 101-148 (Guderian R., Hrsg.).
Zechmeister H.G. (1995). Correlation between altitude and heavy metal deposition in the Alps. *Environmental Pollution* 89, 73-80.
Zechmeister H.G., Grodzinska K., Szarek-Lukaszewska G., (2003). Bryophytes. In: Markert B.A., Breure A.M., Zechmeister H.G. (Eds.), *Bioindicators/Biomonitors* (principles, assessment, concepts). Elsevier, Amsterdam, pp. 329–375.
Zechmeister H.G., Riss A., Hanus-Illnar A. (2004). Biomonitoring of atmospheric heavy metal depositions by mosses in the vicinity of local emission sources. *J. of Atmospheric Chemistry* 49, 461-477.
Zechmeister H.G., Hohenwallner D., Riss A., Hanus-Illnar A. (2005). Estimation of element deposition deriving from road traffic sources by mosses. *Environmental Pollution* 138, 238 - 249.
Zechmeister H.G., Dullinger S., Hohenwallner D., Riss A., Hanus-Illnar A., Scharf S. (2006). Pilot study on road traffic emissions (PAHs, heavy metals) measured by using mosses in a tunnel experiment in Vienna, Austria. *Environmental Science and Pollution Research* 13, 398-405
Zechmeister H.G., Hohenwallner D., Hanus-Illnar A., Hagendorfer H., Roder I., Riss A. (2008). Temporal patterns of metal deposition at various scales in Austria during the last two decades. *Atmospheric Environment* 42, 1301-1309.
Zechmeister H.G., Hohenwallner D., Hanus-Illnar A., Roder I., Riss A. (2009). Schwermetalldepositionen in Österreich – erfasst durch Biomonitoring mit Moosen (Aufsammlung 2005). Umweltbundesamtes Wien, Rept. 0201. ISBN 3-85457-999-3.
Ziegler H. (1988). Weg der Schadstoffe in der Pflanze. In: Hock B. & Elstner E. (Hrsg.) *Schadwirkungen an Pflanzen*, pp. 245-300. 2. Aufl. BI Wissenschaftsverlag Mannheim.

Plant Extracts: A Potential Tool for Controlling Animal Parasitic Nematodes

Pedro Mendoza de Gives*, María Eugenia López Arellano,
Enrique Liébano Hernández and Liliana Aguilar Marcelino
*Centro Nacional de Investigación Disciplinaria en Parasitología Veterinaria, INIFAP
México*

1. Introduction

Many plants play a crucial role in maintaining animal and human life in a natural balance with a tendency to establish an environmental armory among the different biosphere inhabitants. During evolution of living organisms in the biosphere biological interactions with other organisms are established and they affect each other in many ways. Different types of relationships are involved among organisms including parasitism. Heritable strategies of biological adaptation are developed by living organisms to overcome adverse environmental conditions. Plants have developed biochemical mechanisms to defend themselves from biological antagonists that act as their natural enemies (Ryan and Jagendorft, 1995). This principle has led scientists to search for bio-active compounds produced by plants against pathogens (Sheludko, 2010). Since long a number of plants and their metabolites are evaluated against diseases of importance not only in public health (Shah et al., 1987); but also in animal and agricultural production (Githiori et al., 2006). In the present chapter, the importance of using plant extracts as an alternative method of control of animal parasitic nematodes is reviewed from a broad perspective.

2. Use of plants as a source of phyto-medicines

Ancestral cultures worldwide developed, over many centuries, several cures and remedies from plants and plant extracts against many diseases affecting human populations and a traditional medicinal system based on empiric knowledge was established and was improved through time (Hillier and Jewel, 1983). Some devastating infectious diseases *ie.*, malaria, responsible for deaths of thousands of people can be overcome with traditional herbal anti-malarian drugs obtained from South America, Africa and Asia *ie.*, Cinchona (*Cinchona sp.*), Qing hao (*Artemisa annua*), Changshan (*Dichroa febrifuga*), Neem (*Azadirachta indica*), *Cryptolepsis sanguinolenta*) and other plants (Willcox et al., 2005). Researchers around the world have scientifically explored the real effect of many plants used as medicines

---

* Corresponding Author

whose uses are not validated by rigorous scientific experimentation. Many plants are being screened for anti-parasitic effects on animals; since animal behaviour reveals self-medication when animals select and ingest some specific plants (Cousing and Huffman, 2002; Huffman, 2003). In Hawaii, Rodriguez et al (1985) reported an interesting work based on a peculiar behavior of primates in the Hawaiian jungle. Researchers noticed that troops of chimpanzees ate some plants that they had previously selected and which they fed their progeny. Researchers suspected that such plants could contain alkaloids with hallucinogenic effects and probably could act as stimulants as alkaloid drugs do. After in-depth studies of those plants researchers found a group of bio-active compounds which can treat bacterial, fungal and nematodal infections. Hence, the researchers concluded that primate behavior responded to a mechanism of self/cure using selected plants as phyto-medicines (Sumer and Plutkin, 2000). During the last decades the study of medical principles from plants has gained considerable interest and a number of natural bio-active compounds from plant extracts are currently commercially available to cure many diseases.

3. Parasites of veterinary importance

Livestock industry worldwide is severely affected by a number of infectious diseases caused by different kinds of parasites. The present chapter focuses on the use of plant extracts against the group of internal parasites and particularly to helminths known as Gastrointestimal Parasitic Nematodes (GIN); considered to be one of the most economically important group of parasites affecting the animal productivity around the world (Poglaven and Battelli, 2006; Abdel-Ghaffar et al., 2011). The most frequent GIN of ruminants in many countries around the world are: *Haemonchus contortus, Mecistocirrus digitatus, Trichostrongylus colubriformis, T. axei, Bunostomum trigonocephalum, Cooperia curticei, Teladorsagia circumcincta, Nematodirus spp, Trichuris ovis, T. globulosa, Strongyloides papillosus, Gaigeria pachyscelis, Chabertia ovina* and *Oesophagostomum columbianum* (Torres Acosta et al., 2005; Valcárcel Sancho et al., 2009). In this group of parasites the nematodes have a remarkable status as the main pathogens causing severe damage to their hosts. *Haemonchus contortus* and other genera/species of nematodes belonging to the group of trichostrongylids are of major concern because its blood-sucking feeding habits cause anemia that can be so severe resulting in the death of the animals (Macedo Barragán et al., 2009). This group of parasites is widespread in almost all tropical and sub-tropical countries and is considered responsible for deteriorating animal health and productivity.

4. Chemotherapy as the unique method of control

The most common method used to control ruminant helminthiasis is the use of chemical compounds commercially available as anti-helmintic drugs that are regularly administered to animals for deworming; the method is considered simple, safe and cheap (Jackson, 2009). There are several disadvantages in the use of such products such as their adverse effect against beneficial microorganisms in soil once they are eliminated with the feces (Martínez and Cruz, 2009). On the other hand, some anthelmintic compounds can remain as contaminants in animal products destined for human consumption *ie.*, meat, milk, etc. (FAO, 2002). One of the main concerns in the use of anthelmintic drugs for controlling ruminant parasites is the development of anthelmintic resistance in the parasites that decreases the efficacy of the drugs (Sutherland and Leathwick, 2011; Torres-Acosta et al., 2011) and threatens economical sustainability of sheep production (Sargison, 2011). The anthelmintic resistance can reach enormous proportions when parasites develop mutations in their genome against different groups of anthelmintic drugs. Such phenomenon is known as "Multiple anthelmintic resistance" and it is a real threat to the inefficacy of commercially available anthelmintics (Taylor et al., 2009; Saeed et al., 2010). Such situation has motivated workers around the world to look for alternatives to control these parasites. Searching for plant bio-active compounds with medical properties against parasites has gained great interest in order to at least partially replace the use of chemical drugs.

5. Exploring the anti-parasitic properties of plants

A wide range of plants and their products around the world are being explored to look for their possible anthelmintic effects on cestodes and trematodes (Abdel-Ghaffar et al., 2011), and against nematodes (Datsu Kalip et al., 2011). Due to the important economic impact of gastrointestinal parasitic nematodes in the livestock industry around the world, most of the research on plant extracts are being focused on searching bioactive compounds from plants against this important group of parasites. Traditionally, some plants around the world are well known as anti-parasitic plants because they contain substances with anthelmintic effects against parasitic nematodes affecting agricultural crops (Krueger et al., 2009) or animal parasitic nematodes (Galicia Aguilar et al., 2008; López Aroche et al., 2008; De Jesús Gabino et al., 2010). Perhaps, the most known cases of plants with nematicidal properties around the world are garlic (*Allium sativum*) (Iqbal et al., 2001; Qadir et al., 2010) Marigold (*Tagetes erecta*) Krueger et al., 2009; Bhardwaj et al., (2010) and the goosefoot or Epazote (*Chenopodium abrosioides*) (Yadav et al., 2007; Eguale and Mirutse, 2009). Another example is the South African plant *Curtisia dentata* commonly used for ages by rural communities as a remedy to cure a number of diseases caused by bacteria and fungi in either human being or animals (Shai et al., 2008; Dold and Cocks, 2001) and against animal parasitic nematodes (Shai et al., 2009). Nevertheless, every year, the list of new plants with nematocidal *in vitro* and *in vivo* properties against animal parasites is growing as new natural alternatives for replacing (at least partially) the use of chemical drugs (Tables 1 and 2).

Some forage have been evaluated searching for potential bio-active compounds against sheep and goat parasitic nematodes with variable results. However studies must be intensified; since some individual limitations in application have been noticed; *ie.,* toxicity, metabolic disorders and inappropriate applications can cause severe damage and even the death of treated animals (Rahmann and Seip, 2008). Other plants are being investigated as bio-active forages in the control of *Haemonchus contortus* in lambs with good/moderate results. For instance Wormwood (*Artemisia absinthium*) which was offered to lambs for voluntary intake, parasitic burden was reduced almost in 50%. Additionally, faecal egg excretion expressed on a dry matter basis was also reduced by 73% in animals fed with the selected plant (Valderrábano et al., 2010). On the other hand, other plant/plant extracts *ie., Melia azedarach* (Chinaberrry tree, Indian Lilac) have shown promising results in trials that confirmed not only a very good anthelmintic efficiency, but also no side-effects (Akhtar and Riffat, 1984). Some plant extract have shown an extraordinary bio-activity against sheep

Plant	Extract	Target nematode	Anti-nematode Efficiency	Author
Adhatoda vasica	aqueous and ethanolic extracts	*H. contortus, O. circumcincta, Trichostrongylus spp,*	84-89% in vitro hatching egg inhibition	Al-Shaibani et al., 2008
Adhatoda vasica	aqueous and ethanolic extracts	*S. papillosus, Oe.columbianum*	81-85% larval development inhibition	Al-Shaibani et al., 2008
Tagetes erecta	Acetonic extract	*Haemonchus contortus* (L4)	99.7% lethal activity	Galicia Aguilar et al., 2008
Castela tortuosa	Hexanic extract	*Haemonchus contortus* (L4)	95.8% lethal activity	Galicia Aguilar et al., 2008
Prosopis laevigata	Hexanic extract	*Haemonchus contortus* (L3)	81% maximum mortality	López Aroche et al., 2008
Bursera copalifera	Acetonic extract	*Haemonchus contortus* (L3)	66% máximum mortality	López Aroche et al., 2008
Acacia pennatula, Lysiloma latisiliquum, Piscidia piscipula y Leucaena leucocephala	Acetone/water extracts	*H. contortus*	Variable range of larval migration inhibition using different *H. contortus* strains	Calderón Quintanal et al., 2010.
Salvadora persica	Aqueous extract	Strongyline nematodes	99.9% anthelmintic activity	Datsu et al., 2011
Terminalia avicenoides	Aqueous extract	Strongyline nematodes	100 % anthelmintic activity	Datsu et al., 2011

Table 1. In vitro nematocidal effect of different plant extracts against nematodes of livestock importance

Plant	Extract	Target nematode	Animal specie	% efficacy	Author
Artemisia absinthium	Crude Ethanolic extract	*Haemonchus contortus*	sheep	Faecal egg count reduction (FECR) of 90.46%	Tariq et al., 2008
Artemisia absinthium	Crude aqueous extract	*Haemonchus contortus*	sheep	Faecal egg count reduction (FECR) of 80.49%	Tariq et al., 2008
Prosopis laevigata	n-hexanic extract	*Haemonchus contortus*	Jirds	Parasitic burden was reduced in 42.5%	De Jesús Gabino et al., 2010
Parkia biglobosa	Aqueous extract	*Haemonchus, Trichostrongylus, Oesophagostomum* and *Bunostomum* species	Bovine	Produced a high hatching egg inhibition	Soetan et al., 2011
Piper tuberculatum	Oil extract	*Strongyloides venezuelensis*	*Rattus norvegicus*	No in vivo anthelmintic effect	Carvalho et al., 2011

Table 2. In vivo nematocidal effect of different plant extracts against nematodes of livestock importance

parasitic nematodes; *ie.*, supplementing sheep with a *Fumaria parviflora* ethanol extract eliminated fecal eggs and caused 72 and 88% mortality of adult *Haemonchus contortus* and *Trichostrongylus colubriformis*, respectively (Hördegen et al., 2003). These are only a few examples of candidate plant extracts to be used in the control of parasites in sheep and goat farming. Rochfort et al (2008) from Australia published a very complete and extraordinary review about bioactive plants and their impact on animal health and productivity. On the other hand, Diehl et al (2004) published the results of a very interesting research project evaluating eighty six plant extracts from Ivory Coast flora and finding that fifty percent of the evaluated plants had nematocidal activity against *Haemonchus contortus* larvae. Such results showed evidence about the important nematocidal activity of plants from Ivory Coast as potential ethnobotanical tools of control against ruminant parasitic nematodes (Diehl et al., 2004). Some recent reports of nematocidal activity of plant extracts against ruminant parasites in different countries are described as follows: In Pakistan, *Adhatoda vasica* both aqueous and ethanolic extracts exhibit an *in vitro* ovicidal and larvicidal activity ranging between 81-89% against diverse genera/specie of gastrointestinal parasitic nematodes of sheep (Al-Shaibani et al., 2008). In Burkina, Faso, two medicinal plants *Anogeissus leiocarpus* and *Daniellia oliveri* were analyzed to identify their anthelmintic effect against nematodes of sheep abomasum. *A. leiocarpus* and *D. oliveri* showed a maximum lethal activity, between 80 and 100%, respectively, against adult *Haemonchus contortus* (Aldama et al., 2009; Kaboure et al., 2009). Many countries have developed important screening of plant extracts with anthelmintic properties from their native flora with an enormous potential for the control of animal parasitic nematodes with encouraging results. Some countries *i.e.* Brazil, India, South Africa, China and others possesses an extraordinary richness in their medicinal flora and they have currently developed an important industry from plant extracts ably supported by science. Some researchers stand out for their important contributions in this regard: Githiori et al (2006) at the International Livestock Research Institute in Nairobi, Kenya; Iqbal et al (2001) and his group or researchers from the Department of Veterinary Parasitology, University of Agriculture, Faisalabad, Pakistan have developed a solid package of information about a big list of native plants with encouraging results in the control of sheep parasites (Iqbal et al., 2001; 2004; Bachaya et al., 2009).

6. Condensed tannin-rich plants

A number of research works have focused on the anthelmintic effect of tannin rich plants against GIN. This group of bio-active compounds present in selected plant material are being obtained from all over the world, from temperate areas (Athanasiadou et al., 2004; Hoste et al., 2006) as well as from tropical tannin rich fodders (Alonso-Diaz et al., 2010). Interdisciplinary groups of researchers (Hoste et al., 2006; Alonso Díaz et al., 2008; Calderón-Quintal et al., 2010; Martínez-Ortíz-de-Montellano et al., 2010) have developed important research studies on tannin-rich plants in the control of *H. contortus* and other important gastrointestinal nematodes. Most scientific works focused on identifying the bio-active compounds produced by nematocidal plants have reported the presence of different molecules including catechins, condensed tannins, flavonoids and steroids (Oliveira et al., 2009) and polyphenolics (Lorimer et al., 1996); as well as bio-active enzymes such as cystein protease and secondary metabolites such as alkaloids, glycosides and tannins

(Athanasiadou and Kvriazakis, 2004). Further in-depth studies need to be undertaken since even though anti-parasitic properties are being demonstrated, negative effects such as reduction in food intake by animals have been identified and this should be considered before establishing their use as an alternative method of control (Githiori et al., 2006).

In recent studies, researchers are reaching beyond the general knowledge about lethal *in vitro* activity of plants and bio-active compounds derived from selected plants against the most important nematode parasites of ruminants. New efforts are being carried out to find practical applications of plants or plant products in the control of ruminant parasitic nematodes; including ways and means of overcoming limitations in applications to animals (Rahmann and Seipa, 2007). Recently in Laos, reduction in appearance of nematode eggs on goat feces with the Cassava foliage supplement has been demonstrated (Phengvichith and Preston, 2011).

7. Conclusions

The use of chemical anthelmintic drugs for controlling animal parasitic nematodes is rapidly loosing popularity due to a number of disadvantages. Anthelmintic resistance in the parasites is spreading and the inefficacy of chemical anti-parasitic compounds is threatening animal health. New plants with medicinal properties against parasites of ruminants are being investigated around the world with promising results. In the near future natural products obtained from plants extracts seems that likely will become a viable alternative of control of parasitizes of veterinary importance. When plant/plant extracts are being selected for use as anti-parasitic drugs in sheep particular attention should be given to the fact that the bio-active compound could be found in stems, roots, leaves, flowers, fruits or even in the entire plant. This means that obtaining plant extracts is a laborious and complex process. Also, the mode of extraction and the solvent used can determine the success in isolating the expected bioactive compounds; since a wide variety of compounds can be hidden into the structural parts of the plants and the only way they could be isolated is through exploring the use of a range of organic solvents. On the other hand, a rigorous effort to identify possible side effects due to the administration of plant extracts should be established before carrying *in vivo* assays. It is remarkablly important to consider that using plant/plant extracts as a unique method of control is insufficient to control itself the parasitosis in the animals. So, an alternated or combined method with other methods of control should be considered as an integrated method which would lead to reduce the use of chemical anthelmintic drugs.

8. Acknowledgments

Authors whish to express their gratitude to Dr. Felipe Torres Acosta (Autonomous University of Yucatan, Mexico) for his valuable comments on this chapter.

9. References

[1] Abdel-Ghaffar, F., Semmler, M., Khaled, A., Al-Rasheid, S., Strassen, B., Fischer, K., Aksu, G., Klimpel, S., Mehlhorn, H. (2011) The effects of different plant extracts on intestinal cestodes and on trematodes. Parasitology Research (2011) 108:979–984.

[2] Akhtar, M.S. and Riffat, S. 1984. Efficacy of *Melia azedarach* Linn. fruit (Bahain) and Morantel against naturally acquired gastrointestinal nematodes in goats. Pakistan Veterinary Journal, 4: 176-9.

[3] Aldama, K., Belem A. M. Gaston, M., Hamidou, H., Amadou, T., and Sawadogo Laya, S. (2009) In vitro anthelmintic effect of two medicinal plants (*Anogeissus leiocarpus* and *Daniellia oliveri*) on *Haemonchus contortus*, an abosomal nematode of sheep in Burkina Faso. African Journal of Biotechnology 8 (18)4690-4695.

[4] Alonso-Díaz, M.A., Torres-Acosta, J.F.J., Sandoval-Castro, C.A., Capetillo-Leal, C., Brunet, S., Hoste, H. (2008) Effects of four tropical tanniniferous plant extracts on the inhibition of larval migration and the exsheathment process of *Trichostrongylus colubriformis* infective stage. Veterinary Parasitology 153:187–192.

[5] Alonso-Díaz, M.A., Torres-Acosta, J.F.J., Sandoval-Castro, C.A., Hoste, H. (2010) Tannin in tropical tree fodders fed to small ruminants: A friendly foe? Small Ruminant Research 89:164-173.

[6] Al-Shaibani, I.R.M., Phulan, M.S., Arijoand, A., Qureshi, A. (2008) Ovicidal and larvicidal properties of *Adhatoda vasica* (L.) extracts against gastrointestinal nematodes of sheep *in vitro*. Pakistan Veterinary Journal. 28(2): 79-83.

[7] Athanasiadou, S., Kvriazakis. I. (2004) Plant secondary metabolites: antiparasitic effects and their role in ruminant production systems. Proceedings of the Nutrition Society. 63(4):631-639.

[8] Bachaya, H.A., Z. Iqbal, M.N. Khan, A. Jabbar, A.H. Gilani and I.U. Din, 2009. *In vitro* and *In vivo* anthelmintic activity of *Terminalia arjuna* bark. International Journal of Agricultural Biology, 11: 273–278.

[9] Bhardwaj, P, Varshneya, C. and Mittra, S. (2010) Anthelmintic efficacy of *Bauhinia variegata* and *Tagetes patula* against *Haemonchus contortus*. The Indian Veterinary Journal 87(12): 1204-1206.

[10] Calderón-Quintal, J.A., Torres-Acosta, J.F.J., Sandoval-Castro, C.A., Alonso-Díaz, M.A., Hoste, H., Aguilar-Caballero, A. (2010) Adaptation of *Haemonchus contortus* to condensed tannins: can it be possible? Archives of Medical Veterinary 42, 165-171.

[11] Carvalho, O.C., C. Chagas, C.A., Cotinguiba, F., Furlan, M., G. Brito, G.L., Chaves, C.M.F., Stephan, P.M., Bizzo, R.H., Alessandro, F. T., Amarante, F.T.A (2011) The anthelmintic effect of plant extracts on *Haemonchus contortus* and *Strongyloides venezuelensis* Veterinary Parasitology (August 2011) (In Press).

[12] Cousins, D. and Huffman, A.M. (2002) Medicinal properties in the diet of Gorillas: An ethno-pharmacological evaluation. African Study Monographs, 23(2): 65-89.

[13] Datsu Kalip, R., Slyranda Baltini, A., Wycliff, A., Abdulrahaman, F.I. (2011) Preliminary phytochemical screening and *in vitro* anthelmintic effects of aqueous extracts of *Salvadora persica* and *Terminalia avicennoides* against strongyline nematodes of small ruminants in Nigeria. Journal of Animal and Veterinary Advances. 10(4):437-442.

[14] De Jesús Gabino, A.F., Mendoza de Gives, P., Salinas Sánchez, D.O., López Arellano, Ma. E., Liébano Hernández, E., Hernández Velázquez, V.M. and Valladares Cisneros, G. (2010) Anthelmintic effects of *Prosopis laevigata* n-hexanic extract against *Haemonchus contortus* in artificially infected gerbils (*Meriones unguiculatus*) Journal of Helminthology 84:71-75.

[15] Diehl, M.S., Kamanzi Atindehou, K., Téré, H., Betschart, B. (2004) Prospect for anthelminthic plants in the Ivory Coast using ethnobotanical criteria. Journal of Ethnopharmacology 95(2-3)277-284.

[16] Dold, A.R. & Cocks, M.L. (2001). Traditional veterinary medicine in the Alice district of the Eastern Cape Province, South Africa. South African Journal of Science, 97:375-379.

[17] Eguale, T. and Mirutse, G. (2009) In vitro anthelmintic activity of three medicinal plants against *Haemonchus contortus*. International Journal of Green Pharmacy 3(1): 29-34.

[18] FAO (2002) Evaluation of Certain Veterinary Drugs Residues in Food. WHO Technical Report No 911. Fifty-eighth report of the Joint FAO / WHO Expert Committee on Food Additives. Geneva 2002. 62p.

[19] Galicia-Aguilar, H.H., Mendoza de Gives, P., Salinas-Sánchez, D., López-Arellano, Ma. E., Liébano-Hernández, E. (2008) In vitro nematocidal activity of plant extracts of the Mexican flora against *Haemonchus contortus* fourth larval stage. Annals of the New York Academy of Science, 1149:158-160.

[20] Githiori, B.J., Athanasiadou, S. and Thamsborg, M.S. (2006) Use of plants in novel approaches for control of gastrointestinal helminths in livestock with emphasis on small ruminants. Veterinary Parasitology 139(4):308-20.

[21] Hördegen, P., H. Hertzberg, J. Heilmann, W. Langhans and V. Maurer, (2003). The anthelmintic efficacy of five plant products against gastrointestinal trichostrongylids in artificially infected lambs. Veterinary Parasitology, 117: 51-60.

[22] Hoste, H., Jackson, F., Athanasiadou, S., Thamsborg, M.S. and O. Hoskin, O.S. (2006) The effects of tannin-rich plants on parasitic nematodes in ruminants. Trends in Parasitology, 22(6):253-261.

[23] Hillier, S.M. and Jewel, J.A. (1983) Chinese traditional Medicine and Modern Western Medicine: Integration and separation in China. In: *Health Care and Traditional Medicine in China 1800-1982. Boston: Routledge & Kegan Paul, 1983.* 221-241.

[24] Huffman, F. A. (2003) Animal self-medication and ethno-medicine: exploration and exploitation of the medicinal properties of plants. Proceedings of the Nutrition Society 62(2):371-381.

[25] Iqbal, Z., Khalid Nadeem, Q., Khan, M.N., Akhtar, M.S. and Nouman Waraich, F. (2001) In vitro Anthelmintic Activity of *Allium sativum, Zingiber officinale, Curcubita mexicana* and *Ficus religiosa* International Journal of Agricultural Biology, 3(4):454-457.

[26] Iqbal, Z., Lateef, M., Ashraf, M., Jabbar, A. (2004) Anthelmintic activity of *Artemisia brevifolia* in sheep. Journal of Ethnopharmacology, 93(2-3):265-8.

[27] Jackson, F. (2009) Worm control in sheep in the future. Small Ruminant Research. 86:40-45.

[28] Kaboure, A., Belem A. M. G., Gaston, M., Tamboura H. H., Traore, A. and Sawadogo, L. (2009) In vitro anthelmintic effect of two medicinal plants (*Anogeissus leiocarpus* and *Daniellia oliveri*) on *Haemonchus contortus*, an abomasal nematode of sheep in Burkina Faso. African Journal of Biotechnology 8 (18): 4690-4695.

[29] Krueger, R., Dover, E.K., McSorley, R., Wang, K-H. (2009) Marigold (*Tagetes* spp) for nematode management. University of Florida IFAS Extension, Publication ENY-056.

[30] López Aroche, U., Salinas Sánchez, D.O., Mendoza de Gives, P., López Arellano, Ma. E., Liébano Hernández, E., Valladares Cisneros, G., Arias Ataide, D.M. and Hernández Velázquez, V. (2008) In vitro nematicidal effect of medicinal plants from the Sierra de Huautla Biosphere Reserve, Morelos, Mexico against *Haemonchus contortus* infective larvae. Journal of Helminthology 82 (1): 25-31.

[31] Lorimer, D.S., Perry, B.N., Foster, M.L. and Burgess, E. (1996) A Nematode Larval Motility Inhibition Assay for Screening Plant Extracts and Natural Products. Journal of Agriculture Food Chemistry, 1996, *44* (9), pp 2842–2845

[32] Macedo Barragán, R., Arredondo Ruiz, V., Ramírez Rodríguez, J., García Márquez, L.J. (2009) Grazing sheep poisoned by milkweed *Asclepias curassavica* or gastrointestinal nematosis? A case report findings Veterinaria México, 40 (3):275-281.

[33] Martínez, M. I. and Cruz, R.M. (2009) The use of agricultural and livestock chemical products in the cattle-ranching area of Xico, central Veracruz, Mexico, and their possible environmental impact. Acta Zoológica Mexicana 25(3):637-681.

[34] Martínez-Ortíz-de-Montellano, C., Vargas-Magaña, J.J., Canul-Ku, H.L., Miranda-Soberanis, R., Capetillo-Leal, C., Sandoval-Castro, C.A., Hoste, H. and Torres-Acosta, J.F.J. (2010) Effect of a tropical tannin-rich plant *Lysiloma latisiliquum* on adult populations of *Haemonchus contortus* in sheep. Veterinary Parasitology 172(3-4):283-90.

[35] Oliveira, L.M.B., Bevilaqua, C.M.L., Costa, C.T.C., Macedo, I.T.F., Barros, R.S., Rodrigues, A.C.M., Camurca-Vasconcelos, A.L.F., Morais, S.M., Lima, Y.C., Vieira, L.S., Navarro, A.M.C.(2009) Anthelmintic activity of *Cocos nucifera* L. against sheep gastrointestinal nematodes. Veterinary Parasitology 159, 55–59.

[36] Phengvichith, V. and Preston, R.T. (2011) Effect of feeding processed cassava foliage on growth performance and nematode parasite infestation of local goats in Laos. Livestock Research for Rural Development 23(1).

[37] Poglaven, G. and Battelli, G. (2006) An insight into the epidemiology and economic impact of gastro-intestinal nematodes in small ruminants. Parasitologia 48(3):409-413.

[38] Qadir, S., Dixit, K.A., Dixit, P. (2010) Use of medicinal plants to control *Haemonchus contortus* infection in small ruminants. Veterinary World, 3(11)515-518.

[39] Rahmann, G. and Seip, H. (2008) Bioactive forage and phytotherapy to cure and control endo-parasite diseases in sheep and goat farming systems – a review of current scientific knowledge. *In: Landbauforschung Völkenrode*. Bundesforschungsanstalt für Landwirtschaft, pp. 285-295.

[40] Rodriguez, E., Aregullin, A.M., Nishida, T., Wrangham, R., Abramkowski, Z., Finlayson, Towers, H.N.G. (1985) Thiarubrine A, a bioactive constituent of Aspilia (Asteraceae) consumed by wild chimpanzees. Experientia 41: 419.

[41] Ryan, A.C. and Jagendorft, A. (1995) Self defense by plants. Proceedings of the Natural Academy of Sciences. USA 92:4075 Colloquium Paper.

[42] Rochfort, S., Parker, J.A., Dunshea, R.F. (2008) Plant bioactives for ruminant health and productivity. Phytochemistry 69 (2008) 299–322.

[43] Saeed, M., Iqbal, Z., Jabbar, A., Masood, S., Babar, W., Saddiqi, H.A., Yaseen, M., Sarwar, M., Arshad, M. (2010) Multiple anthelmintic resistance and the possible contributory factors in Beetal goats in an irrigated area (Pakistan). Research in Veterinary Science. 88(2):267-72.

[44] Sargison, N.D. (2011)Pharmaceutical control of endoparasitic helminths infections in sheep. Veterinary Clinical North American Food Animal Practice. 27:139-156.

[45] Shah, V., Sunder, R., De Sousa, N.J. (1987) Chonemorphine and Rapanone Antiparasitic Agents from Plant Resources. Journal of Natural Products. 50(4)730-731.

[46] Shai, L.J., Bizimenyera, E.S., Bagla, V., McGaw, L.J., Eloff, J.N. (2009) *Curtisia dentata* (Cornaceae) leaf extracts and isolated compounds inhibit motility of parasitic and free-living nematodes. Onderstepoort Journal of Veterinary Research, 76(2):249-56.

[47] Shai, L.J., McGaw, L.J., Aderogba, M.A., Mdee, L.K., Eloff, J.N. (2008) Four pentacyclic triterpenoids with antifungal and antibacterial activity from *Curtisia dentata* (Burm.f) C.A. Sm. Leaves. Journal of Ethnopharmacology. 26,119(2):238-44.

[48] Soetan, K.O., Lasisi, O.T. and Agboluaje, A.K. (2011) Comparative assessment of *in-vitro* anthelmintic effects of the aqueous extracts of the seeds and leaves of the African locust bean (*Parkia biglobosa*) on bovine nematode eggs. Journal of Cell and Animal Biology Vol. 5 (6), pp. 109-112.

[49] Sheludko, Y.V. (2010) Recent Advances in Plant Biotechnology and Genetic Engineering for Production of Secondary Metabolites. Cytology and Genetics, 44(1): 52–60.

[50] Sumer, J. and Plutkin, M (2000) Chimpazees and self-medication. In: The Natural History of Medicinal Plants Timber Press Inc. Portland, Oregon, USA.

[51] Sutherland, A.I. and Leathwick, M.D. (2011) Anthelmintic resistance in nematode parasites of cattle: a global issue? Trends in Parasitology, 27: 176-181.

[52] Tariq, K.A., Chishti, M.Z., Shawl, A.S. (2008) Anthelmintic activity of extracts of *Artemisia absinthium* against ovine nematodes. Veterinary Parasitology 160(1-2)83-88.

[53] Taylor, M.A., Learmount, J., Lunn, E., Morgan., C. Craig, B.H. (2009) Multiple resistance to anthelmintics in sheep nematodes and comparison of methods used for their detection. Small Ruminant Research 86: 67-70.

[54] Torres-Acosta, J.F.J., Mendoza-de-Gives, P., Aguilar-Caballero, A.J., Cuéllar-Ordaz, J.A. (2011) Anthelmintic resistance in sheep farms: update of the situation in the American continent. Veterinary Parasitology (Accepted for publication).

[55] Torres-Acosta, J.F.J. and Aguilar-Caballero, A.J. (2005). Epidemiología, prevención y control de nematodos gastrointestinales en rumiantes. *En: Rodríguez, V.I. (Editor). Enfermedades de Importancia Económica en Producción Animal. Editorial McGraw-Hill Interamericana / UADY. Pp.143-173.*

[56] Valcárcel-Sancho, F., Rojo-Vázquez, F.A., Olmeda-García, A.S., Arribas-Novillo, B., Márquez-Sopeña, L., Fernando-Pat, N. (2009). Atlas de Parasitología Ovina. Editorial Servet. Zaragoza, España, 137p.

[57] Valderrábano, J., Calvete, C. Uriarte, C. (2010) Effect of feeding bioactive forages on infection and subsequent development of *Haemonchus contortus* in lamb faeces. Veterinary Parasitology, 172 (1-2)89-94.

[58] Willcox, M., Bodeker, G., Rasoanaivo, P. (2005) Traditional Medicinal Plants and Malaria: Volume 4 of the Traditional Herbal Medicine for Modern Times Series. The Journal of Alternative and Complementary Medicine. 11(2): 381-382.
[59] Yadav, N., Vasudeva, N., Sing, S., Sharma, K.S. (2007) Medicinal properties of genus *Chenopodium* Linn. Natural Product Radiance 6(2): 131-134.

7

Understanding Linkages Between Public Participation and Management of Protected Areas – Case Study of Serbia

Jelena Tomićević[1], Ivana Bjedov[1],
Margaret A. Shannon[2] and Dragica Obratov-Petković[1]
[1]*University of Belgrade – Faculty of Forestry, Belgrade,*
[2]*The Rubenstein School of Environment and Natural Resources, 333 George D. Aiken Center, University of Vermont, Burlington, Vermont*
[1]*Serbia*
[2]*USA*

1. Introduction

The issue of participation is an important issue in protected area management. For instance, the IV IUCN World Congress on National Parks and Protected Areas convened in Caracas, Venezuela, called for increased community participation and human equity in decision-making for protected areas in order to improve their management (IUCN, 1993). The term *participation* can be interpreted in very different ways, and therefore it is essential to define it carefully.

Until the 1970s, participation of local people in conservation was often seen as a tool to achieve the local approval to protected area plans, and participation was almost a mere public relations exercise. During the 1980s, participation of the local people was regarded as a mechanism to gain better results in natural resource protection, while in the 1990s, participation has been interpreted more and more as a means to involve local people in protected area management (Pimbert & Pretty, 1997).

It is now widely assumed that participation is required in order to achieve sustainable and effective conservation, particularly in protected areas; that it can bring economic and social benefits to marginalised groups; and that devolution of decision-making will benefit biodiversity (Jeanrenaud, 1999). 'Participatory approaches provide opportunities for the poor to contribute constructively to development' (FAO 1990, p.4; FAO 2001). The FAO People's Participation Programme believes that 'participatory approach is an essential part of any strategy and its call for 'the active involvement and organization of grass roots level of the rural people' (FAO 1990a, p.5).

As sustainability is defined in ecological, economic, and social terms, participation, as a democratic means of decision-making, has been increasingly recognised 'as an essential means and end to the development of the social dimensions of sustainability' (Finger-Stich & Finger, 2003, p.1).

According to Finger-Stich & Finger (2003), 'participation' is defined as, "the voluntary involvement of people who individually or through organised groups deliberate about their respective knowledge, interests, and values while collaboratively defining issues, developing solutions, and taking – or influencing – decisions". Furthermore, defining who can participate will lead to different types of participation processes. Finger-Stich & Finger (2003) distinguished three main types of participation: public participation, representative participation and community participation. This research focuses on community-based participation processes.

Public participation, collaborative management, and community-based management as types of participation may not always be distinct. For example, FAO/ECE/ILO define public participation as: *"a voluntary process whereby people, individually or through organised groups, can exchange information, express opinions and articulate interests, and have the potential to influence decisions or the outcome of the matter at hand"* (FAO/ECE/ILO, 2000, p.9). And Renard (1997) defines collaborative or co-management as: *"a situation in which two or more social actors negotiate, define and guarantee amongst themselves a fair sharing of the management functions, entitlements and responsibilities for a given territory, area or set of natural resources"* (Borrini-Feyerabend et al., 2000, p.1). Co-management (short for collaborative or joint management) – this term has been defined as, *"...durable, verifiable and equitable forms of participation, involving all relevant and legitimate stakeholders in the management and conservation of resources"* (Renard, 1997). 'They may be complementary and evolve into one another over time' (Finger-Stich & Finger, 2003, p. 28). For example, a protected area policy may be drafted in consultation with the general public at the regional and/or national level, then there may be a co-management body to monitor the management of a particular protected area, and it may work in partnership with community-based associations to adapt this management to particular places, activities, and social groups (Finger-Stich & Finger, 2003, p. 28).

In order to understand the meaning of participation, as well as participation processes, the following definitions and understandings are collected from different authors: Participation processes, whatever their type, have the potential to evolve and provide space and opportunities for social learning (Korten, 1990). Participatory theories, such as social forestry (Korten, 1981), emphasize policy-making based on direct citizen participation, ahead of expertise and citizen representative structures. These theories propose a restructuring of institutional arrangements to accommodate greater citizen deliberation. In the field of social forestry, Korten (1980) identified several weaknesses in early traditional community development programmes, which he attributed partly to inappropriate governance structures. He maintained that new arrangements can be achieved through *"innovative social learning (which emphasizes) central facilitation over central control, performance monitoring and self-correction over planning, encourages local initiative and self-control, and reflects a tolerance for the ambiguity and uncertainty inherent in the learning process"* (Korten, 1981, p.613).

"While understanding that all participatory processes entail communicative action, it is useful to recognise that in the situation where problems are being defined and actors are forming or changing their roles, the essence of the participatory process is communicative action. This means that the degree of instrumental or strategic policy development is low since there is not a clear public problem and no organised social interests. Indeed, one can expect this part of the policy process to possibly extend over years as the nature of the public problem is slowly understood and shared understanding emerges through dialogue between the actors" (Shannon, 2003, pp.147-148).

Thus, communicative action leads to a better understanding of the actors, stakeholders and interests and why they are associated with this problem (Finger-Stich & Finger, 2003).

"Participation processes are both a way to manage conflict by seeking compromise between various interests, and they are also a means of developing more creative solutions that would not have emerged without the interaction of stakeholders. The decisions born out of such collaborative thinking and negotiation have the advantage of being the product of all those taking part, and are therefore more likely to be effective. Effective participation is a means and an outcome of collaborative learning" (Finger-Stich & Finger, 2003, p.41).

In general, scholars have agreed about the main points of participation, namely: learning process, communicative action and participation as a means and as well as an outcome of collaborative learning.

One promising overall approach to building cooperation between local people and protected area managers is 'collaborative management' or 'co-management' of protected areas – a partnership whereby various stakeholders agree to share amongst themselves the management functions, rights, and responsibilities for a territory or set of resources under protected area status (Borrini-Feyerabend, 1996).

In recent years, there has been a growing interest in the integrated management of protected areas, which means the ample participation of the local people in the decision-making and management of the area (Ghimire & Pimbert 1997; Orlove & Brush 1996; Shyamshundar 1996; Wells & Brandon 1993).

In our case study, the focus is on the role of public participation in the management of the Special Nature Reserve Zasavica (SNR Zasavica) in Serbia. The aim of this study was to analyse and describe how managers of the SNR Zasavica work with local communities in order to achieve biodiversity conservation.

2. Methods

A qualitative approach to the study was chosen, with triangulation of different data collection methods. Understanding a situation in its entirety and characterization by a number of specific principles like subject orientation, adequacy of theories and methods, reflexivity of the researcher and research are characteristics of qualitative research (Tomićević, 2005). In-depth expert interviews were used in order to collect a great deal of 'rich' information from relatively few people (Veal, 1992). The expert interviews were held during spring 2009 in Belgrade. They included expert from the Institute for Nature Conservation of Serbia, expert from IUCN for SEE and one resource manager of SNR Zasavica. Topics discussed during the interviews were: achievements and development projects that managers of this reserve accomplish. Furthermore, the issue of the participation of local people in the management of protected area was discussed, as well as the way in which local residents are involved in the management of SNR Zasavica. The purpose for expert interviews was not only to provide the personal attitudes towards the Special Nature Reserve Zasavica, but also to obtain the broader understanding of the relationships between different stakeholders. Therefore, this study encompasses both an interest in understanding the specific circumstances of Special Nature Reserve Zasavica and the broader role of public participation in a management of protected areas.

Furthermore, secondary data, i.e. data gathered by research of literature and different case studies which deals with issues of public participation, were considered in the analysis undertaken. Legal, strategy, and other institutional and statutory documents were also analyzed as a basis for understanding the management of protected areas in Serbia.

3. Results and discussion

Managing protected areas is essentially a social process (Lockwood & Kothari, 2006). The traditional approaches to protected areas management are currently being challenged. Indeed, protected areas are undergoing a shift from a preservationist paradigm towards an integrated approach. This process is reflective of social changes. These social changes increased interest in, and demand for, participation in decision-making processes (Tomićević, 2005).

Guidelines for effective participation processes include encouragement of all stakeholders to contribute; opportunities for participation in a manner that best suits the particular understandings, needs and contributions of each participant, and ensuring that participants have access to all relevant information (O' Riorden and O'Riorden, 1993; Moote et al, 1997).

Participation is often promoted by government as a mechanism for giving participant power to influence policy outcomes. However, despite some participatory processes offering opportunities for citizens to express views, and perhaps have an influence at the margin, the core policy agenda and framework may often largely remain under the control of governments (Lockwood & Kothari, 2006).

In general, Balkan Peninsula is one of the richest regions in Europe in terms of biodiversity, but is suffering as a consequence of a decade of conflict followed by political and economic crisis. The fact that most of the protected areas are situated in isolated and poor rural regions makes the situation even worse. Problems are especially noticeable in Serbia, where political, social and economic conflicts combined with conflicts of interests over the use of natural resources by different groups and individuals have resulted in a people moving from rural villages to cities in order to survive (Tomićević, 2005).

In the case of Serbia, protected areas management system has been characterized by a top-down approach. Serbia has a long history of centralized planning for and management of protected areas. Recent research show some improvements towards participatory management approaches in protected areas in Serbia (Tomićević, 2005; Tomićević et al. 2010, Tomićević et al., 2011).

Based upon research in 2004 from the case study from Tara National Park we learned that while local people are generally willing and interested in engaging in participatory management, there are currently no opportunities for the kinds of deliberative discussions regarding management priorities or implementation strategies. The only clear relationship between the local people and Park administration is through direct employment. From the perspective of the Park administration, engaging in collaborative planning with the local people requires support from the State. Regardless of the personal interest of a park manager or the willingness of local people to work with park managers, without adequate resources and commitment, participatory management will not move forward (Tomićević et al., 2010).

The findings of the Tara study indicate the need to strengthen the clarity of nature conservation policy and the missions of the responsible authorities. In addition, in order to promote the involvement of local people and empower the national park management to work with them collaboratively, it is necessary to promote communication among all stakeholders (Tomićević et al., 2010).

Furthermore, research findings from NP Kopaonik in Serbia from 2011 indicate that the poverty of the local people and conflict between the local people and the management authorities are the biggest threat to conservation of plant species and natural resources generally in protected areas (Tomićević et al., 2011).

In general, conservation policies that attempt to keep communities out of the decision-making process and/or benefit sharing are unlikely to be sustainable for a long period. Community support is needed to achieve long-term conservation objectives. There is no substitute for engaging with people. Indeed, public communication and collaboration can significantly enhance conservation objectives and outcomes (Kothari 2006).

Today, there is growing understanding that is necessary to ensure the participation of those stakeholders who have not had a role in the decision making process. According to the expert from the IUCN office for SEE *"Interested public, whether it is an individual or organization, is a significant and unavoidable part in a nature conservation system"* (director of IUCN office for SEE, personal communication, 2009). In our conditions, it requires increased role of local communities, NGOs, educational and scientific - research institutions and the private sector. All stakeholders should take part in shaping and implementing decisions and the implementation of an efficient and permanent dialogue between them (Vukadinović, 2009).

Legislation aimed at biodiversity conservation and the management of protected areas has been developed in Serbia. However, the challenge is to guarantee that the policies work to ensure peoples' livelihoods, to avoid future conflicts and to achieve the best means of protecting nature (Tomićević et al., 2011).

Protected areas in Serbia have experienced considerable weakness and constraints in terms of their management. Most protected areas in Serbia suffer from inadequate funding and have weak institutional and human capacities (Tomićević, 2005, Tomićević et al., 2011). According to the reserve manager of SNR Zasavica the *"main problem in managing is inadequate funding and support from Ministry of Environmental Protection"* (personal communication, 2009).

However, in spite of all difficulties in legal and institutional terms, there are examples where a significant shift in the management of protected areas and public involvement in the process can be noticed. According to expert from the Institute for Nature Conservation of Serbia (public relation manager) and an expert from IUCN office for SEE the SNR Zasavica is an example of good practice, in which the public, in terms of non-governmental organizations, the local population, individuals and various associations take some share in the management of the protected area (personal communication, 2009).

3.1 Special Nature Reserve Zasavica (SNR Zasavica)

SNR Zasavica is located in the southeastern part of Europe, in Serbia, in the area of the southern Vojvodina and northern Mačva, east of the Drina River and south of the Sava River, in the municipality of Sremska Mitrovica and Bogatić (Fig. 1). The Reserve is dominated by water mainly comprising marsh ecosystems with fragments of meadows and forests (Fig. 2).

Fig. 1. Location of SNR Zasavica in Serbia

Fig. 2. Water biotope and marsh ecosystems

The NGO Nature Conservation Movement from Sremska Mitrovica (Pokret gorana Sremska Mitrovica) is the Manager of the Special Nature Reserve Zasavica and this organization is achieving good results mostly thanks to its inventiveness, creativity, management skills and enthusiasm. This non-governmental organization initiated this area to be declared as a SNR in 1997 (UNEP, 2005). Zasavica is a part of a national network of Ramsar sites (wetlands protected according to the Ramsar Convention) (http://www.ramsar.org/pdf/sitelist.pdf), and according to IUCN management categories, it is a Habitat and species management area – category IV. Since 2001 Zasavica is a member of The Europark Federation (http://www.zasavica.org.rs/en/zasavica-lokacija-rezervata/).

Before the proclamation of Zasavica as a Special Nature Reserve, this area was not managed properly. Even more, this area was abandoned and without any kind of protection. Local people and authorities were not realizing the importance of this area in terms of its biological diversity and ecological significance.

Water Management Organization "Sava" Šabac, has used this area for water supply, which led to a significant drying of the very fragile aquatic ecosystem. Because of drainage and regulation, as well as eutrophication processes, this very sensitive ecosystem was in danger of drying up completely (reserve manager, personal communication 2009). In addition, local residents had often used this area as a site for waste disposal, which turned parts of this area into garbage dumps.

Nature Conservation Movement recognized the value and importance of the survival of this area. For twelve years they have realized significant results in maintaining and upgrading this area, individually or in cooperation with the local population, authorities and various institutions. From the very beginning, this NGO launched a whole range of initiatives, programs and projects for the preservation and improvement of this area in a sustainable way (Vukadinović and Tomićević, 2009).

The flora of Zasavica includes invasive species which are in expanding. Invasive plants are introduced species that negatively influence biological diversity of any region. The main activities of the Nature Conservation Movement were: regulation of water level, removal of garbage and shrub and invasive vegetation. On the other hand, important achievements for sustainable economic and social development are part of the project "Zasavica - *Support of local community through sustainable tourism*".

Regulation of water level successfully prevents further drying and desiccation of the water ecosystem. Before the experimental regulation of water level, invasive plant species, especially *Stratitoides adoines*, were widespread and threatened the survival of rare species. Within a period of two years the aquatic eco-system with clean water and without invasive vegetation was re-established (reserve manager, personal communication 2009).

Very rare plant and animal species have been preserved due to ecosystem restoration. Among them is a relict species of insectivore (*Aldrovanda vesiculosa*), which has already disappeared from many European countries, and is on the red list of the flora in Europe. The most important of protected species of animal is the fish -*Umbra crameri* (Simić, 2005). According to Stevanović et al. (1999), there are four species that are critically endangered in Serbia that are found in the Reserve: *Groenlandia densa, Hottonia palustris, Ranunculus lingua, Hippocheris palustris*. For these globally endangered species, Zasavica is the only remaining habitat in Serbia. This indicates the importance of protecting Zasavica in order to preserve the biodiversity of Serbia and Europe (Stanković, 2008).

Various actions for removing of garbage were organized in cooperation with the local population. Cleaning actions were taken several times and all the existing landfill were successfully repaired. One of the actions named *"Nature gave us-and we?"* is aimed to raise awareness of local people and to motivate their involvement in implementing activities (Fig. 3 and 4). Construction of sanitary landfills in the area outside the reserve, enabled local residents to dispose waste in a planned and efficient manner (reserve manager, personal communication 2009).

Fig. 3. The garbage before action

Fig.4. The same area after removal of garbage

As a measure of maintenance of pastures and meadows, removal of shrub vegetation is a regular activity (Simić, 2008). The, diversity of ecosystem is preserved and richness of species successfully conserved in the Reserve (reserve manager, personal communication 2009).

Also, in the forests of the Reserve, in accordance with natural requirements, replacement of hybrid poplar with autochthonous species (oak, ash, willow) was carried out - (Simić, 2008).

These measures, mentioned above, contributed to ecological sustainability of the Reserve. Beside it is also very important sustainable economic and social development of the area. Managers of the Reserve carry out a whole range of programs and projects that will enable all aspects of sustainability (Vukadinović, 2009). One of the most important projects is *"Zasavica- support of local community through sustainable tourism"*. The vision of this project is to ensure protection of biodiversity with development of sustainable tourism - as a form of ecologically rational use (UNEP, 2005).

SNR Zasavica represents a unique entity, very important for preserving biological diversity. Today it represents a very attractive tourist resource. Therefore, the development of the whole area is based on sustainable development and tourism (EAR, 2007).

SNR Zasavica is surrounded by nine villages. Economical and social conditions of the local population are very difficult at this moment (UNEP, 2005). Unemployment is high (over 50%). Therefore, it is necessary to increase local employment and generate other social and cultural benefits. It is very important to motivate local residents to participate in the development of this area (Fig. 5). Through their participation and economic and cultural development, the local community will become aware of the need to protect their natural resources (EAR, 2007).

Fig. 5. Project team and local people – "Helping the community to identify themselves and their opportunities through sustainable tourism."

Therefore, to achieve sustainable development, it is necessary to apply integrated approach in planning and management (Vukadinović and Tomićević, 2009). Also, the successful implementation of various measures of protection, as well as implementation of different development programs, to a large extent depends on the support of the local population support (Zavod za zaštitu prirode, 2007). The aim of this project is to include local communities in active participation. It promotes local actors- farmers, food producers, tourist organizations, NGOs, etc. The main stakeholders of tourist services will be the manager of the Ecotourism Center "Zasavica", Manager of the Reserve and local people (EAR, 2007). Through this project all interested parties will take part in planning, decision making, and development of this area.

The tourism project is ended in the year (2011) and still the final report from IUCN expert is missing but regarding to data from personal communication with park manager of SNR Zasavica we can preliminary conclude that the tourism project has improved livelihoods of local communities (for example according to park manager statistical data shows that in 2002 the number of permanent employed people (local residents) was four and in the year 2011 the number of permanent employed local people is twelve and in the year of 2011 between ten and fifteen locals are temporary employed which are engaged as a tourist guide or in production of traditional food products or handcrafts (personal communication, 2011). Furthermore, during this project in SNR Zasavica are offered some exclusive products such as mangulica meat prepared in different types of sausages, smoked ham and bacon.

Mangulica pigs are descended from wild boar populations and are breaded as traditional animals which still need to be made more profitable for the local communities. Breeding of indigenous animals, such as the mangulica pig is an example of the agricultural practices that need to be preserved and continued as part of the local traditions and promotion of local food products.

4. Conclusion

Public participation is an integral component of protected area management. It is now widely assumed that participation is required as a means of sustaining protected areas. Generally, protected areas are primarily viewed in biological or ecological terms, but they provide numerous functional benefits to humans, and may even be seen as essential to human welfare. Increasingly, they are seen as drivers and providers of social and economic change.

In order to achieve sustainable conservation, state legislators and environmental planners should involve local people in management of protected areas and identify and promote social processes that enable local communities to conserve and enhance biodiversity as a part of their livelihoods. Therefore, protected areas are undergoing a shift from a preservationist paradigm towards an integrated approach.

Management of protected areas in Serbia was characterised by a centralised approach that did not include all stakeholders. This approach in management has brought many problems and misunderstandings between government, non-governmental organisations, private sector and local people.

However, despite all the difficulties, Serbia has started the shift towards greater public involvement in protected area management. An example of good practice in managing of Special Natural Reserve Zasavica shows one of the solutions on how natural resources can be successfully preserved and protected. This proves that despite all economic and political difficulties, a solution for managing of protected areas in a sustainable way can be found. In fact, these circumstances should be seen more as a challenge to create the best possible conditions for successful public participation, rather than as an excuse to avoid any form of participation. The Zasavica example shows that involving the public in planning, decision making and management of protected areas could result ecosystem restoration, species recovery and generation of social and cultural benefits to people and thereby contribute towards sustainable development of the overall regiona where the protected area is situated.

In Serbian protected areas management Zasavica presents specific and unique place which represent a good example of integrated approaches to protected area management compared to other protected areas in Serbia. From this example we learned that protected area management must include benefits for the people living around the protected areas and that financial mechanisms must be developed to sustain both conservation activities and improvements to rural livelihood.

5. References

Borrini-Feyeraband, G. (1996) Collaborative Management of Protected Areas: Tailoring the Approach to the Context. Issues in Social Policy. IUCN, Gland, Switzerland.
Borrini-Feyeraband, G., Farvar, M.T., Nguinguiri, J.C. & Ndangang, V. (2000) Co-management of Natural Resources: Organizing, Negotiating and learning by Doing. GTZ and IUCN, Kasparek Verlag, Heidelberg.
EAR (2007) Studija izvodljivosti - podrška lokalnoj privredi kroz održivi turizam. Sremska Mitrovica.
FAO (1990) Participation in Practice, Lessons from the FAO People's Participation Programme. FAO, Rome.
FAO (2001) SEAGA (Socio-Economic and Gender Analysis) Programme, Field Level Handbook. FAO, Rome.
FAO/ECE/ILO Joint Committee Team of Specialists on Participation in Forestry (2000) Public Participation in Forestry in Europe and North America. ILO, Geneva.
Finger-Stich, A. & Finger, M. (2003) State versus Participation: Natural Resources Management in Europe. IIED and IDS, London, UK.
Ghimire, K. B. & Pimbert, M. P.(1997) Social change and conservation: An overview of issues and concepts. In: Ghimire, K. B. & Pimbert, M. P. (eds.) Social Change and Conservation: Environmental Politics and Impacts of National Parks and Protected Areas. UNRISD and Earthscan, London, pp. 1-45.
IUCN (1993) Parks for Life: Report of the IVth World Congress on National Parks and Protected Areas. IUCN, Gland, Switzerland.
Jeanrenaud, S. (1999): People-oriented conservation: progress to date. In: Stolton, S. & Dudley, (eds.) Partnerships for Protection, New Strategies for Planning and Management for Protected Areas. WWF and IUCN, Earthscan Publication Ltd., London, UK, pp.126-134.
Korten, D.C. (1980) Community organization and rural development a learning process approach. Public Administration Review, September-October: 480-511.
Korten, D.C. (1981) The management of social transformation. Public Administration Review, November-December: 609-618.
Korten, D.C. (1990) Getting to the 21st. Century, Voluntary Action and the Global Agenda. Kumarian Press, West Hartford.
Kothari A (2006) Collaboratively Managed Protected Areas. In: Lockwood M, Worboys G L, Kothari A (eds), Managing protected areas. A global guide. IUCN, Earthscan, London, pp 528-548
Lockwood M, Kothari A (2006) Social Context. In : Lockwood M, Worboys GL, Kothari A (eds), Managing protected areas. A global guide. IUCN, Earthscan, London, pp 41-72
Moote, M.A., McCalaran, M.P. and Chickering, D.K. (1997) Research theory on practice: Applying participatory democracy theory to public land planning, Environment Management, vol 21, no 6, pp 877-889.

O' Riorden, T., O'Riorden, J. (1993) On evalueting public examination of controversial projects, in Foster, H.D. (ed) Advances in Resource Management, Belhaven Press, London

Orlove, B. S. & Brush, S. B. (1996) Anthropology and the conservation of biodiversity. Annual Review of Anthropology 25: 329-352.

Ramsar Convention Wetlands. Available online: http://www.ramsar.org/pdf/sitelist.pdf

Renard, Y. (1997) Collaborative management for conservation. In: Borrini-Feyerabend, G. (ed.) Beyond Fences: Seeking Social Sustainability in Conservation. IUCN, Gland, Switzerland, pp. 65-67.

Shannon, M.A. (2003) Mechanisms for coordination. In: Dubé, Y.C. & Schmithüsen, F. (eds.) Cross-Sectoral Policy Impacts Between Forestry and Other Sectors. FAO Forestry Paper 142. FAO, Rome, pp.145-159.

Shyamsundar, P. (1996) Constraints on socio-buffering around the Mantadia National Park in Madagascar. Environmental Conservation 23: 67–73.

Simić, S. (2008) Izveštaj o realizaciji plana i programa zaštite i razvoja Specijalnog rezervata prirode Zasavica u 2008. godini, Sremska Mitrovica.

Stanković, M. (2008) Međunarodna i nacionalna vrednost biodiverziteta specijalnog rezervata prirode Zasavica. Sokobanja.

Stevanović, V. et al. (1999): Crvena knjiga flore Srbije 1 - iščezli i krajnje ugroženi taksoni (Red List of the Flora of Serbia Vol. 1 – extinct and critically endangered taxa). – Ministry of Environmental Protection, Biological Faculty of the University of Novi Sad, Institute forNature Conservation of Serbia, Belgrade.

Tomićević, J. (2005) Towards Participatory Management: Linking people, Resourses and Management. A Socio-Economic Study of Tara National Park. Freiburg.

Tomićević, J., Shannon, M. A., Vuletić, D. (2010) Developing Local Capacity for Participatory Management of Protected Areas: The Case of Tara National Park. Šumarski list, Vol.134 No.9-10, pp. 503-515.

Tomićević, J., Bjedov, I., Obratov-Petković, D., Milovanović, M. (2011) Exploring the park-people relation: collection of Vaccinium myrtillus l. by local people from Kopaonik National park in Serbia, Environmental management 48 (4):835-846

UNEP (2005) Razvoj metodologije i plan upravljanja turizmom za specijalni rezervat prirode Zasavica. Beograd / Bon.

Veal, A.J., (1992) Research methods for leisure and tourism: a practical guide. Pitman Publishing, London, p. 93–103.

Vukadinović, R. (2009) Učešće javnosti u upravljanju zaštićenim priodnim dobrima. Diplomski rad, Šumarski fakultet, Beograd.

Vukadinović, R., Tomićević, J.(2009) Public participation in management of protected areas - example of Serbia, In: Zlatić, M, Kostadinov, S., Bruk, S.(Eds.) (2009) Global change-challenges for soil management -From degradation trough soil and water conservation to sustainable soil management, Conference Abstract, International conference "Land conservation"- LANDCON 0905, May 26-30, Tara Mountain, Serbia, pp.237

Wells, M. P. & Brandon, K. B. (1993) The principle and practice of buffer zones and local participation in biodiversity conservation. Ambio 22 (2):157-162.

Zavod za zaštitu prirode Srbije (2007) Stanovnici naselja na području zaštoćenih prirodnih dobara Suve planine, Sićevačke i Jelašničke klisure, o zaštiti i razvoju kraja u kome žive. Niš

8

The Materially Finite Global Economy Metered in a Unified Physical Currency

John E. Coulter
Tsinghua University, Beijing
China

1 Introduction

1.1 The economy as a subset of the biosphere

A more comprehensive understanding and more general acceptance of the concept of the biosphere should change the way we think about economics. Economic development is a key goal in most countries and amongst the leaders and citizens, but hides a serious flaw that successes so far have been able to pass on and ignore. The flaw manifests itself, in the early days, as a surprise, in depletion of local raw materials and a build up of unpleasant wastes. The response pattern has been natural and obvious. Other regions wanted the glamour of development, and had unused resources and plenty of space for the wastes and emissions. The passing on of problems originated in the West and has moved to the East. The fact is that much development and profit required some cheating on the environment, and at first it was unconsciously done. But progress has turned full circle, and reached, as in the eloquent and deep lyrics of The Eagle's song, the Last Resort (Henley & Fry, 1976). Environmental scientists who have always been trumped by the success of money makers will be needed to advise on how to manage the biosphere when there is nowhere else easy for resources, when oilwells drill too deep in danger, and when there is no where to hide unwanted emissions to the air, water and soil.

The Global Financial Crisis and Global Environment, including Climate Change, are both hot topics with politicians, academics and in the media, though they are typically treated as unconnected issues. Yet frustration with debates on deficits and bailouts in the trillions leads for calls to get back to the "Real Economy" and the "Real World". It is time to fit economic activity into the biosphere, which is indeed our real world. There needs to be a crossover where activities in money terms are seen to represent (and perhaps misrepresent) the stocks and flows of matter and energy in the environment. At the same time solutions to environmental problems are commonly discussed superficially in terms of budgets, without addressing the technologies needed to be employed to solve problems. Monitoring our state of well-being using dollar terms and worse still, trying to manage and correct problems by means of a "stimulus" or "quantitative easing" is not going to fix our world.

1.2 New science defines energy relevant to economics and puts a physical cost on it

Mention the biosphere to most people and they seem unnerved by this arcane and complex concept, almost as if it was removed from their relevance. Generally, politicians, business people, most academics, and the media do not respond in a way acknowledging we are talking about the world we share and on which we live. This chapter addresses problems with the concept, why it has been overlooked, attempts to show clearly the critical need for awareness, and introduces a new method for monitoring and accounting for what people do to their surroundings, logically and scientifically cumulative from individual actions to global consequences. The method is raw and only newly tested in case studies, but has the advantage of being objectively testable, and removed from the subjective influences imposed by public and private stakeholders in manipulation of financial accounts. Our modern life is mesmerized with manipulated money, and we need to grasp physical reality directly. This new approach estimates that in 2011, 450 exajoules of fossil fuels energy was "released" for "consumption" against the context of 26,000 exajoules of "unavailable" coulombic forces.

If an alien visitor observed our planet from space, and saw the marvelous constructs of infrastructure and cities, vast farmlands and other developments, but was unable to zoom in or understand how people and organizations deal with each other in social contracts of language, documents and money, a fair idea could still be deduced of where on Earth was conducive to quality of life, and where was unpleasant because of natural and man-made problems.

A study is described in this chapter on why monitoring of GDP by financial accounting is increasingly misleading, and what attempts have been made to use non-money approaches to understanding economic activity. This chapter introduces a radical and raw new approach. The initiative is described and conducted as a first step to provide a scientific numeraire that indicates how individuals, groups and regions are impacting on their local surroundings and, in total, on the biosphere.

The problem of monitoring and managing "Spaceship Earth" (Boulding 1966) in money terms is becoming ridiculous. Clean-up costs of environmental disasters are additions to GDP. Financial crashes are addressed by financial experts recommending more money, trillions in the US and hundreds of billions in other default prone nations. In the cacophony of reporting, giving advice, mounting austerity measures and carrying out protests after the 2008 crash, there seems no mention that economists for two centuries have always fallen back on the obvious truth that money is supposed to be used to represent real goods and services. It is not supposed to be an end in itself.

Primitive peoples built up a view of their world founded on food and shelter for survival and dealt in transactions of sharing and then barter. Tokens such as simple shells became a wonderful invention for efficient trading. This evolved into money, leading to centrally administered stamped metal, recognizable to the general public, and conducive to storing, transportation, and small and large market transactions. Almost as soon as invented, money started to distort the world view of some, in official circles, undermining the power to represent reality through wanton increase in money supply, and unofficially through counterfeiting, robbery and any means of cheating possible.

2. When our world was small – Managing the household

The first conscious attempt to monitor and explain what happened with people's activities around us was by Quesnay publishing in 1756 an "Economic Table" of transactions between farms and cottage industries in northern France. The term "economy" then was as novel and as strange as "biosphere" seems now to the public. The term derived from the two Greek words for "household" and "manage" and scientists dealing with our "biosphere" would serve themselves and the public well to raise awareness that the biosphere is the home of 7 billion people, and requires management.

Right from the outset of economic study three centuries ago, Nature was taken for granted – the soil and water were too obvious to discuss. Attention was concentrated on the goods and commodities produced and the services rendered. Quesnay recorded goods and services in francs and plotted the complex transactions between farm produce and the simple outputs of utensil, tools and cloth from the villages.

The first grand attempt to explain the wellbeing of the world, as it was understood at that time, was set out in "The Wealth of Nations" (Smith, 1776) observing, not the whole world, but those parts that were considered important to Europeans of that era. Smith legitimized the accounting of real goods and services in money terms with the image of two hunters agreeing that if the trouble taken in getting a beaver was equivalent to getting two deer, then in money terms a beaver is worth 2 deer. What was not needed to be spelt out was that money is so wonderful for it avoids the unpleasantness of equating one deer with half a beaver. The British Empire expanded all over what we now call the biosphere, and monitored and managed its territories in money terms. The power of money led to exponential developments in resource utilization, financing and leveraging mines and commercial plantations, along with intricate infrastructure in the New World, Africa, and Asia.

Modern students of the biosphere must be aghast that the author of the founding canon on resource utilization (Ricardo, 1817) offered the throwaway line that, of course "water and virgin land are free" because they are abundant. But this was written 60 years before the Oklahoma Land Rush, dramatic evidence that virgin land was indeed free. The sense of superiority that grew out of the industrial revolution engendered a ubiquitous phenomenon that regions of the world still undeveloped were desperate for development. As the mills, factories and foundries of Europe reached limits of local raw materials and encountered downsides of polluting, smokestack industries were enthusiastically welcomed in the New World, where there was plenty of coal and other resources, and virgin land where dumping pollution seemed not to matter.

The economic development theory presumed a linear path towards a better life, technology replacing unpleasant labor, and benefits so obvious that they seemed universal goals. The downside of resource depletion always seemed to have substitutes in other places or through different technologies. The downside of polluting air, water and soils could be confidently addressed, usually by shifting the culprit emitters to further locations where development and new employment were welcome. The cleansing of London of its smog was hailed as a triumph, usually without recognition that the causes were moved to Los Angeles, and now to Beijing and a dozen cities in China.

3. The materially finite planet

The first checks on the blind belief in technology and concerns for broad environment only appeared in the 1960's with clear evidence of farm pollutants in "Silent Spring" (Carson,

1962). This was followed by the philosophical reflection on the first photos from the Apollo space program of the Earth as a closed materially finite system in a vacuum with only energy from the sun and gravity as inputs, set out in the essay, in "The Economics of the Coming Spaceship Earth" (Boulding, 1966). A heightened anxiety of global resource depletion and pollution was stimulated by "The Limits of Growth" (Meadows et al, 1973).

The new environmental awareness called into question the economists assertion of "rational man" which presumed myriad actors in the market place purchasing, producing, investing resulted in a superior outcome to Big Brother central planning. In the context of the Cold War western countries exalted entrepreneurship and faith in new technologies and leapfrog innovations that would overcome immediate problems or resource depletion and dirty industries. There was plenty of evidence to support this faith. Thus each call for sober assessment of global resources and capacity to absorb pollutants was met with various degrees of distain and ridicule, likening nay-sayers to Chicken Little panicking that "the sky is falling" and rebutting depletion concerns with the observation that "the Stone Age did not end because they ran out of stones". In a widely noted public wager, business optimist Julian Simon challenged conservationist Paul Erhlich on the rise or fall of metal prices over the decade from 1980. Through new technologies and more exploration driven by demand, the prices fell.

The thrust of the warnings in The Limits of Growth seemed to be weakened by periods of economic boom. At publication in 1973 the world seemed exhausted and choked, and China took no part in the world economy, preoccupied with its debilitating Cultural Revolution. China changed quickly from 1980 to enter and drive a new level of intensity in development. The 1980 US presidential elections saw candidate Reagan refuting incumbent President Carter's urging to choose smaller cars for less environmental impact, and the following rout underscored to politicians around the world the danger of preaching conservation and humbleness as opposed to bravely dominating the environment.

4. Science and economics as divergent cultures

The sciences and the humanities have been criticized as diverging to the extent that communication breaks down (Snow, 1959). The observation was made that in a gathering of people well educated in the arts, mention of the Second Law of Thermodynamics would likely be met with puzzlement. There has also been a bifurcation between science and economics. The early economists emulated Newtonian mechanics in their modeling of market "forces", parabolic trajectories of demand, and business cycles. As the discipline of economics burgeoned, and enjoyed power as it guided and supported the policies of governments, science developed into myriad specialties. The polymaths made way for very narrowly focused specialists delivering wonderful new discoveries that were not generally understood by the public or economists, but resulting in general progress. Economists impressed by Newton used gravity for their modeling of economic processes, and assessed projections of investment with rates of return precise to two decimal places, when in physics, Heisenberg was revealing the uncertainty principle in even the most concrete of experiments.

The economist forecasts of good news and slewed accounting seem to win over the caution of scientist, and thus assessments of the need to respect ecologies and the diverse biomes, and especially pointing to the material finiteness and interrelatedness of activities on the planet rarely get endorsement over glamour projects promising more to consumers. In government, economic advisors are more influential than science advisors, and have more room to wriggle hinting at potential for short term gains. The 2010 Oscar winning documentary movie, Inside Job, objectively and correctly highlights how influential economists from Ivy League universities unabashedly endorsed reckless deregulation of Wall Street financial services for their personal remuneration. Economics is used to forecast and to influence regulations in the future, and is not subject to rigorous tests of objective repeatability as occurs in science.

Accountants, not scientists tend to leadership in big business. Early this century Enron's core business was selling natural gas, a product where inventories and flows should be easy to confirm. With a Harvard trained financial expert in charge, engineers were belittled for their doggedness and inability to be creative. A culture of ignoring science and presenting financial reports of miraculous results engendered such an aura of respect that simple checking on the natural gas (that is, methane, or CH_4, a very simple molecule that cannot be doctored and substituted for increased profits) dared not be suggested, neither by prudent employees enjoying promises of bonus, by the accountant firm enjoying high billings, or even by government regulators. The charade was untenable, and the "market correction" that economists assure will come into play when needed was severe, leading to a record bankruptcy, layoffs, suicide, and jailings.

We have reached a stage where scientists are unable to articulate convincingly on even what seems indisputable important facts such as the factors influencing climate change, and the desecration of primary forests and grasslands. There are rumblings on Capitol Hill in Washington that the EPA is a "job killer" and should be disbanded. In all human history so far, the biosphere we are now trying to understand and protect has served us well and has endured volcanic eruptions, sun storms, and other giant natural disasters. A generation is growing up in polluted cities blithely accustomed to serious air pollution and undrinkable tap water. The threat of a slow early death seems not nearly as vital as having a job and maximizing short term enjoyments. As long as there are doubts about what impacts human activity has on our surroundings, and especially while the planet seems big enough to process and refreshen even very harmful emissions, scientists will remain the playthings of politicians.

5. Measuring our world without relying on money

Early civilizations measured wealth in weights of grain and gold. Money soon took over and Smith's inquiry into the wealth of nations led 150 years later to the formal estimating of national income and the concept was so powerful that National Accounts and GDP soon became the dominant check on a nation's wellbeing. Criticisms of GDP thinking were met with tinkering at the edges for "green" GDP and "quality of life" and "human development indices".

The oil crises of the 1970's prompted venturing into using energy as a measure of value, not only directly, but also in the "energy cost" of all goods and services. The aluminium industry lends itself to energy analysis simply because it is so energy intensive. As an example, at the height of energy analysis popularity one case study showed that 1.9 ton of zinc concentrate required 18.97 gigajoules to be processed to 1 ton of zinc ingot (Middleton & Paterson, 1977). Yet the approach was overdone and there were many economic activities and commodities where energy input seemed only a minor factor.

One valiant crusader attempting to circumvent the biases caused by money accounting invented the term "emergy", variously described as derived from "energy memory" or "embodied energy" (Odum, 1989). Odum tried to explain economic activity through a language of various energy-related symbols he had created. Given the desire by many well-meaning people to find antidotes for measuring wellbeing in dollars, his quests had serious following but have now mainly been discredited.

The giant step to link economics back to science was made by a respected economist, arguing that economy was modeled as if it was a perpetual motion machine, and specifically that it had no cognizance of the Second Law of Thermodynamics (Georgescu-Roegen, 1971). Published at a time of growing concern about the downsides of economic development, the idea was received with interest and taken up by a new generation of environmentally aware economists.

The biosphere came to prominence with the work of a Georgescu-Roegen protégé, Herman Daly. Daly took the foundational model of macroeconomics, the "circular flow of the economy" and showed that the loops of goods and services in the market ignored Nature as a source of inputs, and as a receiver and recycler of high entropy unwanted outputs to the air, water and soils. Daly's 2-part diagram shows the conventional circular flow as a small loop nested in a very large environment, so generous and forgiving as to be taken for granted, as was the case two centuries ago. The second part of the diagram shows the circular flow at its current size, burning billions of tons of coal etc, drawing down reserves of vital inputs and, highlights the new awareness that on the output side the economy is choking in constrained space for emissions and wastes of many kinds (Daly 1977). But even his tenure as Chief Economist, Environment Division of the World Bank did little to convince World Bank bankers and pro-growth, pro-consumption economists that economies should not blindly strive for higher GDP. A third wave after Georgescu-Roegen attempted to estimate the value of the world's ecosystems and came to the conclusion that the annual service value of the biosphere was about double world GDP (Costanza et al, 1997). This famous article in the prestigious journal, *Nature*, was a milestone in bringing the concept and term, biosphere, to a wider public.

As environmental consciousness gained ground, the analytic tool known variously as Substance Flow Analysis and Material Balance became to be usefully applied. Based on the fact that matter/energy cannot be created or destroyed, the material and energy in major industrial processes could be tracked from insitu reserves through to final product, AND with the inventorying of what wastes had previously been ignored or hidden that must exist to make up the total of what went in. It comes as a surprise to many lay people that a lump of coal, solid and heavy, is primarily made up of carbon atoms, atomic mass 12, but on combustion they form with two atoms of oxygen, each atomic mass 16, so that the invisible, formerly perceived as benign, carbon dioxide floating in the air has an atomic mass of 44.

An advance was made on tidying up energy analysis with the slow picking up of the concept of exergy. In 1956, in Central Europe, Slovinian Zoran Rant made the simple observation that energy cannot be created or destroyed (First Law of Thermodynamics) and so in economic activities we should not say energy is produced or consumed. He coined the term "exergy" meaning useful energy that, in being used, is destroyed. His first step was taken up and promoted by Polish Jan Szargut, but for decades of the Cold War, his publishing, mainly in Polish, went unnoticed in the West. Chinese students, with only narrow options to study abroad, discovered Szargut's work in the Soviet Union and brought it back to China, finding it fitted well with the material inventories used in Chinese accounting under the planned economy (Chen & Chen, 2007). In the West, Stanford University Global Climate and Energy Project have produced excellent, detailed charts of global exergy flows, accumulation and destruction, (Section 8.3 and Figure 10). Szargut's monumental research not only quantified the exergy in fuels but also the "embodied exergy" in anthropogenic products.

6. De-engineering science for a fresh look at the big picture

6.1 Purpose-driven science needs to be de- and re-engineered for broader applications

Much of what is now good and useful science was discovered or evolved for a specific purpose. Carnot encountered the phenomenon now called entropy trying to improve efficiency of steam engines that lift water. Gibbs borrowed the notion to explain why not all reactions are exothermic. The solving of internal combustion engine "knock" by adding lead to the gasoline compound was hailed as a great advance until the downsides of leaded gasoline were encountered. Military urgency has been a great driver of new technologies, with the invention of the nuclear bomb as an extreme example. Radar would not have had a budget to research and develop but for the Battle of Britain. In driving science for a specific purpose, after the goal has been achieved, in subsequent decades with further and broader applications, there may be benefits in revisiting the inventions to fit better into a broader context.

6.2 The energy concept hides more than it reveals

James Joules was a brewer, and wanted to improve efficiency by using the new electric motors instead of steam engines in his brewery. From this purpose in comparing two technologies he evolved the concept of the mechanical equivalent of heat, leading to popularization of the term and concept of energy. Joule's later experiments in finessing his idea are amazing, and he settled on the equivalence factor between rotating a paddle (mechanical) and the resulting rise in temperature to an accuracy of 2 decimal places of the modern tested fact. Yet I dare to suggest that in understanding economies, the global economy, and the biosphere, the concept of energy hides more than it reveals. It certainly confuses many people in positions of power and leads to wrong decisions. First, as mentioned above, energy cannot be produced or consumed. When energy is transformed, it may be in part dissipated, but by definition the energy transformed is the equivalent value to its original state. While all energy is quantified in joules, very different forms of energy are being confused. Economists, business people and politicians use the term "energy" to mean electricity, fossil fuels, and other dynamic forces including wind, rivers, geothermal heat and tides. These forces are different by nature (Urone, 2004).

6.3 Forces

The four forces of totally different characteristics are

1. gravity;
2. the electromagnetic force, in magnets, the heat and light of fire, sunshine, electricity, and "chemical energy";
3. the nuclear force which holds protons and neutrons in a nucleus to define which element an atom is; and
4. the weak force, only encountered in extreme subatomic experiments.

If an intelligent alien was observing the Earth from a distance, and noting the natural and anthropogenic activities in the biosphere, the four forces would be distinguishable. Energy analysis would be an unnecessary complication. Many of the forces are vital to life and to humans, but we cannot change them, and certainly do not pay for them.

i. Start with the Earth's orbiting momentum (giving us seasons) and rotational momentum (for day and night and heat distribution). To be seriously comprehensive in inventorying the energy of the Earth, we need to acknowledge that the Earth's momentum orbiting the sun is 2.66×10^{40} kg/m²/s and the angular momentum in its rotation on its own axis is 7.074×10^{33} kg/m²/s. Whatever the Big Bang was, that imparted those forces, they have evolved to give us what we have got. The point in stating what many might disdainfully comment as superfluous information is that this provides an exercise in acknowledging what energy on Earth we can control or influence (and is a cost) and what is natural and free.
ii. Gravitational force is big in our imagination because it is so obvious and ubiquitous. It is common in our economic modeling but only as a model. We do not pay for gravity.
iii. The force that attracts the protons and neutrons in the nuclei of any atom is so powerful that until the atom was "split" to end World War 2 and then for peaceful electricity generation, each element had to be taken as a "given". To understand the biosphere and the economic analysis we might apply to it, it is best to begin taking atomic structure for granted by putting the nuclear force to one side (i.e. for a simple first step, view any conventional economic processes but not those using nuclear power).
iv. It is the electromagnetic force (emf) which is and can be the agent of intelligent or stupid change in our biosphere. We can light fires and boil water, we can move about, digest, breathe, see and make noises. It is emf that holds an electron in orbital around its proton partner. This is chemistry. This is about fuels, low carbon, what tastes good, and if you feel well.

A lot of emf is beyond our control, starting with sunlight. The emf in photons distils and lifts 1,400 trillion tons of water from the sea each 24 hour rotation to start the water cycle, which is then abetted by the Earth's rotation and the sun's play on air (winds). But in the animals we ascended from, and in primitive communities 100,000 years before recorded history, homonoids were taste testing, moving about, and exercising some mastery over their emf. It is anthropogenic emf that we should understand and monitor in the study of economic activity in the biosphere (Coulter, 2002).

There is hopeless confusion in economics by not arriving at this revelation. For example, we do not pay for hydropower. The emf from the sun lifts millions of tons of water up to the head of river valleys and gravity attracts it back to sea level. The economics is in the cost of building and maintaining the hydropower facility, and that can be quantified in physics and chemistry without accounting in money.

6.4 A false notion

The father of scientific method, Francis Bacon identified four problems with reasoning and the most dangerous was misuse of terms – "names of things which exist, but yet confused and ill-defined, and hastily and irregularly derived from realities" (Bacon, 1620). The said "releasing" of "chemical energy" from carbohydrate or hydrocarbon fuel and pivotal in any comment on the state of the biosphere is just such a brain-crippling notion.

As indicated in 6.1 above, it must have been exciting when scientists discovered how to measure the energy "released" when fuel is combusted. It can now be derived from tables based on previous work showing listings of energy values of bonds of compounds. Totalling the values for bonds listed in the reactants and products and calculating the difference will provide an estimate of energy released. That makes sense to ordinary people, even an economist. Here is the starting total. Here is the final total. The difference is what people call energy (mainly heat). This is a very useful calculation. But there is something strange – the bond energies in the resulting products are greater than in the original reactants. The atoms in the product are locked tighter togther than before, and the increased bonding is equal to the amount of energy released. In specific purpose driven science, all that mattered at first was the resulting number for available energy, and it could be calculated accurately. An engineer looks at a ton of coal and says it can produce 24 gigajoules of heat. That is very useful in comparing where else energy may be harnessed from including hydropower, wind power, tidal power and direct solar power. But now that environmentalists are forced to look at surroundings and the bigger picture, culminating in the materially finite biosphere, it is relevant to ask, where does 24 gigajoules of heat energy in coal come from, especially when in calculations of chemistry, it seems surprisingly the opposite of common sense.

6.5 De-engineering energy calculations

Coal and gasoline have complex chemistry, so the energy released from the simplest hydrocarbon is noted at the molar level, and at the level of a single methane molecule. This is set out in Table 1.

The above data for a mole of methane are readily available from standard textbooks (Atkins & Jones, 2000) and chemistry tables. The data is obtained from calorimetric experiments. In applying chemistry table data to practical everyday problems, a mole of methane (6.02214×10^{23} molecules of CH_4) ought to be converted to the volume in cubic meters and its layman's name of natural gas. It is the frequent failure to convert that exacerbates the divergence of the two cultures, science and economics. There are conversion tables for common fuels. Table 2 is one compilation, noting that it is difficult to find one unified table with the units of fuel and energy in the same and desired format. Obviously the quality of fuel, especially coal, varies and these figures are indicative of the norms.

	compound	bond	no. of atoms	bond energy kJ	totals kJ	kJ	atomic weight	molecular weight	totals
1 mole of methane	CH₄	C-H	4	410	1640		16	16	
(22.4 liters)	O₂	O=O	2	494	988	2628	32	64	80
	CO₂	C=O	2	799	1598		44	44	
	H₂O	H-O	4	460	1840	3438	18	36	80

Energy "released" 810

				eV	eV	eV
one molecule of methane	CH₄	C-H	4	4.2	17	
	O₂	O=O	2	5.1	10.2	27.2
	CO₂	C=O	2	8.3	16.6	
	H₂O	H-O	4	4.8	19.1	35.6

Energy "released" 8.4

Table 1. Conventional energy accounting of methane combustion, 1 mole (810kJ), and below, 1 molecule (8.4eV).

Fuel	Mj/kg
Gasoline	46
Plant Oil	38
Butter	30
Coal	24
Natural gas	20
Bread	15
Wood	14
Potatoes	4

Data from Forinash, 2009.

Table 2. Common fuels and their listed energy outputs, after Forinash, 2009. Actual values depending on qualities, especially carbon content, with coal having the widest variations as anthracite, bituminous coal, or lignite.

It is not normally practical to experimentally investigate the bond energies and energy release for one molecule of a fuel, so calculations simply divide by Avogadro's number and convert joules to eV. For one molecule of methane combined with two molecules of oxygen, the bond energies total about 27eV and the product carbon dioxide and two water molecules total about 35 eV. The 8 eV of energy in the form of heat are "released" because they are not needed. It is pertinent to question, where did the released eV of energy come from and why. This question has not been asked by chemists solely concerned with delivering energy for practical purposes such as heating and driving a car. But from a point of view of understanding the environment, and indeed, the biosphere, this apparent mystery ought to be investigated.

The answer seems to be that the conventional chemists' calculations only deal with the bond energies between atoms in a molecule that are traded. If we look at a simple diatomic molecule, taking O_2 as an example, then the standard test result is 494 kJ of energy per mole required to separate into separate atoms. Thus the matrix of the coulombic forces between two sets of 8 electron-proton pairs is calculated to be that which requires 5.1 eV to separate. In terms of the Second Law of Thermodynamics, and the "available" – "unavailable" nexus that distinguishes entropy from energy, the release of 5.1 eV (available) must come from a far larger base, which afterwards is then totally "unavailable" and in economic terms is waste, useless, or just pollutant. The two atoms each have 8 proton-electron pairs and it is merely a small residual charge of this that forms the molecular bond. In previous study before introduction to Bader's Atoms in Molecule approach (Bader, 1990), the oxygen molecule was depicted as a complex set of 16 proton-electron pairs centered on the two nuclei, with the valence electrons of one nuclei far outside the bond length of 0.121 nanometers, overlapping the other nuclei and attracting the protons in it "like spiders making love" as borrowed from a doctoral dissertation (Coulter, 1993) and shown in Figure 1.

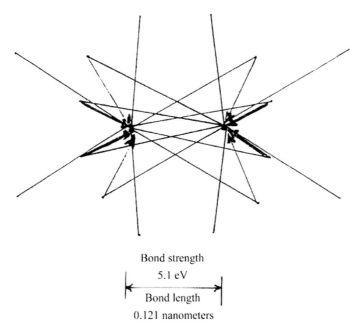

Fig. 1. The oxygen molecule with electrons from one nuclei overlapping the other nuclei, with a complex set of positive and negative charges resulting in a bond strength equal to 5.1 eV.

The phenomenon of entropy inexorably increasing, as observed when any two solids, liquids or gases separated start off with different energies (heat, molecular motion) and are then brought together, can be characterized as simple "even out". Hot and cold even out. High and low pressures even out. The simplest model is the concept of two springs unevenly set, naturally evening out, as shown in Figure 2.

Fig. 2. The core concept of "available" energy evening out (Marshall, 1978)

Appreciation of this phenomenon, and its quantification, is not a trivial point, and has been classed as "one of the most important problems in physical science" (Marshall, 1978).

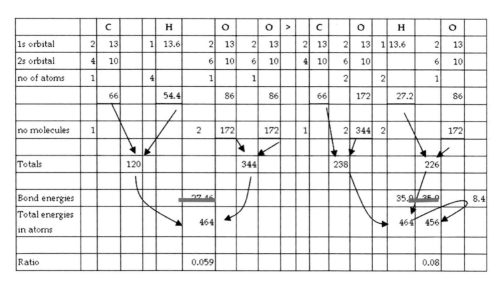

		C		H		O		O	>	C		O		H		O	
1s orbital		2	13	1	13.6	2	13	2	13	2	13	2	13	1	13.6	2	13
2s orbital		4	10			6	10	6	10	4	10	6	10			6	10
no of atoms	1			4		1		1				2		2		1	
			66		54.4		86		86		66		172		27.2		86
no molecules	1					2	172		172	1		2	344	2			172
Totals			120				344					238			226		
Bond energies					27.46										35.0 / 35.0		8.4
Total energies in atoms							464								464	456	
Ratio					0.059										0.08		

Table 3. The aggregate bond energies of reactants is 27.46 eV and of products, 35.86 eV. Taking the hydrogen ionization energy as 13.6 eV and for oxygen and carbon a coulombic force of attraction per proton-electron pair equivalent to 13 eV for inner shell 1s and 10 eV for 2s valence shell, the total energy of the atoms in reactants is 464 eV and, after release of 8.eV, 456 eV in the products. The ratio has evened out from 27.46/464 to 35.86/456, or from 0.059 to 0.079. Thus from this perspective bond energies used in calculating energy released are only a small fraction of the total energy (coulombic forces) within an atom and that this factor must be taken into consideration when we look at energy equations of reactants and products.

The question then focuses on, from what energy is the energy released evening out? It seems fair to conclude that fuels, hydrocarbons and carbohydrates, do have "available" energy from a total amount of coulombic forces within the molecule, and we could estimate the ratio of bond energies to total amount in the reactants and see how that fares in the product. At the time of initial study, the evolving field in physical chemistry pioneered as Atoms in Molecules (Bader, 1990) was unfolding in Canada and not known to this project. In the hydrogen atom the total emf charge existing in the atom in ground state is the coulombic force between the electron and proton and is tabulated as 13.6 eV. It seems that in conventional chemistry it is too difficult and not interesting to estimate the aggregate of coulombic forces in more complex atoms and molecules. For the purpose of suggesting an indicative ratio as available energy evens out to be come unavailable, a very rough estimate of the other two atoms involved in the combustion of methane was made. For carbon and oxygen, it was assumed the two electrons in the 1s inner shell also were attracted to their proton partners with an energy equivalent to 13 eV, and the electrons in the next shell assigned an estimated value of 10 eV.

Subsequently Bader's Atoms in Molecules (AIM) approach became better known, among much controversy. For the purpose of understanding energy released from fuels in the context of entropy change, from a low ratio naturally shifting to a higher ratio in an exothermic reaction, Bader's AIM proves insightful. Previously chemists resorted to telling the public that the shape and form of atoms and molecules was essentially unknowable, the Schrodinger equation for any atom other than the single protoned hydrogen too complex in wave-particle orbitals to picture. Bader radically changed that, applying the observable aspects of electronic charge density. Bader's AIM can contour molecules to show their shape, as in Fig 3, showing a carbon monoxide molecule with the oxygen atom nesting against the carbon atom.

Bader's pioneering work has proven so useful in applications that the main uses now are for determining the shape of complex molecules allowing two new fields of science to emerge and prosper: i.e. material engineering where scientists design the crystal structure of compounds, and drug design to actually "dock" on anomalous biological molecules in animals and humans and "fix" them, particularly successful on rogue enzymes.

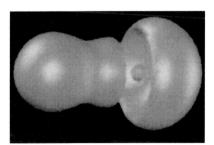

Fig. 3. The radical approach by Bader in his book (1990) and subsequent publications and those of his students allows 3D solid contoured modeling of atoms in molecules. This is a simple carbon monoxide molecule, showing its "shape" according to electron charge density. The blue shape on the left is the electron charge density of a single oxygen atom when bonded to a single carbon atom (the blue mushroom shape on the right).

The application of simply scaling up simple molecules to macroscopically observable objects has not been pursued but offers insights to appreciating molecules and atoms in ionic crystal lattices and then zooming out to observable matter, and conceptually, large volumes, ultimately to representative key chemical reactions in the biosphere.

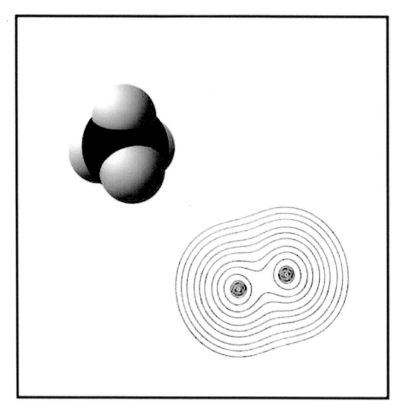

Fig. 4. The side profile of a cube of sides 1 nanometer with a conventional space-filling representation of a methane molecule together with a Bader 2D contour model of the electronic density in an oxygen molecule. The methane model is imaginary, but based on a known C-H bond length of 0.109 nanometers, the molecule is about the right size to fit into the 1 nanometer sided cube. The oxygen molecule outer contour defines 98% of the charge and Bader lists its dimensions accurately as length and width 0.42 and 0.32 nanometers respectively.

In the study (Coulter, 1993) a total of nine theoretical models of atoms in molecules were postulated, scaled up from a single molecule of methane combusting with two oxygen molecules, then a single glucose molecule reacting with six oxygen molecules, through to models of a cubic meter of atom in molecules inventory up to a cubic kilometer that in one year used 237 tons of carbohydrate and 1000 tons of coal. Six actual case studies of real human activity were conducted, beginning with brick making using grass as fuel in 100 cubic meters of space, up to a village in Henan China in 1990 monitoring a population of 1300 people within an imaginary cubic kilometer (100 meters soil depth and 900 meters air) consuming 237 ton of carbohydrate and in their brick kiln and for heating/cooking, 1000 tons of coal. In a simple economic activity, in fact not much change occurred. The change that did occur was an increase in entropy signified by the ratio change of available to unavailable energy, from 1.455228 to 1.455255. The inertness of change was mainly due to the 1 billion tons of clay in the cubic kilometer that did not change.

7. Fitting cubic kilometers into the biosphere

7.1 The concept of the biosphere

It is hard to come to grips with the shape of the biosphere. It can be likened to an appleskin, and it is hard to imagine our world as we know it, with tall mountains and deep oceans and mines, all taking place between the outer and inner layer of an appleskin. This study took the biosphere to include the highest mountains, the deepest wells and the altitudes that airliners fly. To be practical some of the thinning higher atmosphere with its oxygen was excluded. The limits above and below sealevel were set at 12 km. On a model of the Earth the size of a basketball, a biosphere 12 km above and below sealevel would be thinner than half a millimeter. Plus its spherical shape defies modeling – made of glass or plastic it would be too fragile and easily broken. There is an extra difficulty in that the biosphere is not a true sphere. But for practical purposes we can take the given average radius of 6371 km, and subtracting the volume of 12 km down from 12 km up, the volume of the biosphere is 12,241,590,428 cubic kilometers.

7.2 The biosphere as a square prism

The pertinent characteristic to model for understanding of economic development in the biosphere is its material finiteness. To make a model as a square prism, with the y axis proportional to the x and z axes surface area, a glass "aquarium" of depth 24 mm would have to extend 22.5 meters in length and breadth. Thus even without the spherical characteristic of the biosphere, its dimensions are hard to grasp. This fact is exacerbated when attempting to track entropy at the level of Bader's atoms in molecules within cubic nanometers, and then scale up to observable processes and economic activity culminating in the global economy. An approach tested was to keep the y axis as linear scale, from +12 km though sealevel to -12 km, but to track the x axis (that is one side of the "aquarium") on a log 10 scale.

Against these large numbers, the matters vital to us – biomass (all forests and vegetation, animals in land, sea and air) and fossil fuels, hardly show up on the biosphere profile. It is not reasonable to place them on a log scale, and at plain linear scale they are difficult to see. Moreover, they are diversely scattered above and below the Earth's surface, and their chemical interactions can only be tracked by selecting representative processes in representative locations. This simple inventory in cubic kilometers, tons, and in kilometers of the side of a cube are set out in Table 4.

When scaling chemical processes from the molecular level to the macroscopic, as was conducted on an iron ore blast furnace, it was found useful to model the biosphere as a cross-section profile with log 10 scale on horizontal AND the vertical axes. Beginning with the focus on one cubic nanometer at the center, it takes 13 orders of magnitude to reach the altitude or depth of 12 km, 15 orders of magnitude to extend from the seashore through

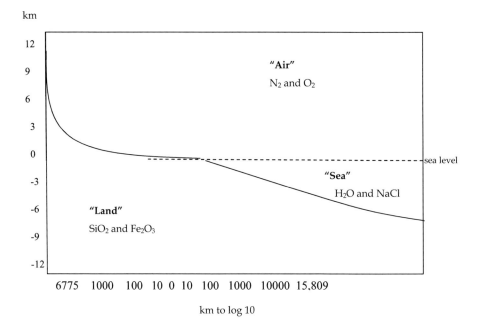

Fig. 5. A profile of the "aquarium" model of the biosphere with the horizontal scale to log 10 stretching from a seashore to land 6,775 km (left) and with sea 15,809 km to the right. As representative key elements, it is extraordinary given the diversity of the biosphere, and its biodiversity, that land, sea and air, roughly equate to solids, liquids and gases and that silicates and iron oxides make up 70% of the Earths crust, the sea is 96.5% H_2O, and air is 99% N_2 and O_2.

	Cubic kilometers	tons	side of cube in km
Total	12,241,590,428		2,305
Air	5,980,916,800	1.00E+06	1,815
Sea	1,350,870,000	1.38E+18	1,105
Land, below sealevel	4,769,925,214	1.19E+19	1,683
Land, above sealevel	104,968,584	3.15E+17	472
fresh water	34,906,900	3.49E+16	327
human constructed buildings	4,000	4.00E+12	16
Biomass	1,841	1.84E+12	12
Coal	702	9.05E+11	9
Oil	200	1.80E+11	6
natural gas	180,000	1.26E+11	56

Table 4. Estimated approximate inventory of major components of biosphere (12 km above and below sealevel).

6,775 km of land, and 16 orders of magnitude to represent the 15,809 km that represents one side of the volumes of the oceans. As will be seen below in tracking Fe_2O_3 through to refined iron, concentrating on the situation of one minimum set of atoms in an iron ore lattice and then lifting the scale ten times, repeatedly, can lead to some aspects of reality otherwise unappreciated. Specifically, the phenomenon of emf can be tracked and why it is the key to economic production costs can be understood.

In Figure 6, the main units of the side of a cube are shown: a nanometer, centimeter, meter, kilometer and then outside that, extrapolation to wide open spaces up to and higher than the highest mountain (8.848 x 10^{12} nanometers and the deepest sea (1.091 x 10^{13} nanometers). But this would seem limited in application except for the introduction of the concept of a "nanometer scanner".

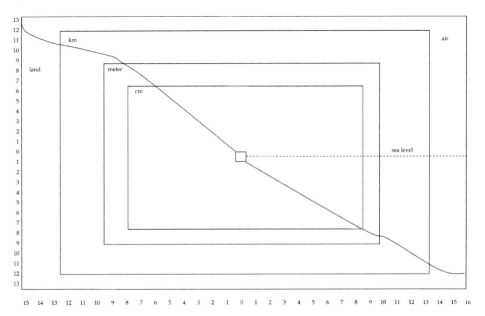

Fig. 6. Log10 scale matrix of profile of the biosphere's volume, from outer perimeter Y=24km, X=22,585km down to 1 km, 1 meter, 1 cc and 1 nanometer. Note annotations of km at 10^{12}, meter at 10^9, cc at 10^7 in units of nanometers.

7.3 The nanometer scanner

In computer war games now, the hunter looks ahead at objects in focus with peripheral objects unclear. Should something to the side attract interest and the hunter turns there, the item of interest becomes clear and can be zoomed in on. This concept similarly allows analysts to study one key cubic nanometer of matter and energy in detail and then move out to broader resolutions with some intimacy of what is going on in fine detail, and able to stop and check wherever is needed. Figure 7 shows a screen with 3 orders of magnitude. Should the attention shift to one side, then one of the 10 square nanometers would enlarge to show 1 nanometer.

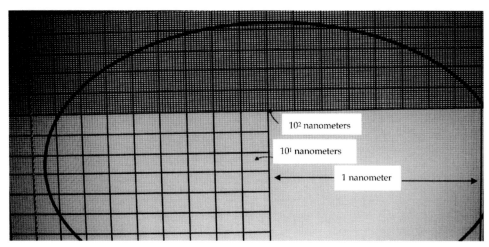

Fig. 7. Three orders of magnitude visible on the screen, with 1 square nanometer as the focus of attention. Shifting attention to the left would allow zooming in to show one of the 10 nanometer squares in detail.

8. Iron refining case study

8.1 Iron ore at the molecular level

Several millennia ago it is supposed that primitive people discovered a certain rock beside a hot fire oozed metal which was useful for tools. The basic technical process of iron refining (requiring forced oxygen onto the ore to heat it, now known to be a temperature of 1600°C) is at least a thousand years old. With the advent of modern techniques in chemistry the lattice crystal of Fe_2O_3 is now known and can be represented accurately. If Bader's Atoms in Molecules theory (Bader 1990) can be applied, the shape of the atoms in the lattice can be plotted with a high degree of accuracy and bring intelligent understanding to the bonding of atoms and what it takes to separate them. What previously was just blind fierce heating and causing atoms to oscillate wildly till they were released from the bond can now be reconsidered as a challenge to accurately targeting atoms with a mild head-on jolt to cause the iron to release from the oxygen. This is analogous to intelligently potting billiard balls with a queue, as opposed to violently and blindly shaking the table hoping that the balls may fall in. This would be a major advance in industrial refining. Insight into the possibilities provoked a senior iron industry executive to exclaim, "That would be the holy grail!" (Coulter, 2007). A schema of a set of iron and oxygen atoms nesting in an iron ore face can be presented (Pivovarov, 1997) and on scale shown within a 1 nanometer sided-cube. The object is to dislodge the oxygen molecules so that the iron is refined. In current technologies carbon atoms from the blast furnace process using carbon atoms from coke or calcium carbonate are used, hitting the oxygen atom at random angles, and given the intensity of action due to fierce heat, dislodging and bonding with a much higher bond strength than that previously existed between the iron and oxygen atoms. The oxygen atom is "married off" and carried away and eventually only iron atoms are left as molten pig iron. This is shown in Figure 8.

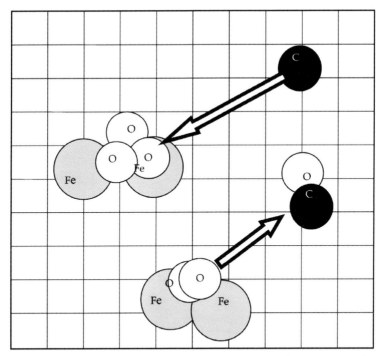

Fig. 8. Iron atoms (gray) are bonded to oxygen atoms (white) in the iron ore Fe_2O_3. A carbon atom (dark gray) strikes an oxygen atom (white) in the upper diagram and because of its high speed (and momentum) dislodges the oxygen atom, bonding it as a carbon monoxide molecule with a bond energy more than double that of Fe-O. The iron ore matrix is modeled after Pivovarov, 1997. The perimeter square denotes sides of 1 nanometer and the atom diameters are to scale. Bader's theory is set to improve details of understanding and lead to more efficient technologies

8.2 Entropy of a blast furnace

The iron industry in China provides perhaps the best single example of how global industry functions and can be quantified in science without having to resort to accounting in money. In 2010 China produced about 600 million tons of crude steel, about 40% of global output. China's blast furnaces range from very large to small county-level technologies, and several typical medium sized operations were studied. In an application of Substance Flow Analysis in the refining of iron in a blast furnace in North China, survey showed that the matter inputted, including 1.75 tons of iron ore, 0.75 tons of coke and 0.10 tons of methane gas, totaled 4.99 tons, and that through mass balance accounting the outputs were all recognized. Producing 1 ton of iron ore also produces 750 kg of carbon dioxide. The entropy increased from 17,879,562 units to 18,365,986 units (2.7%) Entropy calculations were based on molar entropy tables (Atkins & Jones, 2002).

A table of tonnages and another table of moles and molar energy (together in Table 5) was produced, and also a volumetric schema of inputs and outputs was deduced (Figure 9).

INPUTS

		mol wt gm	Tons	Mols	density in kg per cm	volume in cc	one side in cc
hematite	Fe_2O_3	159.7	1.75	10,958	2500	700,000	89
coke	C	12	0.75	62,500	610	1,229,508	107
limestone	$CaCO_3$	100.08	0.25	2,498	1394	179,340	56
methane	CH_4	16	0.10	6,126		100,000	46
oxygen	$2O_2$	64	0.39	12,251	1.43	272,727,273	648
air for	O_2	32	0.39	12,251			
heating	N_2	28.02	1.36	48,392		2,832,000,000	1415
			4.99			3,106,936,121	

inputs

tons		mol wt gm	Mols	molar entropy J/K	total entropy J/K
Fe_2O_3	1.75	159.7	10,958	87	957,733
C	0.75	12	62,500	20	1,250,000
$CaCO_3$	0.25	100.08	2,498	93	232,064
CH_4	0.1	16	6,126	186	1,140,951
$2O_2$	0.39	64	12,251	205	2,513,203
O_2	0.39	32	12,251	205	2,513,203
N_2	1.36	28.02	48,392	192	9,272,407
					17,879,562

OUTPUTS

	tons	density in kg per cm	volume in cc	one side in cc
Fe	1.00	7,850	127,389	50
CO_2	0.75	1.98	378,787,879	724
CaO	1.00	1,394	717,360	90
CO_2	0.27	1.98	136,363,636	515
$2H_2O$	0.22	1,000	220,518	60
O_2	0.39			
N_2	1.36		2,832,000,000	1415
	4.99		3,348,216,782	

outputs

	mol wt gm	mols	molar entropy J/K	total entropy J/K	tonnes
Fe	55.85	17,905	27	488,451	1.00
CO_2	44	17,045	214	3,643,295	0.75
CaO	56.08	17,832	40	708,809	1.00
CO_2	44	6,136	214	1,311,586	0.27
$2H_2O$	36	6,126	70	428,234	0.22
O_2	32	12,188	205	2,513,203	0.39
N_2	28.02	49,375	192	9,272,407	1.36
				18,365,986	

Table 5. The Substance Flow Analysis of an iron refining process showing that to refine 1 ton of pig iron requires a mass balance in and out of 4. tons of matter, and that the entropy increased from 17,879,562 units to 18,365986.

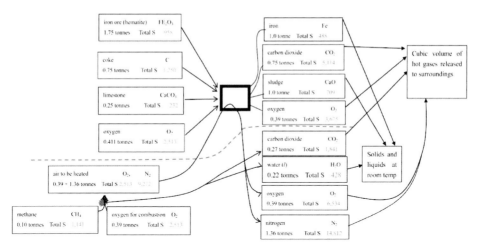

Fig. 9. A "Black Box" approach to measuring entropy in an out of a blast furnace. Assumptions. All data is for 1 tonne of iron produced in 1 minute. Matter measured in mols by dividing weight by molecular weight. Total entropy means molar entropy in kJ/K times number of mols.

Blast furnace chemistry is complex and varies. It possibly requires about 13 steps involving coke and limestone and a designated range of temperatures. Economics focuses first on the iron and steel output for sale and secondly what costs are involved in inputs. With the global output now around 1.5 billion tons in 2011, in a closed, materially finite biosphere, it is imperative to acknowledge down at the basic level of a cubic nanometer, while separating a few atoms of iron into a nearly pure iron lattice, the electromagnetic forces released out of fossil fuel is occurring by changing the ratio of bond energies to total atomic charges to a higher number, making them unavailable for future use. The ratio when considering one methane molecule was estimated to change from 0.059 to 0.079. At the macroscopic level of refining one ton of pig iron, the same phenomenon measured in an entirely different way result is an estimated to increase entropy by 2.7%. In an entire village making bricks over one year, that ratio of bond energies in molecules to their total base charge was calculated to increase from 1.455228 to 1.455255.

8.3 Global extrapolation

Whether at the scale of a nanometer or a kilometer, human manipulation of emf results in an increasing ratio of unavailable columbic forces. Entropy increases. At the global scale, so far problems have been localized. The Earth is still a big place for human activity, and defining the 24 km skin as the biosphere, we have almost 5 billion cubic kilometers of solid "land" – silicates and other rocks, of which we have barely, literally, "scratched the surface". Underground are 200 zetajoules of methane clathrate (Hermann, 2006) formed during the accretion of the planet that have not been tapped except for one small facility in Russia.

The concept of biosphere adds both a vertical dimension of space and a characteristic of material finiteness to what has previously been conceived of as a "flat earth" even by the most modern economists, strategic analysts and world leaders. From primitive hunters through to the peasants of Quesnays's pre-industrial revolution France, right up to modern national planners in various countries, there is a prevalence to conceiving of "landuse" maps in two dimensions. The maps on the walls of the world's great planners, in business, diplomacy and military, are usually flat Mercator projections. Trade requires transport, and the camels of the Silk Road and the galleons of Europe's western seaboard were the media of world trade for many centuries, and no one thought of anything but moving across the surface of the Earth. There was a gradual cognizance of the wealth in fertile soils, then water wells and finally mines and oil wells. But the further from the surface the more difficult economic exploitation became and recent troubles in deep sea oil wells demonstrate the limits of current technologies. And the vertical dimension above the surface was not considered until air travel and then air pollution became imprinted on our consciousness.

What we thought we already knew, and was so obvious as to ignore, now requires a studious and inquisitive visit in the world of the biosphere. The points below may seem puerile and obvious, but there is a purpose in enunciating them:

i. On Earth, 2D space is a trap for incoming sunlight, rain, if any, and air. This is obvious for farmland, and two storied farm fields would be a problem. The trap feature is something of value.
ii. Land, sea and air have radically different viscosities, porosity and visible transparency. People live and most economic processes are centered on the land surface. Even air travel, fishing and mining activities are centered for economic transactions on the land surface.
iii. Land, namely rock, is the antithesis of fluids. It is "solid as a rock". Only when weathered and crumbled and mixed with organic matter is it soil. We don't like quicksand.
iv. Water flows. On land it flows attracted by gravity. In the sea it flows by temperature and the spinning of the Earth, and manifests the moon's gravity in tides.
v. Air is invisible, and has a mere density of about 1.4kg/cubic meter. Only under significant pressure changes from temperature (weather) can we feel the wind, and occasionally very strong winds. It constantly delivers us oxygen, and carries emissions, including our own, away, and delivers carbon dioxide to plants.

In the era since 2008, when hundreds of billions of dollars can be conjured, not even by printing bills, but by a click on a computer screen to dilute what participants in markets think goods and services are worth, it is a reasonable check on reality to survey the physical world using science. China till 1993 published National Accounts in tonnages of all major commodities. This provided a perspective that can be calculated in exergy produced and consumed (destroyed by being used up). While financial National Accounts boast a GDP grossing all goods and services produced without acknowledging from what source they were drawn, the exergy accounts show in China for 1993 that the per capita output was 9.88 giajoules, but that the per capita input of exergy (used up) was 53.3 gigajoules (Chen and Chen, 2006).

One contribution to bridging the gap between surreal financial reporting and the physical world is in the emerging discipline of environmental informatics. In the past several decades new technologies have produced advanced sensors to monitor a vast range of physical conditions from simple temperature and moisture to many pollutants in air, water and soil, and through wireless technologies and powerful computing, integrate the information into a real time picture of a farm, river, lake, city's air quality, or a whole region. In some fields this approach has gone global.

Environmental informatics can be applied to the refining of iron from an iron ore molecule to the global emission situation. Reconciling the simple facts with Figure 6 is challenging and insightful. If the central cubic nanometer is solid iron ore crystal at the face, then moving to the next cell and zooming out to a space where there are a 1000 cubic nanometers in a blast furnace, with high speed gas, mainly carbon atoms hitting the crystal face and ricocheting off with an oxygen atom, we can then conduct an imaginary track through centimeters and meters to hundreds of meters away from the chimney. It is theoretically plausible to zoom out to the biosphere and observe CO in the atmosphere, as shown in Figure 10. While it would be too much to track each cubic nanometer, should any challenge be made as to the veracity of the observation, links in the chain of tracking can be traced, zoomed in on and discussed.

Scientists may feel uncomfortable at this brash shorthand, but it has more objectivity than many assumptions made by economists. The bare truth is that microeconomics does not add up to macroeconomics, even in theory. In microeconomics individuals and enterprises conduct their activities based on their market perceptions, and markets may be sensitive enough to correct in hours, but maybe in decades. To some extent some wars are market corrections, adjusting for conflict over resources, often land or oil, or even just water. There is no logical connection between how grassroots microeconomics can be aggregated to a nation's GDP; how around 200 GDPs are added to estimate global GDP is a work of rough fiction, and gives little insight into the wellbeing and positive and negative contributions to the state of our planet.

When energy accountants conclude that the global burning of fossil fuels in 2011, liquids, natural gas and coal, was 450 exajoules, there are two observations that can be made on the basis of the content of this chapter: first, the aggregate can be checked and verified as a reasonable number, and can be disaggregated and argued about and discrepancies identified, and either corrected or explained. The disaggregation can zoom down to where the number of atoms must balance, at least in theory, and approximately, in practice. The exercise is much more a scientific method than estimates at global GDP, and more meaningful than global accounts in money terms of energy usage. The second observation is that that energy was not just "released" from nowhere in order to drive generators and transport and blast furnaces and heating. It is the numerator begging us to identify the denominator. In the case of methane, as a representative fossil fuel, the 8 eV of energy released came from a denominator of 464 eV, so that the ratio increased from 0.059 to 0.079. In the case of burning 100 kg of methane to refine a ton of iron, using a different approach to calculation, entropy increased 2.7%. A different approach again was taken looking at a village making bricks over a year, where change was also measured against the backdrop of a whole cubic kilometer of the biosphere.

Whatever the methodology, some attempt should be made to quantify the increase in entropy at the global scale, thus providing some guidance on how to alleviate degradation of the biosphere. Since the 450 exajoules of fossil fuels consumed in 2011 consist of liquids, gas and coal, the ratios at the molecular level would be one useful indicator. Despite the complexities of liquid fossil fuels and lignite and anthracite coals, their carbon composition all is in the vicinity of 70-90% and more exact calculation can be made, but the ratio is roughly that as of the representative methane combustion model. In rough terms, total energy from fuels annually is 0.5 zetajoules "released" from 26 zetajoules locked up in the coulombic forces of proton-electron pairs in carbon, hydrogen and oxygen atoms.

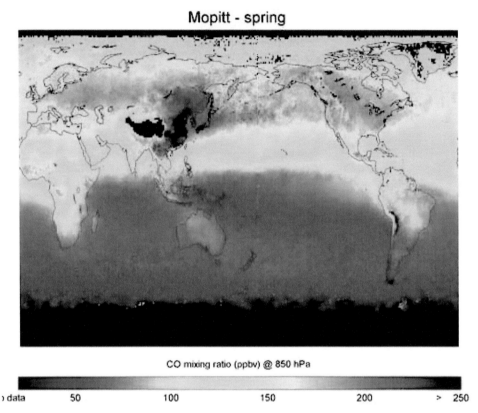

Fig. 10. Composite satellite image of the world showing heavy CO from China and drifting across Korea and Japan, even to US. Very dark pixels over Tibet are simply no reading because the data is from a lower altitude equivalent to air pressure of 850 hPa.

In this frame of mind, it is insightful to review the global exergy chart published by Stanford University Global Climate and Energy Project (Exergy Flow Chart, 2009). Dispelling the confusion in the term, energy, exergy can be rigorously quantified as 1) stored, 2) flowing, or 3) destroyed (used up and "unavailable" for further use). On the chart, the storage and flow of fossil fuels can be summarized as in Table 6.

Stanford University Global Climate and Energy Project exergy chart extends over 2 A3 pages and has extremely well researched detail integrated into a comprehensive system covering the entire biosphere. It begins with an input flow of 162 x 10^{15} joules per second

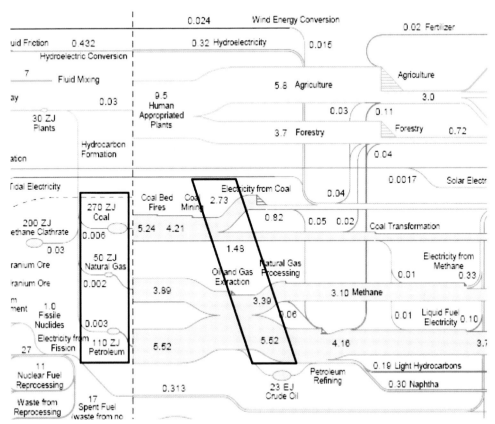

Fig. 11. Part of Stanford University Global Climate and Energy Project exergy chart showing fossil fuel reservoirs (in rectangle) and flows (in rhomboid). This exergy flow totaling 402 exajoules per year (2008) can be related to the 450 exajoules of energy "released" in 2011. For insight into the impact on the biosphere, we should present the result as energy "released" (or exergy flow) against the context of making 26 zetajoules in carbon, hydrogen and oxygen atoms now locked up and "unavailable".

	stored	flow		
	10^{21}J	10^{12}J/s	10^{18}J/year	
human appropriated	30	5.8	182.9	agriculture
plants		3.7	116.7	forestry
coal	270	2.37	74.7	electricity
		1.48	46.7	heating, refining
gas	50	3.39	106.9	gas
petroleum	110	5.52	174.1	Liquid
SUBTOTAL		12.76	402.5	

Table 6. Exergy stored and flowing from plant and fossil fuel. Data copied from Stanford University Global Climate and Energy Project exergy chart. Extract in Figure 10.

(162 petawatts) solar radiation. Of this 15PW are converted through photosynthesis and at any one time 30×10^{21} joules are stored biologically. Underground coal, gas and petroleum are in reservoirs of 270, 50 and 110×10^{21} joules respectively. The flows used by humans are listed and quantified in the table as in order, agriculture, forests, electricity, heat and refining, gas utilities, and liquid petroleum. The total of 402.5 exajoules in 2008 estimates is commensurate with the 450 exajoules estimated for 2011 from a consensus of international sources.

9. What kinds of energy count?

9.1 Global energy

Section 6.3 above makes a point of warning to take care in discussing global energy. Most participants would be thrown into confusion if an objective, broadminded scientist (perhaps from another planet) began with assessment of planet Earth's orbital momentum ($2.66 kg/m^2/sec$ followed by 44 zeroes) and angular momentum (needs 33 zeroes). Even more mundane energy line items such as incoming solar radiation (162 petajoules per second), or the parts of that which lift 1,400 trillion tons of water into the sky each 24 hours (41 petajoules per second) or warm the surface to livable conditions (37 petajoules per second) make human harnessing of energy quite puny in comparison. Clearly it is critical to be cognizant when quantifying economic activity against the biosphere of what is truly free energy, and in fact beyond human control, and what part can be influenced, harnessed, managed by humans. This latter, classed as under anthropogenic influence, is what ought to be monitored and analysed.

9.2 Anthropogenic electromagnetic forces

Of the four kinds of forces (section 6.3) the nuclear force holding protons and neutrons together in a nuclei to give an element its identity is not of relevance. We cannot economically turn lead (82 protons) into gold (79 protons). The last 60 years of development of harnessing the nuclear force for generating electricity are an added complexity and in this initial study is not discussed. The force of gravity is ubiquitous and though it plays a crucial part in economic processes is not a cost, and cannot be paid for. If we wish to influence gravity, we must harness electromagnetic forces to change the status, such as lifting using electricity or gasoline engines. The fourth force recognized in physics (the weak force only encountered in particle physics) is not relevant to this chapter.

Thus it becomes clear how confusing are the pronouncements by economists, business people, public administrators up to and including world leaders, and yes, some scientists, on the topic of "energy". The concept of exergy is far more concise, and apart from the nuclear power industry is entirely related to electromagnetic forces. Moreover, to address economic management issues, natural exergy flows that are truly free should not be included in costs. Hydropower, wind turbines, tidal power generators all have costs in manufacture of facilities, and in service and parts maintenance, but not in the flow of natural exergy.

The Szargut approach (Szargut, 1988) to exergy analysis introduces with clarity the concept of "embodied exergy" in matter, which has as its starting point a reference environment.

The question is posed, how much exergy is required to transform that matter in initial condition to a subsequent condition? As an example, iron ore lying insitu is assigned a value of 13.54 kJ/mol, possibly the exergy cost of realizing it is there. Then to transform into pure iron, the exergy cost embodied in the final product is 374.8 kJ/mol (Ranz-Villarino, Valero & Cebellero, 1998, Table 1).

10. Conclusions

10.1 Useful trends

The trend of economic and financial analysis has not resulted in the market correction needed. Analysts do not know where to look for answers, and therefore recommend remedies still stuck in the old language of solving global economic and environmental problems with more money, stimulus packages, quantitative easing and bailouts. There is an imagined linear trend of improved livelihood, and any recession is treated as unexpected and temporary, and at the worst, a hint of a double dip before recovery takes hold again. This approach is destined to catastrophic correction.

The real and actual trend occurring in the biosphere is an inexorable increase in entropy. Put another way humans are destroying exergy at a rate much higher than the flows we can harness. Realization of this trend immediately directs thinking into exploring how can the flows and stores we do not have the technology not to harness be tapped. Clearly they are not easy, but some of them have not even been considered.

The term "energy" confuses analysis of how to manage our biosphere. We have amounts of energy from the sun, and from the orbiting, spinning Earth, and its core that are unimaginably large. We need to study, firstly, what is useful energy, and secondly how much of that is free (eg wind) and what do we have to do to harness that which is not free. In short, the economic cost can be linked to anthropogenic exergy, as is addressed by Stanford University Global Climate and Energy Project.

The trend exposed in dollar terms, that the value of ecological services per year is about double global GDP (Costanza et al, 1997) can be explored in terms of science without relying on money, which is a refreshing relief when governments are severely watering down currencies. In this chapter, case studies were described that began at the scale of the simplest hydrocarbon molecule (methane) in a cubic nanometer and scaled up through various studies to address global fuel exergy flow and destruction. The key trend which needs to be explored and developed into strong science is that anthropogenic exergy, or what we popularly think of as energy we can harness, must come from somewhere, and after it is used (destroyed) it leaves a base of unusable energy. In lay terms coulombic bonds in some molecules easily release "energy" (they can explode, or at least react spontaneously, even if slowly, as in sugars we eat). The resultant product molecules are more "locked up" with, for example, the coulombic force between a single oxygen and carbon atom three times as strong as that between two oxygen atoms. That is why CO is toxic and O_2 is the breath of life. No way can we breath CO hoping the oxygen atom will split off and match up with another oxygen atom. What we perceive conventionally as usable energy has been addressed in this chapter as a numerator that has a scientifically calculable denominator. All our actions drive the ratio of the pair to a higher number (entropy inexorably increases). The ratio is presented as an estimate or at least intimated as a number yet to be understood in Table 7.

case study	unit of measure	Product-reactant	numerator	denominator	change	ratio %

computed from laboratory tests

case study	unit of measure	Product-reactant	numerator	denominator	change	ratio %
single methane molecule	electronvolt			464	27.26/464	0.059
		35.90-27.46	8.4	456	35.90/456	0.079

actual laboratory tests

case study	unit of measure	Product-reactant	numerator	denominator	change	ratio %
mole of methane	kilojoule					0.059
		3438-2628	810			0.079

actual laboratory tests

case study	unit of measure	Product-reactant	numerator	denominator	change	ratio %
kilogram of methane	megajoule					0.059
			20			0.079

typical boiler

case study	unit of measure	Product-reactant	numerator	denominator	change	ratio %
ton of coal	gigajoules					
			24			?

village brickmakers

case study	unit of measure	Product-reactant	numerator	denominator	change	ratio %
237 tons of carbohydrate and 1000 tons of coal	"released" to "locked up" coulombic forces					1.455255
						1.455255

typical blast furnace

case study	unit of measure	Product-reactant	numerator	denominator	change	ratio %
0.7 ton coal 0.1 ton methane	entropy in joules/degree		17,879,562	18,365,986		2.7

global statistics

case study	unit of measure	Product-reactant	numerator	denominator	change	ratio %
all fossil fuels burned 2011	exajoules		450	26,000? (estimate)		?

Table 7. A summary of 7 scales of exergy destruction starting from the combustion of one molecule of the simple fuel, methane, up to global fossil fuels estimated to be "consumed" in 2011. At different scales different units are used, and different methods of quantification are applied, but the universal point is that there is exergy destruction (mainly heat released) as a numerator over a denominator which in the past has not been acknowledged or even conceptualized. The denominator for global fossil fuels is simply an estimate commensurate with computations at lower scales.

11. Acknowledgement

The direction of Professor Shi Lei, Environmental Science and Engineering Department, Tsinghua University, especially in iron and aluminium Substance Flow Analysis has been an inspiration in linking chemistry at the molecular level to macroscales.

12. References

Atkins, P, Jones, L. (2002) *Chemistry: Molecules, Matter and Change*. 4th ed. Freeman, New York

Bacon, F. (1620). *The New Organon(Cambridge Texts in the History of Philosophy)* Editors Jardine, L & Silverthorne, M. , Cambridge, London.

Bader, R. (1990) *Atoms in Molecules: A Quantum Theory*. Oxford University Press, Oxford, UK

Boulding, K. (1966) The Economics of the Coming Spaceship Earth, paper presented Sixth Resources for the Future Forum on Environmental Quality in a Growing Economy in Washington, D.C.

Chen, G Q, Chen, B. (2007). Resource Analysis of the Chinese society 1980–2002 based on exergy. *Energy Policy*, 2007, 35: 2038-2050

Costanza, R, d'Arge, R, de Groot, R, Farber, S, Grasso, M, Hannon, B, Limburg, K, Naeem, S, O'Neill, R V, Paruelo, J, Raskin, R G, Sutton, P, van den Belt, M. (1997). The value of the world's ecosystem services and natural capital. *Nature*, 1997, 387(6630): 15 doi:10.1038/387253a0

Coulter, J. (1993). Entropy analysis of economic activity. http://books.google.com/books/about/Entropy_analysis_of_an_economic_activity.html?id=RKeHNAAACAAJ

Coulter, J. (2002). The Chemistry of Markets. *Ecological Economics*. 41(2002) April 2002 pp.1-3

Coulter, J. (2007) http://www.chinadaily.com.cn/opinion/2007-10/17/content_6182376.htm

Daly, H. (1977). *Steady State Economy*, Freeman, San Francisco.

Exergy Flow Chart, Stanford University (2009) . http://gcep.stanford.edu/

Forinash, K (2009) http://homepages.ius.edu/kforinas/ClassRefs/Storage/Fuel.htm

Georgescu-Roegen, N. (1971). *The Entropy Law and the Economic Process*. Harvard University Press, 0-674-25781-2, Massachusetts

Henley, D & Fry, G. http://www.lyrics007.com/Eagles%20Lyrics/The%20Last%20Resort%20Lyrics.html

Hermann, W. Quantifying global exergy resources. *Energy*. 31. pp.1685-1702.

Marshall, A, (1978) *Biophysical Chemistry*. John Wiley & Sons, New York

Odum, H T. (1970). *Environment, Power and Society*. Wiley Interscience, New York

Pivovarov, S. (1997) Letter to the editor: Surface structure and site density of the oxide-solution interface. *Jounal of Colloid and Interface Science*, 1997, 196: 321–323

Ranz-Villarino, L, Valero, A, Cebollero, M. (1998). Applications of Szargut's methodology to an exergetic account of Earth's mineral capital. Paper delivered at Seminar on Contemporary Problems in Thermal Engineering, Silesia, Poland, 1998

Ricardo, D. (1969) *On the Principles of Political Economy and Taxation*. Dent, London, (first published 1817)

Snow, C.P., (1993 reprint) *The Two Cultures*, Cambridge University Press, London.

Szargut, J, Morris D, Stewart F. (1988). *Exergy Analysis of Thermal, Chemical, and Metallurgical Processes*. Monograph.

Szargut, J. (2003).Anthropegenic and natural exergy losses (exergy balance of the Earth's surface and atmosphere). *Energy*, 2003, 28: 1047-1054

Szargut, J. (2005) *Exergy Method: Technical and Ecological Applications*. Southhampton: WIT Press,

Urone, P. (2004) *College Physics* (2nd edition), Brooks/Cole Publishing Company, ISBN 0-534-37688-6, reprinted in the People's Republic of China.

Part 3

Biosphere Reserves

9

Biosphere Reserves

Murat Özyavuz
Namık Kemal University, Faculty of Agriculture, Department of Landscape Architecture
Turkey

1. Introduction

Protected areas are essential for biodiversity conservation. They are the cornerstones of virtually all national and international conservation strategies, set aside to maintain functioning natural ecosystems, to act as refuges for species and to maintain ecological processes that cannot survive in most intensely managed landscapes and seascapes. Protected areas act as benchmarks against which we understand human interactions with the natural world. Today they are often the only hope we have of stopping many threatened or endemic species from becoming extinct (Dudley, 2008).

Most protected areas exist in natural or near-natural ecosystems, or are being restored to such a state, although there are exceptions. Many contain major features of earth history and earth processes while others document the subtle interplay between human activity and nature in cultural landscapes. Larger and more natural protected areas also provide space for evolution and future ecological adaptation and restoration, both increasingly important under conditions of rapid climate change.

The term "protected area" is therefore shorthand for a sometimes bewildering array of land and water designations, of which some of the best known are *national park*, *nature reserve*, *wilderness area*, *wildlife management area* and *landscape protected area* but can also include such approaches as *community conserved areas*. More importantly, the term embraces a wide range of different management approaches, from highly protected sites where few if any people are allowed to enter, through parks where the emphasis is on conservation but visitors are welcome, to much less restrictive approaches where conservation is integrated into the traditional (and sometimes not so traditional) human lifestyles or even takes place alongside limited sustainable resource extraction. Some protected areas ban activities like food collecting, hunting or extraction of natural resources while for others it is an accepted and even a necessary part of management. The approaches taken in terrestrial, inland water and marine protected areas may also differ significantly and these differences are spelled out later in the guidelines (Dudley, 2008). Today roughly a tenth of the world's land surface is under some form of protected area. Over the last 40 years the global protected area estate has increased from an area the size of the United Kingdom to an area the size of South America. However, significant challenges remain. Many protected areas are not yet fully implemented or managed. Protected areas continue to be established, and received a boost in 2004 when the Convention on Biological Diversity (CBD) agreed an ambitious *Programme of Work on Protected Areas*, based on the key outcomes from the Vth IUCN World Parks

Congress,1 which aims to complete ecologically-representative protected area systems around the world and has almost a hundred time limited targets.

The IUCN protected area management categories are a global framework, recognized by the Convention on Biological Diversity, for categorizing the variety of protected area management types. Squeezing the almost infinite array of approaches into six categories can never be more than an approximation. But the depth of interest and the passion of the debate surrounding the revision of these categories show that for many conservationists, and others, they represent a critical overarching framework that helps to shape the management and the priorities of protected areas around the world. The first effort to clarify terminology was made in 1933, at the International Conference for the Protection of Fauna and Flora, in London. This set out four protected area categories: *national park*; *strict nature reserve*; *fauna and flora reserve*; and *reserve with prohibition for hunting and collecting*. In 1942, the Western Hemisphere Convention on Nature Protection and Wildlife Preservation also incorporated four types: *national park*; *national reserve*; *nature monument*; and *strict wilderness reserve* (Holdgate 1999). In 1978, Ten categories were proposed, defined mainly by management objective, all of which were considered important, with no category inherently more valuable than another: scientific reserve, *national park, natural monument/national landmark, nature conservation reserve, protected landscape, resource reserve, anthropological reserve, multiple-use management area, biosphere reserve and world heritage site (natural)*. In 1994, IUCN and the World Conservation Monitoring Centre are definition new protected area categories: *strict protection/strict nature reserve and wilderness area, ecosystem conservation and protection, conservation of natural features, conservation through active management, landscape/seascape conservation and recreation, sustainable use of natural resources*. This is the most important changes in the definitions that the biosphere reserves and world heritage sites. These categories are not discrete management categories but international designations generally overlain on other categories.

The concept of biosphere reserves as originated by a task force of UNESCO's Man and the Biosphere (MAB) Programme in 1974, The biosphere reserve network was launched in 1976 and, as of June 2011, has grown to include 580 reserves in 114 countries. The network is a key component in MAB's objective of achieving a sustainable balance between the sometimes-conflicting goals of conserving biological diversity, promoting economic development, and maintaining associated cultural values. Biosphere reserves are sites where this objective is tested, refined, demonstrated and implemented. In this chapter, which is one of the most important protected areas in the approach to the information given in relation to the biosphere reserve.

2. Protected areas and management categories

The original intent of the IUCN Protected Area Management Categories system was to create a common understanding of protected areas, both within and between countries. This is set out in the introduction to the Guidelines by the then Chair of CNPPA (Commission on National Parks and Protected Areas, now known as the World Commission on Protected Areas), P.H.C. (Bing) Lucas who wrote: "*These guidelines have a special significance as they are intended for everyone involved in protected areas, providing a common language by which managers, planners, researchers, politicians and citizens groups in all countries can exchange information and views*" (The International Union For Conservation of Nature [IUCN], 1994).

Though its Commission on National Parks and Protected Areas (CNPPA), IUCN has given international guidance on the categorisation of protected areas for nearly a quarter of a century. The purposes of this advice have been (The International Union For Conservation of Nature [IUCN], 1994):

- to alert governments to the importance to the importance of protected areas;
- to encourage governments to develop systems of protected areas with management aims tailored to national and local circumstances
- to reduce the confusion which has arisen from the adoption of many different terms to describe different kinds of protected areas;
- to provide international standards to help global and regional accounting and comparisons between countries;
- to provide a framework for the collection, handling and dissemination of data about protected areas; and
- generally to improve communication and understanding between all those engaged in conservation.

As a first step, the General Assembly of IUCN defined the term "national park" in 1969. Much pioneer work was done by Dr Ray Dasmann from which emerged a preliminary categories system published by IUCN in 1973. In 1978, IUCN published the CNPPA report on Categories, Objectives and Criteria for Protected Areas, which was prepared by the CNPPA Committee on Criteria and Nomenclature chaired by Dr Kenton Miller. This proposed these ten categories. Ten categories were proposed, defined mainly by management objective, all of which were considered important, with no category inherently more valuable than another (Dudley, 2008):

Group A: Categories for which CNNPA will take special responsibility

i. Scientific Reserve/Strict Nature Reserve
ii. National Park
iii. Natural Monument/Natural Landmark
iv. Nature Conservation Reserve/Managed Nature Reserve/Wildlife Sanctuary
v. Protected Landscape

Group B: Other categories of importance to IUCN, but not exclusively in the scope of CNNPA

vi. Resource Reserve
vii. Natural Biotic Area/Anthropological Reserve
viii. Multiple Use Management Area/Managed Resource Area

Group C: Categories that are part of international programmes.

ix. **Biosphere reserve**
x. World Heritage Site (natural)

This system of categories has been widely used. It has been incorporated in some national legislation, used in dialogue between the world's protected area mangers, and has formed the organisational structure of the UN List of National Parks and Protected Areas.

Nonetheless, experience has shown that the 1978 categories system is in need of review and updating. The differences between certain categories are not always clear, and the treatment of marine conservation needs strengthening. Categories IX and X are not discrete management categories but international designations generally overlain on other categories. Some of the criteria have been found to be in need of rather more flexible interpretation to meet the varying conditions around the world. Finally, the language used to describe some of the concept underlying the categorisation needs updating, reflecting new understanding of the, natural environment, and of human interactions with it, which have emerged over recent years (The International Union For Conservation of Nature [IUCN], 1994).

In 1984 CNNPA established a task force to update the categories. This reported in 1990, advising that a new system be built around the 1978 categories I–V, whilst abandoning categories VI–X (Eidsvik, 1990). CNPPA referred this to the 1992 World Parks Congress in Caracas, Venezuela. A three-day workshop there proposed maintaining a category that would be close to what had previously been category VIII for protected areas where sustainable use of natural resources was an objective. The Congress supported this and in January 1994, the IUCN General Assembly meeting in Buenos Aires approved the new system. Guidelines were published by IUCN and the World Conservation Monitoring Centre later that year. These set out a definition of a "protected area" – *An area of land and/or sea especially dedicated to the protection and maintenance of biological diversity, and of natural and associated cultural resources, and managed through legal or other effective means* – and six categories (Dudley, 2008):

Areas managed mainly for:

i. Strict protection [1a] Strict nature reserve and [1b] Wilderness area
ii. Ecosystem conservation and protection (i.e., National Park)
iii. Conservation of natural features (i.e. Natural Monuments)
iv. Conservation through active management (i.e. Habitat/species management area)
v. Landscape/seascape conservation and recreation (i.e. Protected landscape/seascape)
vi. Sustainable use of natural resources (i.e. Managed resource protected areas)

The 1994 guidelines (Guidelines for Protected Area Management Categories) are based on key principles: the basis of categorization is by primary management objective; assignment to a category is not a commentary on management effectiveness; the categories system is international; national names for protected areas may vary; all categories are important; and a gradation of human intervention is implied (Dudley, 2008).

2.1 Biosphere reserve

Biosphere reserves are designed to deal with one of the most important questions the World faces today: How can we reconcile conservation of biodiversity and biological resources with their sustainable use? An effective Biosphere reserve involves natural and social scientists; conservation and development groups; management authorities and local communities, all working together on this complex issue.

The concept of biosphere reserves as originated by a Task Force of UNESCO's Man and the Biosphere (MAB) Programme in 1974, The biosphere reserve network was launched in 1976 and, as of June 2011, had grown to include 580 reserves in 114 countries. The network is a key component in MAB's objective of achieving a sustainable balance between the sometimes-conflicting goals of conserving biological diversity, promoting economic development, and maintaining associated cultural values. Biosphere reserves are sites where this objective is tested, refined, demonstrated and implemented (United Nations Educational, Scientific and Cultural Organization [UNESCO], 1996). Landmarks in the history of the Man and Biosphere Programme are given below Table 1 (Pocono Environmental Education Center and The Center for Russian Nature Conservation/Tides Center, 2001)

1970: UNESCO General Conference. Delegates support the formation of the Man and Biosphere program, intended to ensure harmonious coexistence between rural populations and the environment from which they derive their subsistence.
1971: First meeting of the MAB International Coordinating Committee. The official birthday of the MAB program sees the establishment of 13 (later 14) research programs aimed at evaluating human impact on various natural systems.
1972: UN Conference on Human Environment, Stockholm. Recommendations are made for establishing a global network of protected areas to conserve representative examples of ecosystems around the world.
1976: UNESCO meeting on the designation of Biosphere Reserves. A total of 59 biosphere reserves are established in eight countries.
1983: First International Biosphere Reserve Conference, Minsk. Delegates develop a detailed action plan for biosphere reserves, including proposals for research, monitoring, training, education, and local participation.
1992: Earth Summit, Rio de Janeiro. The Convention on Biological Diversity (CBD) introduced the "ecosystem approach" promoting holistic analysis of environmental as well as social and economic problems. The Agenda 21 Action Plan advocates conserving local environments and involving the local population in conservation and development.
1995: UNESCO Seville Conference. The "Seville Strategy" is adopted, emphasizing the importance of sustainable development in biosphere reserve creation and management. The Statuary Framework is drawn up regarding the nomination, approval, networking, periodic revision, and withdrawal of biosphere reserves by UNESCO.
2000: Seville +5 Meeting, Pamplona. Delegates review the implementation of the Seville Strategy over the past five years on the international level.
2008: Madrid Action Plan for Biosphere Reserves (2008-2013) *Madrid Action Plan* was agreed at the 3rd World Congress of Biosphere Reserves which was held in Madrid in February 2008. It builds on the Seville Strategy and aims to capitalize on the strategic advantages of the Seville instruments and raise biosphere reserves to be the principal internationally-designated areas dedicated to sustainable development in the 21st century

Table 1. History of Biosphere Reserve Program

In 1983, UNESCO and UNEP jointly convened the First International Biosphere Reserve Congress in Minsk (Belarus), in cooperation with FAO and IUCN. The Congress's activities gave rise in 1984 to an 'Action Plan for Biosphere Reserves.' which was formally endorsed by the UNESCO General Conference and by the Governing Council of UNEP. While much of this Action Plan remains valid today, the context in which biosphere reserves operate has changed considerably as was shown by the UNCED process and, in particular, the Convention on Biological Diversity. The Convention was signed at the 'Earth Summit' in Rio de Janeiro in June 1992, entered into force in December 1993 and has now been ratified by more than 100 countries. The major objectives of the Convention are: conservation of biological diversity; sustainable use of its components; and fair and equitable sharing of benefits arising from the utilization of genetic resources. Biosphere reserves promote this integrated approach and are thus well placed to contribute to the implementation of the Convention (UNESCO, 1996).

In the decade since the Minsk Congress, thinking about protected areas as a whole and about the biosphere reserves has been developing along parallel lines. Most importantly, the link between conservation of biodiversity and the development needs of local communities – a central component of the biosphere reserve approach – is now recognized as a key feature of the successful management of most national parks, nature reserves and other protected areas. At the Fourth World Congress on National Parks and Protected Areas, held in Caracas, Venezuela, in February 1992, the world's protected-area planners and managers adopted many of the ideas (community involvement, the links between conservation and development, the importance of international collaboration) that are essential aspects of biosphere reserves. The Congress also approved a resolution in support of biosphere reserves.

There have also been important innovations in the management of biosphere reserves themselves. New methodologies for involving stakeholders in decision-making processes and resolving conflicts have been developed, and increased attention has been given to the need to use regional approaches. New kinds of biosphere reserves, such as cluster and transboundary reserves, have been devised, and many biosphere reserves have evolved considerably, from a primary focus on conservation to a greater integration of conservation and development through increasing cooperation among stakeholders. And new international networks, fuelled by technological advances, including more powerful computers and the Internet, have greatly facilitated communication and cooperation between biosphere reserves in different countries (UNESCO, 1996).

In this context, the Executive Board of UNESCO decided in 1991 to establish an Advisory Committee for Biosphere Reserves, This Advisory Committee considered that it was time to evaluate the effectiveness of the 1984 Action Plan, to analyze its implementation, and to develop a strategy for biosphere reserves as we move into the 21st century. To this end, and in accordance with Resolution 27/C/2.3 of the General Conference, UNESCO organized the International Conference on Biosphere Reserves at the invitation of the Spanish authorities in Seville (Spain) from 20 to 25 March 1995. This Conference was attended by some 400 experts from 102 countries and 15 international and regional organizations. The Conference was organized to enable an evaluation of the experience in implementing the 1984 Action Plan, a reflection on the role for biosphere reserves in the context of the 21st century (which gave rise to the vision statement) and the elaboration of a draft Statutory Framework for the World Network. The Conference drew up the Seville Strategy, which is presented below. The International Coordinating Council of the Man and the Biosphere (MAB) Programme, meeting for its 13th session (12-16 June 1995) gave its strong support to the Seville Strategy (UNESCO, 1996).

In sum, biosphere reserves should preserve and generate natural and cultural values through management that is scientifically correct, culturally creative and operationally sustainable. The World Network of Biosphere Reserves, as implemented through the Seville Strategy, is thus an integrating tool which can help to create greater solidarity among peoples and nations of the world.

In addition, goals and objectives for biosphere reserves as shortly defined in the Seville Strategy. These objectives were defined in the context of broad goals and recommendations at three organizational levels: international, national and individual reserves. The objectives for individual reserves are described in Table 2 (UNESCO, 2010).

GOALS	OBJECTIVES
Utilize Biosphere Reserves as models of land management and of approaches to sustainable development	Secure the support and involvement of local people
	Ensure better harmonization and interaction among the different biosphere reserve zones
	Integrate biosphere reserves into regional planning
Use Biosphere Reserves for research, monitoring, education and training	Improve knowledge of the interactions between humans and the biosphere
	Improve monitoring activities
	Improve education, public awareness and involvement
Implement the Biosphere Reserve Concept	Improve training for specialists and managers
	Integrate the functions of biosphere reserves
	Strengthen the World Network of Biosphere Reserves

Table 2. Goals and objectives for biosphere reserves as defined in the Seville Strategy

The decisions made at conference in Seville in detail the goals and objectives are given below (UNESCO, 1996).

GOAL 1. Use Biosphere Reserves to Conserve Natural and Cultural Diversity
    Objective 1.1. Improve the coverage of natural and cultural biodiversity by means of the World Network of Biosphere Reserves.
    Objective 1.2. Integrate biosphere reserves into conservation planning.

GOAL 2. Utilize Biosphere Reserves as Models of Land Management and of Approaches to Sustainable Development
    Objective 2.1. Secure the support and involvement of local people

Objective 2.2. Ensure better harmonization and interaction among the different biosphere reserve zones.
Objective 2.3. Integrate biosphere reserves into regional planning.

GOAL 3. Use Biosphere Reserves for Research, Monitoring, Education and Training
Objective 3.1. Improve knowledge of the interactions between humans and the biosphere.
Objective 3.2. Improve monitoring activities.
Objective 3.3. Improve education, public awareness and involvement.
Objective 3.4. Improve training for specialists and managers.

GOAL 4. Implement the Biosphere Reserve Concept
Objective 4.1. Integrate the functions of biosphere reserves.
Objective 4.2. Strengthen the World Network of Biosphere Reserves.

The Statutory Framework sets out the rules for governing the functioning of the World Network, giving a formal definition, a set of functions and criteria, and a designation procedure. In particular, it sets out a periodic review of biosphere reserves designated over eleven years ago to bring them up to the revised standards and criteria. At the request of the MAB ICC, a "Seville +5" international meeting of experts was organized in October 2000 in Pamplona, Spain. Aimed at taking stock of the first five years of implementation of the Seville Strategy, the meeting highlighted the following points (Pocono Environmental Education Center and The Center for Russian Nature Conservation/Tides Center, 2001):

Use biosphere reserves to implement the Convention on Biological Diversity

The biosphere reserve concept encapsulates the three major concerns of the Convention on Biological Diversity (conservation of biological diversity, sustainable use and sharing of benefits at the local, national and international levels). Moreover, this Convention has now adopted the "Ecosystem Approach" which is close to the principles of biosphere reserves.

Develop biosphere reserves in a wide variety of environmental, economic, and cultural situations

Since 1995, an increasing number of biosphere reserves worldwide founded in coastal marine areas. Biosphere reserves have been established in previously un-represented regions noted for their exceptional biodiversity, and are also increasingly being established near or around important urban centers.

Explore and demonstrate approaches to sustainable development on a regional scale

Since 1995, the new biosphere reserve nominations demonstrate a clear rise in the size and complexity of decision making in biosphere reserves. In some cases, these recent biosphere reserves correspond to entire "bioregions" spanning several administrative areas. As a result, interest in creating transboundary biosphere reserves has grown. The Seville +5 meeting produced specific recommendations for transboundary biosphere reserves.

Bring together all interested groups in a partnership approach

In this process of creating larger sites, the biosphere reserve often consists of multiple units, with several core areas and buffer zones within a large transition area. The coordination mechanisms of such sites are often innovative, consisting of a consortium or a committee on which all stakeholders are represented, usually with a rotating chair. New funding mechanisms are also being tried.

Strengthen regional networks as components of the World Network of Biosphere Reserves

Computer technology increases the ease of informal communications and access to information and has enhanced cooperation in regional groupings. Various sub-networks include ArabMAB, AfriMAB, IberoMAB (19 countries of Latin America plus Portugal and Spain), EABRN (East Asian Biosphere Reserves Network) and EuroMAB (including North America). A new MAB network for South and Central Asia was recently formed.

Madrid Action Plan is one of the latest studies on Biosphere Reserves. *The Madrid Action Plan* was agreed at the 3rd World Congress of Biosphere Reserves which was held in Madrid in February 2008. It builds on the Seville Strategy and aims to capitalize on the strategic advantages of the Seville instruments and raise biosphere reserves to be the principal internationally-designated areas dedicated to sustainable development in the 21st century. The biosphere reserve (BR) concept has proved its value beyond protected areas and is increasingly embraced by scientists, planners, policy makers and local communities to bring a variety of knowledge, scientific investigations and experiences to link biodiversity conservation and socio-economic development for human well-being. Thus the focus is on developing models for global, national and local sustainability, and for biosphere reserves to serve as learning sites for policy professionals, decision-makers, research and scientific communities, management practitioners and stakeholder communities to work together to translate global principles of sustainable development into locally relevant praxis. Individual biosphere reserves remain under the jurisdiction of the States where they are situated, which take the measures they deem necessary to improve the functioning of the individual sites (UNESCO, 2008).

2.1.1 The Biosphere Reserve concept

Biosphere reserves are areas of terrestrial and coastal/marine ecosystems or a combination thereof which are internationally recognized within the framework of UNESCO's Programme on Man and The Biosphere (MAB) (Statutory Framework of the World Network of Biosphere reserves). Reserves are nominated by national governments; each reserve must meet a minimal set of criteria and adhere to a minimal set of conditions before being admitted to the Network (UNESCO, 1996).

Biosphere reserves are 'living laboratories for sustainable development' and represent learning centers for environmental and human adaptability. Biosphere reserves are the only sites under the UN systems that specifically call for conservation and sustainable development to proceed along mutually supportive paths. Such mutuality requires cultural sensitivity, scientific expertise, and consensus-driven policy and decision-making (UNESCO, 2010).

Biosphere Reserves are special entities (sites) for both the people and the nature and are living examples of how human beings and nature can co-exist while respecting each other's needs. These reserves contain genetic elements evolved over millions of years that hold the key to future adaptations and survival. The high degree of diversity and endemism and associated traditional farming systems and knowledge held by the people in these reserves are the product of centuries of human innovation and experimentation. These sites have Global importance, having tremendous potential for future economic development, especially as a result of emerging new trends in Biotechnology (Government of India Ministry of Environment and Forests, 2007).

Biosphere reserves are areas that are recognized by the UNESCO's program on MAB. One of the primary objectives of MAB is to achieve a sustainable balance between the goals of conserving biological diversity, promoting economic development, and maintaining associated cultural values.

Characteristics of Biosphere Reserves

The characteristic features of Biosphere reserves are (Government of India Ministry of Environment and Forests, 2007):

1. Each Biosphere Reserves are protected areas of land and/or coastal environments wherein people are an integral component of the system. Together, they constitute a worldwide network linked by International understanding for exchange of scientific information.
2. The network of BRs includes significant examples of biomes throughout the world.
3. Each BR includes one or more of the following categories
    i. BRs are representative examples of natural biomes.
    ii. BRs conserve unique communities of biodiversity or areas with unusual natural features of exceptional interest. It is recognized that these representative areas may also contain unique features of landscapes, ecosystems and genetic variations e.g. one population of a globally rare species; their representativeness and uniqueness may both be characteristics of an area.
    iii. BRs have examples of harmonious landscapes resulting from traditional patterns of land-use.
    iv. BRs have examples of modified or degraded ecosystems capable of being restored to more natural conditions.
    v. BRs generally have a non-manipulative core area, in combination with areas in which baseline measurements, experimental and manipulative research, education and training is carried out. Where these areas are not contiguous, they can be associated in a cluster.

Biosphere reserves are areas of terrestrial and coastal/marine ecosystems. Reserves are nominated by national governments; each reserve must meet a minimal set of criteria and adhere to a minimal set of conditions before being admitted to the World Network. Each biosphere reserve is intended to fulfill three complementary functions: 1) a conservation function, to preserve genetic resources, species, ecosystems and landscapes; 2) a development function, to foster sustainable economic and human development; and, 3) a logistic support function, to support demonstration projects, environmental education and training, and research and monitoring related to local, national and global issues of conservation and sustainable development (UNESCO, 2010) (Figure1).

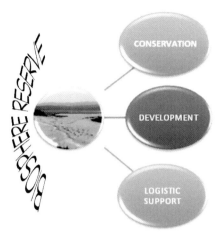

Fig. 1. Biosphere reserves functions

The main principal tasks of biosphere reserves are as detailed below (The Federal Ministry for The Environment, 2007):

Conservation of natural systems and biodiversity;

- Preservation and restoration of natural and near-natural ecosystems
- Development and preservation of diverse cultural landscapes
- Sustainable use of the natural resources of water, soil and air
- Safeguarding genetic resources and the diversity of ecosystems

Economic sustainability;

- Use of environmentally sound Technologies
- Establishment of regional value chains
- Production and marketing of sustainability produced products.
- Environmentally and socially compatible tourism.
- Ecologically adapted and site-appropriate land use, with due regard for a regions distinctive characteristics

Education for sustainable development, research and monitoring:

- Development of the skills to shape a sustainable future
- Public relations work, practice-based information and education
- Promotion of sustainable lifestyles
- Long-term ecosystem research and environmental monitoring
- Advancement of interdisciplinary research in ecological, social and economic disciplines

Structure and design of Biosphere Reserves

In order to undertake complementary activities of biodiversity conservation and development of sustainable management aspects, Biosphere Reserves are demarcated into three inter-related zones. These are natural or core zone manipulation or buffer zone and a transition zone outside the buffer zone (Anonymous, 2007) (Figure 2).

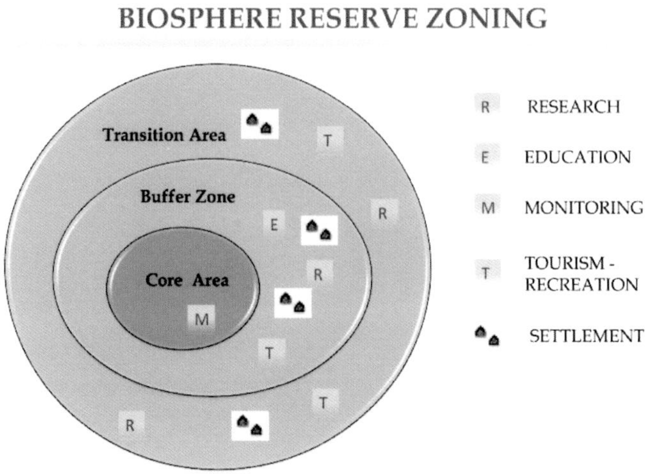

BIOSPHERE RESERVE ZONING

Fig. 2. Biosphere reserves zoning system

The Core Zone

The core zone is kept absolutely undisturbed. It must contain suitable habitat for numerous plant and animal species, including higher order predators and may contain centres of endemism. Core areas often conserve the wild relatives of economic species and also represent important genetic reservoirs. The core zones also contain places of exceptional scientific interest. A core zone secures legal protection and management and research activities that do not affect natural processes and wildlife are allowed. Strict nature reserves and wilderness portions of the area are designated as core areas of BR. The core zone is to be kept free from all human pressures external to the system.

The Buffer Zone

In the Buffer Zone, which adjoins or surrounds core zone, uses and activities are managed in ways that protect the core zone. These uses and activities include restoration, demonstration sites for enhancing value addition to the resources, limited recreation, tourism, fishing and grazing, which are permitted to reduce its effect on core zone. Research and educational activities are to be encouraged. Human activities, if natural within BR, are likely to be permitted to continue if these do not adversely affect the ecological diversity.

The Transition Zone

The Transition Zone is the outermost part of a Biosphere Reserve. This is usually not delimited one and is a zone of cooperation where conservation, knowledge and management skills are applied and uses are managed in harmony with the purpose of the Biosphere Reserve. This includes settlements, crop lands, managed forests and area for intensive recreation, and other economic uses characteristic of the region. In Buffer Zone and the Transition Zones, manipulative macro-management practices are used. Experimental research areas are used for understanding the patterns and processes in the ecosystem. Modified or degraded landscapes are included as rehabilitation areas to restore the ecology in a way that it returns to sustainable productivity.

Although originally envisioned as a series of concentric rings, the three zones have been implemented in many different ways in order to meet local needs and conditions. In fact, one of the greatest strengths of the biosphere reserve concept has been the flexibility and creativity with which it has been carried out in various situations. Some countries have enacted legislation specifically to establish biosphere reserves. In many others, the core areas and buffer zones are designated (in whole or in part) as protected areas under national law. A large number of biosphere reserves simultaneously belong to other national systems of protected areas (such as national parks or nature reserves) and/or other international networks (such as World Heritage or Ramsar sites). Ownership arrangements may vary, too. The core areas of biosphere reserves are mostly public land, but can also be privately owned, or belong to non-governmental organizations. In many cases, the buffer zone is in private or community ownership and this is generally the case for the transition area. The Seville Strategy for Biosphere Reserves reflects this wide range of circumstances (UNESCO, 2010).

At first glance, much of the biosphere reserves and other protected status may not be seen a difference. At first glance, much of the biosphere reserves and other protected status may not be seen a difference. But the nature of the biosphere reserves and other protected areas management objectives and the expected functions of these areas are quite different from each other. In fact, within the biosphere reserves and the IUCN classification independent of each other in fields such as nature conservation areas must be thinking. In some cases fields are independent of each other as in the examples are designed mostly with the biosphere reserve approach Occasionally, there also exist examples in nature conservation areas are being adapted to the existing nature conservation area, covering a wider area as a biosphere reserve is determined. The differences between the biosphere reserves and other conservation areas are given in Table 3 (Bioret, 2001).

The coordinator of the Biosphere Reserve is not the direct manager of the territory concerned: he/she merely coordinates or facilitates. One of the main problems encountered in Biosphere Reserves is the need for visibility of the management structure and adequate recognition of the coordinator. The role of the Biosphere Reserve coordinator is to moderate and communicate the different aspirations and needs of each partner around a 'common territory project' (a project which balances consideration of the environment, economy and equity of a specific area) with which all stakeholders can identify themselves (resource users, professional groups, local populations, government agencies, elected officials, scientists etc.). Hence a Biosphere Reserve coordinator must ensure (Bioret, 2001):

- Identification of the main conservation and development issues and potentialities, both at the scale of the territory concerned and at the scale of the wider biogeographically region. Certain conservation or development priorities and even sustainable development experiments may be envisaged.
- Identification of the main management issues concerning human interaction with nature using the ecosystem approach. Different types of interaction can be highlighted including
  - negative interactions: divergence of interests;
  - neutral interactions and;
  - positive interactions: convergence of interests.
  - resolving conflicts throughout mediation processes.

- Setting up working groups devoted to the common concerns of the main groups of actors.
- Organization of thematic workshops and training sessions.
- Promotion of results of successful experiments.
- Carrying out the periodic review of the Biosphere Reserve using a multidisciplinary approach.

Protected Areas	Biosphere Reserves
One type of land *a single category of land, usually relatively small in size and managed for a single purpose (e.g. nature conservation)*	A mosaic of different types of land *several categories of land, generally managed for different purposes (conservation, development, etc)*
One type of objective and function *conservation*	Overlapping of different types of objectives and functions *conservation, development, logistical support*
One main category of interests – natural – landscape – cultural – historical	Multitude of interests *often conflicting: farming, forestry, fisheries, tourism, science, local and national government*
One manager *well identified, directly in charge of the management of the territory*	Several managers *working more or less independently without consultation*
Simple zonation	Complex zonation *three zones, transition area without demarcated outer limit*
Protection through regulation	Various means of protection *Regulation limited to the core areas, existence of management agreements or contracts*
Management plan *single planning scenario applied to a well-defined land area*	Guide to Biosphere Reserve coordination *harmonization of different planning scenarios for different areas in line with Biosphere Reserve concept; emphasis on local Participation*
Single ecosystem approach *populations, ecosystem functioning*	Landscape approach *complex of ecosystems*
Manager	Coordinator

Table 3. TitleThe differences between the biosphere reserves and other protected areas (Bioret, 2001)

This approach can be realized by setting up a management guide for the Biosphere Reserve Territory. Here, a GIS can prove to be a relevant and efficient tool for the Biosphere Reserve Coordinator, since it can be used to set up, structure and continuously update a database for the Biosphere Reserve, and provide an excellent basis for decision-making by facilitating the elaboration of various zoning scenarios. The maps produced using a GIS can also help in discussions and consultations with the local communities and the various stakeholders (Bioret, 2001).

2.1.2 Ecotourism and Biosphere Reserves

The concept of ecotourism and the concept of Biosphere Reserves go together in many ways. Some people feel that although tourism is and should be an important activity in Biosphere Reserves, especially in the core areas, the only responsible tourism in these areas is ecologically conscious and sustainable tourism, thus "ecotourism". Others have pointed out, however, that there are some distinctive and possibly diffi cult aspects to ecotourism in Biosphere Reserves that must be considered if ecotourism is to contribute positively to all functions of a Biosphere Reserve -in the core area, the buffer zones, and the transition or cooperation zone ,and if the Biosphere Reserve is to contribute in a strong and positive way to the ecotourism experience. For it is that positive experience, which requires a high level of knowledge and preparation from both leaders and participants, that will give the ecotourists the emotional and inspirational stimulus that Biosphere Reserves can provide, and develop increased understanding and perspective so that they will take henceforth an active personal responsibility toward the natural environment. If it can achieve this understanding and commitment, ecotourism can help Biosphere Reserves fulfill their responsibilities to demonstrate a balanced relationship between humans and Nature (Canada MAB Committee, 2002).

A decade ago, a perceptive report of the Canadian Environmental Advisory Council called "Ecotourism in Canada" (Scace, et.al., 1992) drew attention to characteristics of ecotourism in protected and managed areas that had been identified by Prof. James Butler (Butler, 1990). Although the situations and institutions are always evolving, the characteristics of ecotourism in protected and environmentally managed areas listed by Butler are pertinent today (Canada MAB Committee, 2002):

- Ecotourism must promote positive environmental ethics - fostering environmentally conscious behavior in its participants;
- Ecotourism does not degrade the natural resource or interfere with natural environmental processes;
- Ecotourism concentrates on intrinsic rather than extrinsic values. Facilities and services never become attractions in their own right;
- Ecotourism is biocentric rather than homocentric in philosophy. Ecotourists enter the environment accepting it on its terms, not expecting it to be modified for their convenience;
- Ecotourism must benefit the wildlife and natural environment. The environment and ecological functions remain essentially undisturbed by the tourism;
- Ecotourism provides a first-hand experience with the natural environment. movies and zoological parks do not constitute an ecotourism experience;
- Ecotourism has an "expectation of gratification" that is measured in terms of education and/or appreciation rather than in thrill-seeking or physical achievement;
- Ecotourism has a high knowledge-based and experience-based dimension.

Biosphere reserve planning work to be done to be successful eco-tourism activities are given below (Roots, 2002).

General comments

1. To be successful within Biosphere Reserves, the ecotourism activities and tourist experience should be compatible with all the characteristics of ecotourism in protected

and managed areas as outlined by Butler (previously referred to) and in addition should also help the tourist to understand, take part in, and contribute to the three main functions of a Biosphere Reserve:
   a. Conservation of:
      - Ecosystems
      - Ecological functions
      - Landscape, hydrology/coastal features, and habitats
      - Historical and local cultural attributes
   b. Improvement of Social, economic and cultural development and sustainability;
   c. Provision of sites and opportunities for research, monitoring, education, communication and international contact and Exchange (the logistic function)
2. Ecotourism in Biosphere Reserves must incorporate visits to and explanations to visitors of the role of the Core Areas, Buffer Zones, and Transition or Co-operation zones in a way that shows to the visitors that these, together, demonstrate balance between human activities and Nature.
3. Biosphere Reserves, in general, can support the preamble to the Declaration on Ecotourism and help to implement the spirit of all the main clauses. But the priority for Biosphere Reserves is Nature and the future, not immediate commercial business.
4. In turn, ecotourism, properly informed and conducted, can support the functions of Biosphere Reserves.

Characteristics of Biosphere Reserves that is important to ecotourism

There are some important characteristics of Biosphere Reserves that should be kept in mind as they relate to tourism in general and to ecotourism (Roots, 2002):

1. Biosphere Reserves are, above all, selected places where relatively undisturbed Nature and active economic and social development, with all its cultural overtones, exist side by side in the same ecological setting. It is the study and knowledge of interaction and inter-dependence of these, and the lessons learned from that study, that characterize a Biosphere Reserve and its role as an exemplary of sustainable development.

Tourists must be able to visualize and understand this interaction, and learn from it. Tour operators must understand and support this purpose. Otherwise, the tourist visit will not be "ecotourism" in a Biosphere Reserve.

2. Biosphere Reserves are not government structures or entities. They typically include land of which some parts are owned and managed by governments, and other parts owned by industry, private citizens and institutions, all of whom have voluntarily co-operated for a common purpose of ecological protection, sustained social and economic development, and learning about long-term management of living resources and ecological productivity. No ownership or legal rights are in any way changed by the creation of a Biosphere Reserve.

Therefore Biosphere Reserves, as complete entities, are not subject to special regulations. They are subject to and must be governed by all the regulations that apply to their component parts. Tourists must be helped to understand that Biosphere Reserves are not special protected "parks", but places where ecological responsibility is exercised through ordinary land rights and regulations.

3. Biosphere Reserves are fundamentally places of research and monitoring in both the natural and social sciences. This means that all activities, including all aspects of tourism itself, can and should be scientifically studied and monitored; and the effects of such activities on the biosphere, natural resources, local and regional economy, cultural integrity and expression should also be studied and monitored.

Biosphere Reserves may offer an unique opportunity to observe and measure the impacts of different numbers or groupings, timing, movements, etc., of tourists. Thus they can provide assessment or models of the "tourist carrying capacity" of landscapes and ecosystems. Such studies require investment, data-gathering and persons knowledgeable in the theoretical and applied social and behavioral sciences to be associated with the ecotourism activity. The network of Biosphere Reserves, within a country or internationally, can facilitate comparative studies.

4. Biosphere Reserves are areas that will be managed, studied, and monitored over a long period of time - at least over several decades. They are throughout much of the world prime places to observe and record long-term environmental and ecological change, including the effects of regional and global changes due to natural causes as well as to changing human activities and technologies.

Ecotourism in Biosphere Reserves can be an excellent way of educating people about global change or changes in the local Nature and the environment over time. At the same time extreme care must be taken to ensure that tourism activities do not disrupt the irreplaceable role of Biosphere Reserves as long-term ecological benchmarks and monitors.

3. Conclusion

Biosphere reserves are part of an international program implemented by the United Nations Educational, Scientific and Cultural Organization since 1970 for conservation of biological diversity and sustainable land management using a participatory approach. The main goal of this program is effectiveness of the participatory approach for sustainable resources management. Conservation of biological diversity, training, monitoring, information sharing, and experience exchanging at a global level are fundamental elements of this program (Özyavuz et al. 2006; Özyavuz and Yazgan, 2010). According to the Madrid Declaration (UNESCO, 2008), the biosphere reserves, to be considered learning sites of sustainable development, should be territories where the relationship between conservation of biodiversity and local cultural and socioeconomic development has been reconciled. In other words, biosphere reserves are expected to be territories where all the four above mentioned functions are being permanently fulfilled. Empirical evidence gained by our field survey suggests that association of the biosphere reserve with the administration of protected area, either the protected landscape area or the national park, does not seem to be an ideal institutional model for the biosphere reserve existence. At least, such a model is hardly to guarantee fulfilling of the fourth function of the biosphere reserve – promotion of sustainable development (Kusova et al., 2008). Though the biosphere-reserve approach, established for the protection and sustainability of biodiversity and natural resources, is a new concept for our country, there are many protected areas which could incorporate into the international network system.

4. References

Bioret, F. (2001). Biosphere Reserve Manager or Coordinator?, *Parks*, Conclusion from EuroMAB IUCN, Vol. 11, No. 1, 26-28 p., UK.

Butler, J. (1990). A protected Areas Vision for Canada. *Canadian Environmental Advisory Council*, 88 pp., Ottawa.

Canada MAB Commitee, (2002). Ecotourism and Sustainable Development in Biosphere Reserves: Experiences and Prospects, *Workshop Summary Reports*, Canada MAB Committee, 54 pp., Canada.

Dudley, N. (Editor) (2008). Guidelines for Applying Protected Area Management Categories, *International Union for Conservation of Nature and Natural Resources*, 86 pp., ISBN 978-2-8317-1086-0, Gland, Switzerland:

Eidsvik, H. (1990). A Framework for Classifying Terrestrial and Marine Protected Areas. Based on the Work of the CNPPA Task Force on Classification, *IUCN/CNPPA*, Unpublisged.

Government of India Ministry of Environment and Forests, (2007). Protection, Development, Maintenance and Research in Biosphere Reserves in India, *Guidelines and Proformae*, Government of India Ministry of Environment and Forests, 35 pp., New Delhi.

IUCN. (1994). Guidelines for Protected Area Management Categories, *IUCN Commission on National Parks and Protected Areas with the assistance of the World Conservation Monitoring Centre*, pp. 5-7, ISBN 2-8317-0201-1, Switzerland and Cambridge, UK.

Kušová, D., Těšitel, J., Bartoš, M. (2008). Biosphere reserves - learning sites of sustainable development?, *Silva Gabreta*, vol. 14(3), 221-234.

Özyavuz, M., Korkut, A.B., Etli, B. (2006). The concept of Biosphere Reserves and Biosphere Reserve Area Studies in Turkey, *Journal of Environmental Protection and Ecology*, 3, 638-647.

Özyavuz, M., Yazgan, M.E. (2010). Planning of İğneada Longos (Flooded) Forests as a Biosphere Reserves, *Journal of Coastal Research*, 26(6), 1104-1111.

Pocono Environmental Education Center and The Center for Russian Nature Conservation/Tides Center, (2001). Biosphere Reserves in Russia, *Russian Conservation News, Pocono Environmental Education Center and The Center for Russian Nature Conservation*, No: 27, 43pp., ISSN 1026-6380, Moscow, Russia.

Roots, F. (2002). Some Factors in the Relationship Between Tourism and Biosphere Reserves, Ecotourism and Sustainable Development in Biosphere Reserves: Experiences and Prospects, *Workshop Summary Reports*, Canada MAB Committee, 54 pp., Canada.

Scace, R.C. et al. 1991. Ecotourism in Canada. Vol. 1. *Unpublished report for Canadian Environmental Advisory Council*, Canada.

The Federal Ministry for The Environment, (2007). UNESCO Biosphere Reserves, Model Regions of World Standing, *the Federal Ministry for The Environment, Nature Conservation and Nuclear Safety*, 40 pp., Berlin, Germany.

UNESCO, (1996). Biosphere reserves: The Seville Strategy and the Statutory Framework of the World Network, 19 p., *UNESCO*, Paris.

UNESCO, (2008). Madrid Action Plan for Biosphere Reserves (2008-2013), 32 pp., *UNESCO*, Paris.

UNESCO, (2010). Lessons from Biosphere Reserves in the Asia-Pacific Region, and a Way Forward, Regional Science Bureau for Asia&the Pacific, 76 pp. *UNESCO Office*, Jakarta.

10

The Importance of Biosphere Reserve in Nature Protection and the Situation in Turkey

Sibel Saricam[1] and Umit Erdem[2]
[1]*Osmangazi University Faculty of Agriculture, Eskisehir*
[2]*Ege University Center for Environmental Studies, Izmir*
Turkey

1. Introduction

"Preservation" implies keeping things as they are, and "protection" implies keeping outside interference at bay. "Reservation" involving refuges, reserves and sanctuaries, aims to avoid use or exploitaion of an area and 'zoning' or 'segregation' seeks multipurpose use by careful management of resources. "Conservation" in its modern sense is a broader concept, comprehanding all of these limited approaches and more (Evans, 1997). According to another description, the concept of conservation is a set of precautions which are needed for the sustainability of life and are developed in order to prevent the extinction of future resources by determining optimum usage methods (Gul& Sahin, 2010).

Today environmental problems have reached to an alarming level and this has drawn strong attention to the subject of "protection and sustainable usage of environmental resources" which has been on the agenda for a long period of time. Besides being a common problem for all the nations of the world, this problem has become the common policy of their governments.

Irresponsible use of natural resources and destruction of natural factors on the Earth are not something new. Therefore, the precautions that are taken to protect nature are not new as well (Yucel, 1995). The notion of modern protection started to develop at the beginning of 1800s and 'Nature Protection' was accepted as a discipline in 1900s (Zeydanli, 2008). However, by the 21st century efforts for the protection of nature have a significant rising tendency in global scale (Yalinkilic&Arpa, 2005). Although the actions taken for protecting nature vary partially from country to country, the common problem is that the protected areas cannot be managed effectively due to several reasons, including the variation in the physical size of the areas. The main reason of the problem is the disagreement between the human use in the area and the conservation of natural and cultural values. It has been realized that the reason of the failure in the efforts of protection in protected areas was at the point of avoiding the local residents. Therefore, integration of the local people with the protection efforts is required. Over the last few decades, the participation-paradigm has grown in research, policy, and practice of natural resource management, biodiversity conservation, and stewardship of ecosystem services

(Schultz et al., 2010). One of the points of view emerged from the participation of local communities to the management process of the protected areas is the holistic approach. For most of the local communities, the most important purpose of the protected area management is to provide the continuity of their lifestyles and cultural heritage. The protection of biological diversity is a component of the mentioned purpose, it does not constitute a purpose on its own. Within this context, "Human and the Biosphere" Programme of UNESCO is extremely important because it is the first environmental program which targets the relationship of the human and the nature. In 1984, MAB ICC approved the concept "Human is a part of biosphere reserve" in "Action Plan for Biosphere Reserve Areas". At the same time, while describing a biosphere reserve area and its management, the importance of local residents and their socio-economic development is included in the description (Price et al., 1998). Participation and cooperation have been increasingly emphasized in the discussions concerning biosphere reserves (Kleeman&Welp, 2008). In fact, this cooperation becomes more meaningful in "transboundary biosphere reserves". "Transboundary Biosphere Reserve Areas" have support in functions like promoting peace, protecting and managing resources and the environment, preserving and enhancing cultural values and especially protecting transboundary people, in addition to the three functions of classic biosphere reserve areas. Moreover, sharing the same "Transboundary Biosphere Reserve Area" can be helpful for building up trust and peace among neighbouring countries (Corn, 1993).

As can be seen, the role of Biosphere reserves in nature protection is unique and important.

2. What is a biosphere reserve?

Biosphere reserves are sites established by countries and recognized under UNESCO's "Human and the Biosphere" (MAB) Programme to promote sustainable development based on local community efforts and sound science. As places that seek to reconcile conservation of biological and cultural diversity and economic and social development through partnerships between people and nature, they are ideal to test and demonstrate innovative approaches to sustainable development from local to international scales. For achieving the three interconnected functions: conservation, development and logistic support, each biosphere reserve must contain three connected zones (Unesco, 2011).

The zonation scheme, with the three zones, is the landmark and the identity of biosphere reserves and is recognized as such (Cibien&Jardin, 2007). These zones are called "the core zone, the buffer zone and the transition/development zone" (Fig. 1). The functions of these zones are going to be explained below.

Core zone can be one or more in number. It is a securely protected site for conserving biological diversity, monitoring minimally disturbed ecosystems and undertaking non-destructive research and other low-impact uses (such as education); buffer zone is the part which usally surrounds or adjoins the core area, and is used for co-operative activities compatible with sound ecological practices, including environmental education, recreation, ecotourism and applied and basic research; and a flexible transition/development zone, or area of co-operation, is the part which may contain a diversity of agricultural activities, settlements and other uses and in which local communities, management agencies, scientists, non-governmental organizations, cultural groups, economic interests and other stakeholders work together to manage and sustainably develop the area's resources (Unesco, 1996).

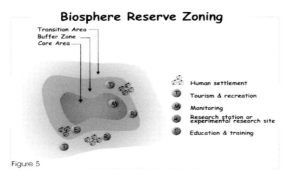

Fig. 1. Biosphere reserve zoning scheme

3. The difference of biosphere reserve in the concept of nature protection

Protected areas exist in almost all countries. Most of these areas were determined throughout 20th century and announced officially and this is accepted as "the largest scaled, logical land use change made by human" throughout history (Dudley et al., 2005).

Today the International Union for Conservation of Nature (IUCN) determines the fundamental policies and systems for conserving nature and IUCN defines a protected area as "an area of land and/or sea especially dedicated to the protection of biological diversity, and of natural and associated cultural resources, and managed through legal or other effective means". According to "1978-Protected Area Classification of IUCN" Biosphere reserves were in the status of protected areas, later on in 1994 they were removed from this status, but were transformed into a state of area which has international importance including other categories, as well (IUCN, 1994). At first sight, there seems to be slight differences between biosphere reserve areas and other areas with conservation status. However, management purposes and expected functions of these areas are rather different (Zal et al., 2006). The differences are indicated in Table 1.

Nature conservation is often understood to occur only within the limited boundaries of protected areas, managed by government agencies. These are conceived as islands of conservation where any form of human intervention is considered harmful for conservation. In contrast to this model, yet complementing its very cause, are thousands of 'unofficial' protected areas across the globe, managed and sustained by ordinary people. In fact, indigenous, mobile, and local communities have played a critical role in conserving a diversity of natural environments and species for millennia, for various economic, cultural, spiritual and aesthetic purposes (Pathak et al, n.d). At this point, the difference of biosphere areas becomes clear with the fact that they do not ignore "human" factor. Today, national parks in many countries increasingly reflect the objectives of biosphere reserves--more cooperation among adjacent land managers, more local involvement, more emphasis on the role of research and public education (Gregg, 1989).

Protected Areas	Biosphere Reserves
- One type of land, a single category of land, usually relatively small in size and managed for a single purpose (e.g. nature conservation)	-A mosaic of different types of land several categories of lands, generally managed for different purposes (e.g. conservation, development, etc.)
-One type of purpose and function Conservation	-Harmonization of different types of purposes and functions Conservation, development, logistical support
-One main category of interests Natural, Landscape,Cultural,Historical	-Multitude of interests often conflicting: farming, forestry, fisheries, tourism, science, local and national government
-One manager well identified, directly in charge of the management of the territory	-Several managers working more or less independently without consultation
-Simple zonation	-Complex zonation three zones, transition/development area without demarcated outer limit
-Protection through regulation	-Various means of protection Regulation limited to the core areas, existence of management agreements or contracts
-Management plan single planning scenario applied to a well-defined land area	-Guide to Biosphere Reserve coordination harmonisation of different planning scenarios for different areas in line with Biosphere Reserve concept; emphasis on local participation
-Single ecosystem approach populations, ecosystem functioning	-Landscape approach complex of ecosystems
Manager	Coordinator

Table 1. Differences between Protected Areas and Biosphere Reserve (Bioret 2001)

4. Biosphere reserve action in Turkey

Turkey is located at the point where the continents Asia, Europe and Africa meet. Therefore, it is endowed with a rich biodiversity which makes it unique on European and World scale

(Keskin&Sarac, 2008). Furthermore, since Anatolia has hosted numerous civilizations for ages, it has a unique and rich cultural heritage.

Forest Law and Land Hunting Law are the first laws related with the nature conservation that came into force in 1937. However, "Nature Conservation and National Park" concept which firstly entered the agenda of Turkey in 1948 by scientific studies could exist in a law with the revised Forest Law in 1956 and turned out to be an application (Kaplan, 2003).

In Table 2. the status of protected natural and cultural resources/areas in Turkey are shown in accordance with their current legal basis. With a new regulation of law on August 17, 2011, the authority related with the determination, registration, announcement and management of National Parks, Natural Sites, Nature Protection Areas, Natural Monuments, Natural Parks, Wetland Areas, Private Nature Protection Areas and other areas with protection status is conveyed to Special Environmental Protection Agency of Ministry of Environment (Special Environmental Protection Agency, 2011).

Protected Areas	Legal Basis
National Park, Nature Protection Area, Natural Monument and Natural Park	Natural Parks Law
Wildlife Protection Area, Wildlife Development Area	Land Hunting Law
Protection Forests, Gene Protection Forests, Seed Stands, and Inner Forest Resting Places	Forest Law
Natural Sites	Cultural and Natural Heritage Protection Law
Aquaculture Production Areas	Aquaculture Law
World Heritage Areas, Private Nature Protection Areas, Emerald Network Areas, Wetland Areas/Ramsar Areas, Biosphere Reserve and Nature 2000	International Agreements

Table 2. Legal Structure of Protected Areas In Turkey

In spite of all these protection status, unfortunately the reality of losing natural and cultural diversity all around the world and especially in developing countries is also valid for Turkey as well. Main reason of this situation is the disagreement between the rules and the human use in the protected area. However, if biosphere reserve area approach is applied to the protected area and its surroundings, it seems to be an effective solution in protection by balancing the usage.

Camili (Macahel) Biosphere Reserve Area which is officially approved in 2005 is the one and only Biosphere Reserve Area in Turkey for now. On the other hand, "Project of Protection and Sustainable Development of Biological Diversity and Natural Resources in Yildiz Mountains" is being carried out with biosphere reserve approach by General Directorate of

Forestry in coordination with General Directorate of Nature Protection and National Parks. In addition, studies pertaining to discussion of some areas with biosphere reserve approach are made and continuing to be made in scientific platforms.

Moreover, Cetinkaya (2002), Altan et al., (2004), Zal et al., (2006), Ozyavuz (2010), Saricam&Erdem (2010) have made several studies related with the planning of several areas as biosphere reserve areas in our country.

5. A potential biosphere reserve: Karaburun Peninsula

Karaburun Peninsula, has an area of 436 km² between meridians of 38° 21 N-38° 41 N latitude and 26° 21 E- 26° 39 E longitude. With the islands around the peninsula, the total area is 442 km² (Fig 2).

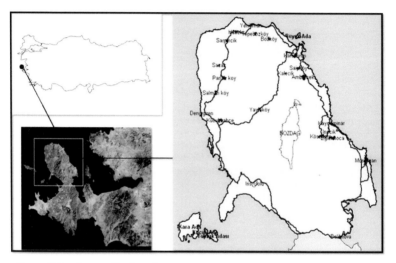

Fig. 2. Location of Karaburun Peninsula

Karaburun Peninsula is one of the rare areas which has protected its natural and cultural characteristics relatively compared with other Aegean coasts because of its rocky coasts and difficulties in transportation. It is the living and breeding ground of Mediterranean Monk Seal (*Monachus monachus*) and Audouin's Gull (*Larus audouinii*) which are included in IUCN Red List. This situation makes the area important on international scale. At the same time, all the islands of the area and untouched coasts are important for the existence of raptors and sea birds (Eken et al., 2006). Forest, maquias and frigana vegetation formations which represent Mediterranean climate geography exist in the area.

Karaburun Peninsula in which Turkish-Greek population lived together in part has traces of the interaction of this population. This cultural fact makes the history of the peninsula and the traditional usage types of the peninsula important. Fundamental human activity is agriculture in the peninsula although the agricultural areas are limited because of its rough topographic structure. Fishery and livestock are listed as the second and the third activities in the area. Tourism sector is not a common means for a living.

Coastal areas of the peninsula are 1st, 2nd and 3rd Level Natural Sites. 1st Level Natural Sites are the areas which must be protected as they are in terms of public weal because they have universal values in terms of scientific protection and they have interesting properties and beauties which are rare. 2nd Level Natural Sites are defined as areas which can be put into service by considering public weal besides protection and development of the natural structure. 3rd Level Natural Sites are defined as areas which can be opened to housing by considering the region's potential and land use features in the way of protection and development of the natural structure (Gul&Sahin, 2010). 61 km² of the peninsula is in the 1st Level Natural Sites Category, an area of 3,9 km² is in the 2nd Level Natural Sites Category and 5,3 km² of the peninsula is considered in the 3rd Level Natural Sites Category. Black Island in Gerence Bay was registered as Wildlife Protection Area in 1994 besides being in the 1st Level Natural Sites Category.

Wildlife Protection Areas are defined as the areas which have wildlife values and which are absolutely protected with their plants and animal species and their continuity is provided (Ministry of Forest and Water Works, 2011). Ministry of Agriculture and Rural Affairs General Directorate of Protection Control prohibited the fishery activities down to 20m depth from the coast, in the area between Ardic Headland and Ege University Control Station in Mordogan District (except the sedimentation fishery area which is active in Ayibaligi Location) (Sad-Afag, 2011). (Figure 3)

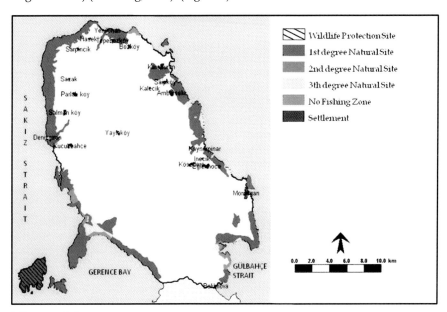

Fig. 3. Protected Areas

Although the area is protected with "protection status" like "Natural Site, Wildlife Protection Site" and with a special regulation for Mediterranean Monk Seals which is called "No Fishing Zone", threats against nature and cultural structure of the area are increasingly continuing.

The most important threat of all is the "population migration". Local residents of the peninsula, especially the young people leave where they live and go to big cities for better facilities in business and a better life. This causes the area to be abandoned and brings the loss of local values. Another important problem for the area is the unplanned second house building activity. These houses are only used in summer and the increase in number threatens the wildlife and the flora of the peninsula. This activity is dense especially on the eastern coast but it is rare on the western coast. This observation seems to be the messenger of a future increase in housing. According to a research; it is understood that there were about 3.500 second houses on the peninsula in 1990. In 2002 the number of the houses increased up to 10.000. After applying Trend Analysis Method to the increase between 1990 and 2002, it is estimated that the number of second houses will be about 74.000 in 2020 (Nurlu et al., 2002). Despite having limited plantation areas because of the rough land formation of the peninsula, agricultural activities are still the primary means of living here. Unfortunately, the increase in number of second houses means a big problem for the agriculture sector. A lot of fertile fields that are cultivable are under the risk of second house invasion.

In addition, serious attempts are needed to re-discover the viticulture and olive production for the economy of the peninsula because the areas that are used to grow quality grapes are no longer in use. Today these areas are about to be lost completely. Since these areas are not in agricultural use, the danger of second house invasion is on the way. Another threatening factor for the flora of the area is the uncontrolled goat feeding. Due to having a few meadows which are not enough for feeding goats, these animals can cause much destruction to the land and threaten the native plants and animal species. On the other hand, marble and mercury mining activities are continuing on the peninsula. Consequently, the waste products coming out of the mines are damaging the agricultural areas and the flora, together with the mountainous habitat. Similarly, wilful fire raising for having lands for agriculture or for building second houses is threatening the natural flora and the wildlife of the peninsula. Moreover, increasing fish farming activities in the bays of the peninsula which have inefficient water circulation is not only another threatening factor for the coastal and underwater flora and fauna of the peninsula but also a warning for the man-kind to resque the currently clean coasts and sea before becoming polluted in the future.

In 2007, in a written notice of Ministry of Environment and Forest, it is declared that "Due to the necessity of protection, fish farms cannot be established in closed bays and gulfs which have delicate area qualifications". Despite this fact, fish farming activities are increasing.

Additionally, Mediterranean Monk Seals attack fishermen's nets just to feed themselves but this instinctive natural behaviour causes fishermen to harm the seals.

Consequently, Karaburun Peninsula carries the potential of being a biosphere reserve area because it is an area where human interference to nature is minimal, it is an important place in terms of bio-diversity, it has facilities to be a model for sustainable development efforts in regional scale and its area is big enough to perform the three fundamental functions of a biosphere reserve (Table 3).

Karaburun Peninsula	
Protection	It is a living and breeding area for Mediterranean Monk Seals which are important in international scale. At the same time, it is an important area for several bird and plant species which are important in International and European scale.
Development	It is an area that agriculture, livestock, fishery and tourism activities can be held. It is an area open to its traditional production, in the area daffodil and hyacinth agriculture and olive production can be made. Viticulture which is now almost extinct can be re-discovered .
Logistic Support	It enables local people to become conscious by scientific socio-economic research and observations.

Table 3. Potential function capacity of Karaburun Peninsula biosphere reserve area

5.1 Materials and method

In the research; 1/25.000 scaled Topographical Map Sections of Karaburun Peninsula, Sections of Environmental Landscape Revision Plan and Aster Satellite images with 15 m resolution, dated May 2005 are used. Moreover, Quickbird satellite image provided by Google Earth has been evaluated in certain points in order to enable the decision process in the phase of interpretation of Aster Satellite image. Seal Observation Records (1997-2002) which belong to Karaburun Peninsula by Underwater Research Society-Mediterranean Monk Seal Research Group, birdwatching records kept by Ege University Birdwatching Community and shared by Turkish National Bird Databank, Kusbank (2003-2004) are among our resource materials.

In the research, in the process of data transfer to the computers and its assessment; software GeoMedia Professional 4.0, Map Info 7.0 of GIS and in the stage of satellite image processing Image-Analyst and Microstation 95 software are used. In land studies topographical data is collected via GPS.

In biosphere reserve zoning; wildlife (Mediterranean Monk Seals and birds), vegetation, current land use and protected areas are taken as the base.

By using the Mediterranean Monk Seal observation record data of "Community of Underwater Research-Mediterranean Monk Seal Research Group", two maps "Seal Observation Map" and "The Map of Breeding Caves and Cavities for Mediterranean Monk Seals" are developed. The records of Karaburun Peninsula Bird Atlas Studies are divided into categories according to SPEC (Species of European Conservation Concern). Based on these records, the Map indicating the breeding areas of Audouin's Gull which is included in Spec 1 category and important in global scale is formed; mapping according to Spec categories is realized and Maps related with the Existence of Raptors and The Richness of Species in Bird Existence are developed. Vegetation formation and vegetation intensity of the research area is composed with the studies on Aster satellite images of May 2005 and on land observations.

In the determination of vegetation intensity, NDVI (Normalized Difference Vegetation Index) which is the assessment of plant reflections in nature is applied. By considering the land covering rate of vegetation with controlled classification, five different classifications are made as Very Intense (100%-80%), Intense (80%-60%), Poorly Intense (60%-40%), Very Poorly Intense (40%-20%) and Bare Rock (20%-0%) and Vegetation Intensity Map is formed.

Furthermore, Vegetation Formation Map is formed by determining six different groups of areas as Forest, Afforestation Area, Maquias-Forest, Maquias, Frigana and Bare Area. By the help of thematic maps formed in GIS environment, regions were determined by using single layer, multi layer inquiries and overlaying.

5.2 Findings

5.2.1 Land use

Inspite of its extremely rough topography, agricultural activities have always maintained its priority in the economic structure of Karaburun Peninsula. Especially grape and olive production has been important in agricultural activities since ancient ages. Selecting these two products which have the best harmony with the soil and climate conditions and building terraces on the slopes for the prevention of erosion and minimization of soil loss have always increased the importance of agricultural activities, especially viticulture and olive production, in the peninsula. After the population exchange between Greece and Turkey (1923), by the departure of Greeks from the peninsula, the gap in the agricultural production is tried to be filled by the settlement of immigrants but this could not prevent the quick collapse of the agricultural structure (Isik, 2002). Vineyard areas which used to be 65.500 da in 1926 are very limited narrow areas today (Isik, 2002). Most of the vineyard areas today lie idle. The origin of worldwide known "Sultana Seedless Grapes" is Karaburun Peninsula. Besides "Sultana Grape", "Hurma Olive" is an olive species peculiar to Karaburun. Olive oil production in small scale local factories is the most essential and still up-to-date alternative means of earning a living for the local people (Erdem et al., 2001). Other important agricultural products of the peninsula are artichoke and tangerine. The fact that artichoke occupies 77% of the total vegetation area implies the importance of this product for the peninsula. Another agricultural activity which has been important in the recent years is daffodil cultivation. Grown daffodil and hyacinth are directly sent to Istanbul and Ankara and make a great economic contribution to Karaburun (Karaburun Municipality, 2011).

Since the meadowy areas are limited in Karaburun Peninsula, livestock hasn't developed much. In contrast, the number of goats is high related with the rough land of the peninsula. Yayla Village which is in the middle of the peninsula has the most livestock activities. Again the village has the most of goats; dairy and cheese production are the main sources of income (Isik, 2002).

Fishery is among important sources of income for the residents of Karaburun. The most famous fish of the region are grey mullets and red mullets. However, because of the fishery activities of high capacity fishing vessels coming from the Black Sea (some of which poach fish very close to the coast), fishermen of the region who fish amateurishly suffer much (Karaburun Municipality, 2011). In addition, fish farms are located to coves and bays of the peninsula.

Aegean and Mediterranean Coasts are very attractive in terms of tourism and especially in summer season they are intensely used. Although Karaburun is on the Aegean Coast, it hasn't been popular touristically. Just because of this reason, Karaburun has protected its natural features until now.

Narrow and crooked roads to Karaburun Peninsula are the most important factor that delays the potential of the area in terms of tourism. Also Karaburun Peninsula serves to domestic tourism more than foreign travel. Domestic tourism mostly consists of the usage of people who come from and around İzmir.

5.2.2 Wildlife

Mediterranean Monk Seals prefer quiet and deserted coasts which people cannot reach easily and no housing exists. Their living area must be away from human activities. Caves and cavities which function as breeding and/or sheltering areas are suitable for them (Kirac&Guclusoy, 2008). After the assessment of the data related with the observation records and living-breeding areas of Mediterranean Monk Seals; it has been understood that the areas Mediterranean Monk Seals have been seen are surprisingly the eastern coasts of the peninsula where the human-settlement is dense (Fig. 4). Since coastal caves and cavities are mostly in the eastern side of the peninsula and indispensable for breeding of Mediterranean Monk Seals, these animals prefer the eastern coasts that housing and second house settlement is dense.

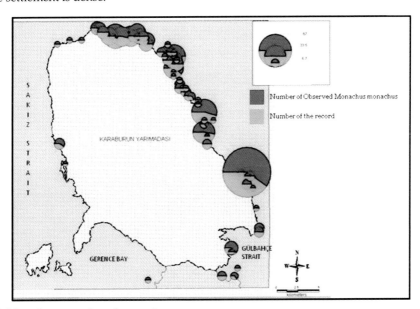

Fig. 4. *Monachus monachus* observation map

It is certain that there are not many observation records of the western part of the peninsula but this does not mean that western coastal areas are not used by seals or these parts are unimportant. There are not so many houses and settlements in the western coast. Therefore, the observation records are formed by the statements of comparatively few residents of the region or the fishermen. In other words, less population means less observation reports. On the contrary, cliffs in the western coasts of the peninsula include appropriate places for the seals to stroll around and have a rest. There are eleven breeding caves and cavities in the peninsula. In three of these caves, definite breeding records have been seen. Two of them are in the northern coast of the peninsula and both are in 1st Level Natural Sites. The other one which is the most important breeding cave is in "Mordoğan-Ayıbalığı Location" and away from protection with intensive human use (Fig.5). In order to protect this cave, the sea region which surrounds the cave is announced as "No Fishing Zone" by the General Directorate of Protection and Control. Even if no breeding record is seen in other caves and cavities, they have utmost importance because of being potential breeding caves. These caves are located in the 1st and 2nd Level Natural Sites. For breeding, feeding its pups and having a rest Mediterranean Monk Seal uses the land and for feeding and copulating it uses the sea. In providing protection to the living and reproduction areas, the sea is as important as the coastal land area. However, for a living creature which covers a distance of about 40 kilometres per day like the Mediterranean Monk Seal (Kirac&Guclusoy, 2008), it is hard to determine protection areas in the sea. On the other hand, the study about the diving behaviours of Mediterranean Monk Seals which is made by Dendrinos et al., (2007) has been a reference while determining the protection areas inside the sea. The study indicates that a Mediterranean Monk Seal can dive down to 40 m in the Aegean Sea even if it has diving records of 123 m.

Seals prefer quiet and restful cliffs and coasts away from human. However, second housing especially near the breeding caves of the seals (Fig.6), swimming activities of people where the breeding caves exist and the fishery activities cause the seals to leave these areas which are vital to them.

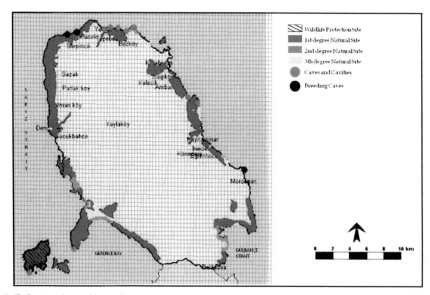

Fig. 5. Relationship of breeding caves and protected sites

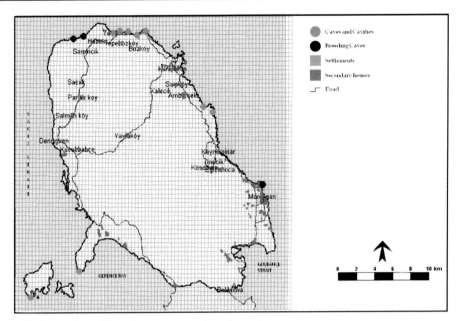

Fig. 6. Relationship of breeding caves and settlements

In accordance with the Bird Atlas studies carried out in the peninsula, 27 families and 67 species were detected. However, these studies were conducted with limited facilities by Aegean Wildlife Conservation Society so it is estimated that the real numbers are more than the reported. Audoin's Gulls (*Larus audouinii*) within Spec 1 category which has a protection priority in global scale are seen in palisades and offshore islands in colonies. In ovulation areas, they prefer medium flora. The ovulation period begins from the second half of April and continues until the beginning of May. In the first two weeks of July, pups hatch (Trakus, 2011). Big Island near Karaburun settlement and Blask Island in the southwest of the peninsula are known as the breeding areas of Audoin's Gulls (Fig.7).

The area is important for birds of prey at the same (Fig.8). Wild species such as Golden Eagle *(Aquila chrysaetos)*, Short-Toes Snake Eagle *(Circaetus gallicus)*, Eurasian Sparrowhawk *(Accipiter nisus)*, Common Buzzard *(Buteo buteo)*, Long Legged Buzzard *(Buteo rufinus)*, Western March Harrier *(Circus aeruginosus)* exist in the area. Besides this, coordinate information pertaining to Eleonora's Falcon *(Falco eleonorae)* is not given and it is not mapped in order not to pose a threat. However, it is stated by specialists that breeding records belonging to Eleonora's Falcon are no more seen in the area.

Richness of Species and diversity in a certain area is the indication of a healthy environment. Diversity of Species in terms of bird existence is divided into four categories between 1 and 22. According to the evaluation of the results, Iris Lagoon wetland region and Bozkoy and its surroundings are the areas that 15-22 and 10-15 species are seen (Fig.9). Iris Lagoon wetland and its surroundings where birds of Spec 2 and Spec 3 categories are intensely observed are very important. In other words, these areas are the priority areas for birds (Fig.10).

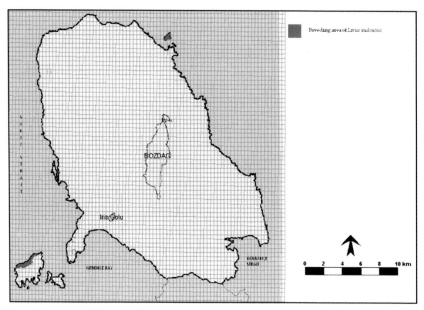

Fig. 7. Breeding area of *Larus audouinii*

The vicinity of Karaburun settlement to Big Island, intensity of human activities especially in summer and boat cruises to Black Island threaten the living and breeding areas of Audoin's Gulls.

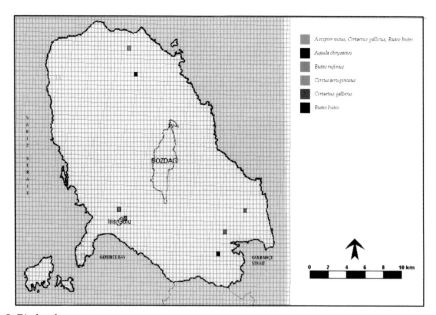

Fig. 8. Birds of prey

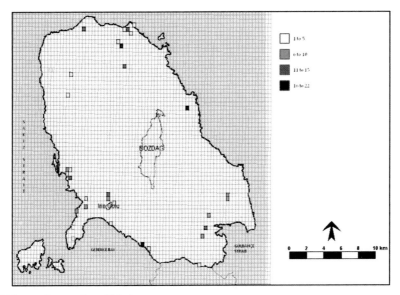

Fig. 9. Richness in species of birds

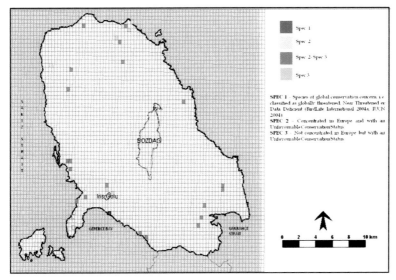

Fig. 10. Species of European Conservation Concern or SPEC

5.2.3 Vegetation

The area has the properties of typical Mediterranean vegetation geography. Vegetation formation is classified as Forest, Maquias, Maquias-Forest, Frigana, Meadow and Bare Area. The most spread vegetation formation in the area according to vegetation maps and field observations is "frigana" (Fig.11). Thorny Burnet *(Sarcopoterium spinosum)*, Cistus species

(Cistus sp.) and Thymus species *(Origanus onites* ve *Thymbra spica)* are among the dominant species of frigana formation. Forest areas consist of natural forests and forests acquired with afforestation. Natural forests are formed of Mediterranean Pine *(Pinus brutia)* and exist in separated small groups for several reasons. In the afforestation areas, 80% East Mediterranean Pine *(Pinus brutia)*, 10% Stone Pines *(Pinus pinea)* and 10% Cypresses *(Cupressus sp.)* are used. Another afforestation activity in the area is the olive afforestation. This afforestation activity is conducted under the control of the Ministry of Environment and Provincial Department of Environment and Forestry with the allocation of lands which belong to the Treasury to business concerns and the people.

What's more, Mastic *(Pistacia lentiscus)*, Terebinth *(Pistacia terebintus)*, Common Myrtle *(Myrtus communis)*, Spanish Broom *(Spartium junceum)*, Prickly Juniper *(Juniperus oxycedrus var 'Macrocarpa')*, Greek Strawberry Tree *(Arbutus andrachne)* and Strawberry Tree *(Arbutus unedo)* can be listed among dominant species in maquias vegetation. In addition, there is Sweetbay *(Laurus nobilis)* in natural forests. They usually exist in natural forests on the northwest hillside of Bozdag. Natural existence of Sweetbay in an area means that the area has a good microclimate.

In the region, Lake İris is the area which has wetland ecosystems properties. It is formed of reedy and marshy area. The area and its surroundings are very important especially for the birds.

The vegetation intensity is divided into five groups as 0%-20% Bare rock, 20%-40% Very Poorly Intense, 40%-60% Poorly Intense, 60%-80% Intense and 80%-100% Very Intense. After the mapping and the land observations, the dominant group is understood to be "Intense" vegetation by 60%-80% (Fig.12).

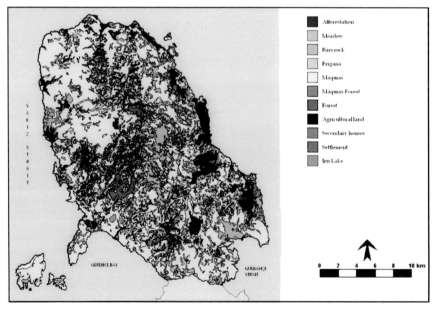

Fig. 11. Vegetation types

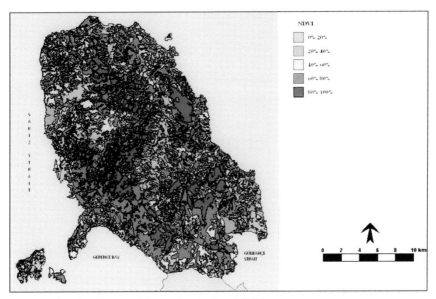

Fig. 12. Normalised Difference Vegetation Index (NDVI)

6. Conclusion

The natural and cultural values of the peninsula are sometimes destroyed although it is important in national and local scale. Unfortunately, the current protection status in the peninsula is not effective enough in terms of "protection". This condition has both social and economic dimensions. The major reason of the failure in protection is that, the peninsula couldn't reach a sufficient degree in economic development. As mentioned in the previous sections, "biosphere reserve area" formula has the "protection" and "development" components in itself and is the best solution for Karaburun Peninsula. First of all, it is obvious that the species and their habitats which need to be protected urgently must be rescued from human use and its pressure. With this in mind, for the protection of Mediterranean Monk Seals which are important in international scale, breeding caves and cavities are thought to be included in the "Core Area". Since Mediterranean Monk Seals use the coasts and the sea together, it is a must to exclude the human-activity both from the coastal region and the sea eco-system while determining the "Core Area". Fortunately, most of the breeding caves and cavities are in Natural Sites and this is an advantage in determining the "Core Area". The location of two breeding caves in the Northern part of the peninsula is in the 1st Level Natural Site. Additionally, there is no second house settlement around the area. These two factors make it easy to take these definite breeding caves under protection. For the protection regions that are inside the sea, the data that "Mediterranean Monk Seals dive down to 40m in the Aegean Sea" is taken into consideration and in the "Depth Map", 40m of depth contour is taken as the basis. In that case, "Posidonia" which spreads down to 40m depth, and supplies oxygen to stocks of the Mediterranean and the layer where the juvenile fish complete their growth is going to be taken under protection. On the other hand, the breeding cave in "Ayıbalığı" location which has vital importance for Mediterranean Monk

Seals is surrounded with intensive human use and does not have any protection status other than being a "No Fishing Zone". So, this zone which is determined by the General Directorate of Protection and Control is taken as the basis in determining the "Core Area". Other caves and cavities which are the potential reproduction areas of Mediterranean Monk Seals are included in the Core Area, although definite breeding records haven't been found yet. These caves and cavities are taken into borders of 200mx200m and the core areas which cover these caves and cavities are not connected to eachother. If these areas are connected to eachother, the dimensions of the area become larger. This situation causes difficulties in application because in this case no permission is going to be given for human use.

Similarly, if too many and wide core areas are formed in the sea, fishermen can react negatively and more threat can occur for the seals. Therefore, small core areas around the caves are thought to be more effective in protection.

Another important point is that, after the assessment of the data obtained from bird watch, the reproduction places of bird species of Spec1 and Spec2 which are important in global and European scales are thought to be included in the core areas. Of course the reproduction areas of Audoin's Gulls (*Larus audouinii*) which is in Spec 1 category are taken into the core areas, in order to provide sustainability of the species. These areas already have the Natural Site and Wildlife Protection Area Status which enabled us to label these areas as the "Core Areas". Most of the areas where Spec 2 category birds are observed have intensive human activity. Thus these areas aren't included in the "Core Areas".

15-22 and 10-15 different bird species are observed in "Bozköy" and "Lake İris", the areas which have richness in species of birds. Therefore, these areas are thought to be taken into consideration as "Core Areas" but since Bozköy and the surroundings have human settlements, it is excluded. Lake İris is accepted as a Core Area because it is very important for the birds and it is the only wetland of the peninsula. Moreover, no land use intersects with this area. All these features are important for this lake to be a Core Area.

In the area, existence of natural forests is continuously decreasing. Thus the natural forest and the dense textured maquias in the western part of "Bozdag" are considered as one of the "Core areas". Since Lake İris and its surrounding wetland is near "Bozdag", all of these mentioned areas are taken as one "Core Area". In fact this area has no protection status, so necessary attempts must be made to protect this region by law.

The reason why the number of core areas in Karaburun Peninsula is quite a lot originates from the obligation of protection of species which prefer different habitats and the obligation of protection of ecosystems which are under danger of extinction, instead of protection of a single species or a single habitat. In addition, due to much distance among these areas, independent and unconnected core areas are determined. "Buffer zone" which is surrounding the core area consists of the coastal areas (in order to control the housing), areas with very dense and dense textured maquias which have connection with the forests and maquias of the core area, destroyed sections of the natural forests, afforestation areas, habitats which are important for birds, areas where agriculture is not intense and Bozdag. In this area, a controlled usage is going to be applied. Bozdag is going to provide opportunities for activities such as climbing, walking, camping, bird watching, nature observation and reforestation of destroyed forest areas.

On the other hand, another equally important subject is the development of the peninsula.

In fact, the peninsula is an area which provides opportunities for the development of sustainable area usage activities such as agriculture, fishery and tourism. Regeneration of traditional area usage types in the area can prevent the migration of local population to metropolises. Production of "Hurma olives" which can be eaten directly after they are grown on tree, depending on the climatic features of the area, can be developed. The economy of the area can be regenerated by developing "sultani seedless grapes and wines". In its history, the peninsula used to be an important center of trade with all these features.

The abandoned Greek villages in the area can be restored and evaluated within ecologic tourism concept. For providing the local people with income, any kind of alternative tourism model that is appropriate for the region can be planned, promoted and developed.

"Transition/development zone" which surrounds the buffer zone is an area where sustainable usage activities are going to be tried and developed. "Transition/development zone" includes the areas with intense agricultural activities and secondary housing. Waste vineyard areas between Kucukbahce and Sarpincik villages are left in the transition/development area in order to regenerate the local economy. Abandoned Greek villages are left in the transition/development area as well, in order to develop the eco-touristic activities.

Consequently, in Karaburun Peninsula recommended Biosphere Reserve Area Zoning is as follows: Core zone 8%, Buffer zone 48% and Transition/development Zone 44%.

It is satisfactory that these ratios are in a magnitude which is not going to create any disagreement between protection and usage.

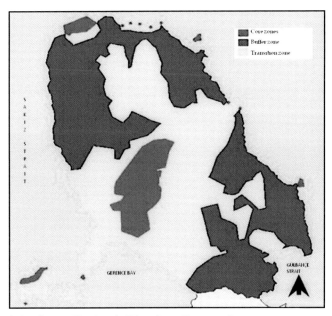

Fig. 13. Zones of Karaburun Peninsula Biopshere Reserve Area

7. References

Altan, T., Artar, M., Atik, M. &Cetinkaya, G. (2004). *Cukurova Delta Biosphere Reserve Management Plan*, LIFE- Çukurova Delta Biosphere Reserve Project, Cukurova University, ISBN 975487118-3, Adana. *(In Turkish)*

Bioret, F. (2001). Biosphere Reserve manager or coordinator?, *Parks, The international journal for protected area managers*, Vol 11, No 1, pp.26-28, ISSN 0960-233X

Cetinkaya, G.& Altan, T. (2002). Koprulu Canyon National Park should be reorganized as a biosphere reserve, *Proceedings of Turkish Costs and Marine Sites IV. National Conference*, pp. 163-171, ISBN 975-429-199-3, Izmir- Turkey, November 5-8, 2002. *(In Turkish)*

Cibien, C. & Jardin, M. (2007). How does the zonation of a biosphere reserve contribute to sustainable development? *In Proceedings of EuroMAB 2007*, Zal, N.&Lise, Y. (Ed(s)), pp. 37-47, Turkish National Commission for UNESCO, ISBN 978-975-94457-4-4, Antalya, Turkey

Corn, M.L. (1993). Biosphere Reserves: Overview, *In: National Council for Science and the Environment*, 05.09.2011, Available from: http://cnie.org/NLE/CRSreports/Biodiversity/biodv-34.cfm

Dendrinos, P., Karamanlidis, A.A., Androukaki, E& McConnell, B. J. (2007). Diving development and behaviour of a rehabilitated monk seal (Monachus monachus). *Marine Mammal Science*, Vol 23, Issue 2, pp. 387–397, ISSN 1748-7692

Dudley, N., Mulongoy, K.J., Cohen, S., Stolton, S., Barber , C.V.&Gidda, S.B. (2005). *Towards Effective Protected Area Systems Action Guide to Implement The Convention Biological Diversity Programme of Work on Protected Areas*, (Turkish Translation:Dr. S. Kalem), Technical Series No: 18, WWF, ISBN: 978-605-60247-0-2, Ankara *(In Turkish)*

Eken, G., Bozdogan, M., Isfendiyaroglu, S., Kilic, D.T.& Lise, Y. (Ed(s)). (2006). *Important Natural Areas of Turkey*, Vol. I., Nature Society, ISBN 978-975-98901-3-1, Ankara. *(In Turkish)*

Erdem, U., Nurlu, E., Yılmaz, O.&Veryeri, O.(2001). Natural Structure and Agricultural Areas: A Case Study of Karaburun Peninsula, Turkey, *In. CIHEAM*, 07.09.2011, Available from http://ressources.ciheam.org/om/pdf/a53/03001743.pdf

Evans, D. (1997). *A History of Nature Conservation in Britain*, 2 nd edition, Taylor&Francis, ISBN 0-203-74827, NewYork.

Gregg, W.P. (1989). On Wilderness, National Parks and Biosphere Reserves, *Proceedings of the Symposium on Biosphere Reserves*, Fourth World Wilderness Congress, Colorado USA, pp. 33-41, 14-17 September 1987 Retrieved from http://pdf.usaid.gov/pdf_docs/PNABE321.pdf

Gul, A. & Sahin, C. (2010). Analysis of current status of natural sit areas in Turkey (Case studies in Bodrum Peninsula), *Proceedings of 3th National Blacksea Forestry Congress*, Faculty of Forestry Artvin Coruh University, Artvin, Vol: IV pp: 1564-1574, Artvin, May 20-22, 2010, Retrieved from http://www.artvin.edu.tr/karok3/IV.Cilt/(1564-1574).pdf *(In Turkish)*

Isik, S. (2002), *Historical Geography of Karaburun Peninsula*, Publication of Ege University Faculty of Letters:120, ISBN 9754835438, Izmir *(In Turkish)*

IUCN, (1994). *Guideliness for protected areas management categories*, CNPPA with the assistance of WXMC., X+261pp, IUCN, Gland, Switzerland and Cambridge, ISBN 2-9317-0201-1, UK.

Kaplan, S. (2003). Nature Conservation Practies and Laws In Turkey, Journal of The World of Public Management, No:16, pp.29-33. Retrieved from http://kamyon.politics.ankara.edu.tr/dergi/belgeler/kydd/61.pdf *(In Turkish)*

Karaburun Municipality. (September, 2011). Karaburun socio-economic profile, In: *Karaburun Municipality*, 09.09.2011, Available from http://www.karaburun.bel.tr/kara/english/sosyoekonomieng.htm

Keskin, A.T.&Sarac, B. (2008). Nature Conservation and Planning, *"Planning" Journal of the Chamber of City Planners*, No 42, pp. 73-87, ISSN: 1300-7319 *(In Turkish)*

Kirac, C.O.& Guclusoy,H. (2008). *Foca and Mediterranean Monk Seal; Conservation and Monitoring of the Mediterranean Monk Seal (Monachus monachus) In Foça Special Environment Protection Area,* Ministry of Environment and Forest – Publication of Special Environment Protection Area, ISBN 978-605-89540-0-7, Ankara *(In Turkish)*

Kleeman, S. & Welp, M. (2008). Participatory and Integrated Management of Biosphere Reserves – Lessons from Case Studies and a Global Survey, *GAIA*, Vol.17,Supplement 1, pp. 161-168, ISSN 0940-5550

Ministry of Forest and Water Works. (September, 2011). Legislation on Wildlife Protection and Wildlife Development Areas, *In: Ministry of Forest and Water Works,* 09.09.2011, Available from http://www2.cevreorman.gov.tr/yasa/yonetmelik.asp *(In Turkish)*

Nurlu, E., Gokcekus, H., Yilmaz, O.& Erdem, U. (2003). Sustainable Development and Land Use Changes in Karaburun Peninsula, *Local Resources and Global Trades Environments and Agriculture in the Mediterranean Region*,Camarda, D.&Grassini, L. (Ed(s)), pp 221-230, Bari-CHEAM-IAMB, ISBN 2-85352-278-4, Bari

Ozyavuz, M.&Yazgan, M.E. (2010). Planning of Igneada longos (flooded) forests as a biosphere reserve. *Journal of Coastal Research*, Vol.26, No.6, pp.1104–1111, ISSN 0749-0208.

Pathak, N., Balasinorwala, T., Kothari, A.& Bushley,B.R. (no date). People in Conservation Community Conserved Areas in India, *In: IUCN*, 06.09.2011, Available from: http://cmsdata.iucn.org/downloads/cca_india_brochure__2_.pdf

Price, M. F., MacDonald, F.& Nuttall, I. (1998). Review of UK Biosphere Reserves, *In: UK MAB Committee,* 05.09.2011, Available from: http://www.ukmab.net/uploads/file/Review_of_UK_Biosphere_Reserves_1999.pdf

Sad-Afag, (September 2011). Conservation and Advocacy, Karaburun Activities, In: Underwater Research Society, Mediterranean Seal Research Groupe, 11.09.2011, Available from: http://www.sadafag.org/english/index.php?bolum=somut-basarilar

Saricam Yigiter, S.& Erdem, Ü. (2010). Planning of Izmir-Karaburun Peninsula as Biosphere Reserve Area, Ekoloji, Vol. 17, No: 77, pp.42-50, ISSN 1300-1361 *(In Turkish)*

Schultz, L., Duit, A.&Folke, C. (2010). Participation, Adaptive Co-management, and Management Performance in the World Network of Biosphere Reserves, *World Development*, Vol.39 No.4, pp. 662-671, ISSN: 0305-750X

Special Environmental Protection Agency (2011). Duties of the General Directorate of Protection of Natural Assets. In: General Directorate of Protection of Natural Assets, 20.09.2011, Available from: http://www.ozelcevre.gov.tr/icerik-15-Gorev-ve-Yetkiler.html *(In Turkish)*

Trakus (September, 2011). Audouins Gull (Larus audounii), *In: Anonym Birds of Turkey*, 08.09.2011, Available from http://www.trakus.org/kods_bird/uye/?fsx=2fsdl17@d&tur=Ada%20mart%FDs%FD *(In Turkish)*

UNESCO, (1996). Biosphere Reserves: The Seville Strategy and the Statutory Framework of the World Network. *In: Unesco Man and Bisphere Programme,* Paris, France, 18p., Available from: http://www.unesco.org/mab/doc/brs/Strategy.pdf

Unesco, (August, 2011). Biosphere Reserves-Learning Sites for Sustainable Development, *In: Unesco Man and Bisphere Programme,* 19.08.2011, Available from: http://www.unesco.org/new/en/natural-sciences/environment/ecological-sciences/biosphere-reserves/

Yalinkilic, M.K. & Arpa, N. (2005). Ecotourism and The Protected Areas In Turkey, Proceedings of Protected Natural Areas Symposium, Suleyman Demirel University, Isparta, pp. 3-13. *(In Turkish)*

Yucel, M. (1995). *Nature Conservation Areas and Planning,* Publication of Cukurova University Faculty of Agriculture, General Publication Number:104, Textbook Publication Number:9 Adana-Turkey *(In Turkish)*

Zal, N., Eczacıbaşı G.B.,&Karauz Er, E.S. (2006). *Planning of Lower Meric Valley Flood Plain As A Biosphere Reserve,* Ministry of Environment and Forest Publication Number:289, Central Anatolia Forestry Research Institute Publication Number:69, Ankara

Zeydanlı, U. (2008). Conservation Management Practice, In: *Forest and Biodiversity,* Ülgen, H.& Zeydanlı, U. (Ed(s).), pp. 117-140, Donmez Ofset, ISBN 978-605-89908-0-7, Ankara. Retrieved from http://www.dkm.org.tr/tr/orman_icindekiler.html *(In Turkish)*

11

Wealth of Flora and Vegetation in the La Campana-Peñuelas Biosphere Reserve, Valparaiso Region, Chile

Enrique Hauenstein
Catholic University of Temuco, Faculty of Natural Resources, Temuco
Chile

1. Introduction

Chile has a National System for State-Protected Wilderness Areas (Sistema Nacional de Áreas Silvestres Protegidas del Estado – SNASPE), which consists of three categories: National Parks (PN), National Reserves (RN) and Natural Monuments (MN). It presently contains 95 units, covering in total 19% of the country's territory. SNASPE has become a fundamental pillar not only for safeguarding an important part of Chile's natural heritage but also for protecting and valuing our cultural heritage, particularly where it is integrated in the areas which make up this system (Oltremari, 2002). By 1985, UNESCO had designated 7 Biosphere Reserves in Chile: Lauca, Fray Jorge, Juan Fernández, La Campana-Peñuelas, Araucarias, Laguna San Rafael and Torres del Paine. To these were later added Cabo de Hornos [Cape Horn], Bosques Templados Lluviosos de los Andes [Wet Temperate Andean Forests], and recently Laguna del Laja-Nevados de Chillán.

The Valparaiso Region contains La Campana National Park, declared a Biosphere Reserve in February of 1985 jointly with Lago Peñuelas National Reserve due to their high biological and ecosystem diversity, representative of the mediterranean environments characteristic of this region (Weber, 1986; CONAF, 1992, 2008; Elórtegui & Moreira, 2002). In 1834, during his stay in Chile, the great naturalist Charles Darwin climbed to the peak of La Campana where he was amazed by the presence of forests of palm trees (*Jubaea chilensis*), and especially by recording a specimen at 1,350 m above sea level (Elórtegui & Moreira, 2002).

The central zone of Chile, also called the mesomorphic or mediterranean zone, extends approximately from 32° to 37° S (Pisano, 1956). It has a mediterranean climate, which Koeppen (1931, 1948) classifies as a "warm-temperate climate with sufficient humidity", in the subdivision "winter rains and prolonged dry season", characterised by regular periods of rain in winter and a strongly marked dry season which may extend from six to eight months. This corresponds to other mediterranean zones around the world such as in California and southern Europe in the Northern Hemisphere, and Australia and southern Africa in the Southern Hemisphere (Grau 1992; Arroyo et al., 1995). The climatic conditions mean that the vegetation in these regions has specially adapted characteristics, such as the presence of sclerophyllous leaves, lignotubers and a great capacity for water-use efficiency (Money & Kumerow, 1971; Araya & Avila, 1981; Avila et al., 1981). According to

Marticorena et al. (1995), the central zone of Chile is a focus where endemic species are concentrated, with great wealth and diversity of flora. Mittermeier et al. (1998) indicate the presence of 1,800 species of endemic plants for this area, leading it to be considered as one of the world's 25 hotspots, requiring priority protection (WWF & IUCN, 1997; Myers et al., 2000).

In this context, flora and vegetation studies are essential basic elements for developing proposals for the conservation and management of species and ecosystems, or defining priority areas (Cavieres et al., 2001; Teillier et al., 2005). The object of the present study is to contribute to knowledge of the flora and vegetation of the La Campana-Peñuelas Biosphere Reserve.

2. Methods and results

2.1 Study area

This Biosphere Reserve includes La Campana National Park (32°55' to 33°01'S; 71°01' to 71°09'W) and Lago Peñuelas National Reserve (33°07' to 33°13'S; 71°24' to 71°34'W). It is important to indicate that by means of offer of the National Forest Corporation of Chile, the year 2008 extended his surface in near 14 times, going on from 17,095 to 238,216 ha (CONAF, 2008). The altitude ranges from ± 350 to ± 2,222 masl, and the highest peaks are "El Roble" (2,222 masl) and "La Campana" (1,920 masl). Some sectors of RN Lago Peñuelas contain plantations of exotic species (*Pinus radiata* D. Don and *Eucalyptus globulus* Labill.). Another part corresponds to Peñuelas lake, a reservoir covering 1,600 ha which supplies water to the cities of Valparaiso and Viña del Mar, and forms a wetland which is important for migratory,

Fig. 1. Map of location of La Campaña-Peñuelas Biosphere Reserve in the Region of Valparaiso, Chile.

occasional or resident birds; more than 125 species of water, land and shore birds have been recorded in the lake (Strang, 1983) (figure 1, photos 1, 2, 3). The climate of the area is temperate-mediterranean, with sufficient humidity, winter rains and a prolonged dry season. The average annual temperature is 13.5°C; the average maximum is 17.1°C and the average minimum is 9.4°C; the temperature occasionally falls to 0°C, with frosts between May and September. Precipitation is seasonal (656 mm/year), from the end of May to August, with a markedly dry summer of six to eight months from October to March (Di Castri & Hajek, 1976; Luebert & Pliscoff, 2006).

Photo 1. Place of access to the Reserve for the sector of the Peñuelas lake.

Photo 2. Place of access to the Reserve for the sector of Ocoa's Palms.

Photo 3. Sight of the Hill the Oak, major summit of La Campana-Peñuelas Biosphere Reserve.

2.2 Methodology

The study of the flora and vegetation of the Reserve was carried out by field activities and bibliographic review. Field trips were done in February and November 2001, with intensive collection and 27 phytosociological surveys in nine sampling stations in the Peñuelas National Reserve. The area of each inventory was 4 m² for herbaceous vegetation, 25 m² for shrubs and 100 m² for forest species, within areas greater than the minimum area (Steubing et al., 2002). Aquatic and marsh vegetation were also included on the shores of Peñuelas lake. The phytosociological tables were processed according to the methodology proposed by Braun-Blanquet (1964, 1979), as explained in Ramírez & Westermeier (1976). The bibliographic review was based principally on the following works: Looser (1927, 1944), Rundel & Weisser (1975), Rodríguez (1979, 1982), ICSA (1980), Villaseñor (1980, 1986), Villaseñor & Serey (1980-1981), Balduzzi et al. (1981, 1982), Rodríguez & Calderón (1982), CONAF (1986, 1994), Zöllner et al. (1995), Elórtegui & Moreira (2002), Novoa et al. (2006), CONAF (2008) and Hauenstein et al. (2009).

The identification, nomenclature and geographical origin of each species was based on Muñoz (1966), Navas (1973, 1976, 1979), Hoffmann (1978, 1991), Marticorena & Quezada (1985), Matthei (1995), Hoffmann et al. (1998) and Marticorena & Rodríguez (1995, 2001, 2003, 2005). The International Plant Names Index (IPNI 2011) was used to update the scientific names and abbreviations. Life forms were determined according to the scheme proposed by Ellenberg & Mueller-Dombois (1966) and the state of conservation of the species was determined considering the proposals of Benoit (1989), updated by a meeting of experts in September 1997 (Baeza et al., 1998; Belmonte et al., 1998; Ravenna et al., 1998), and de Novoa et al. (2006) for the Orchidaceae. The degree of human disturbance of the site was determined on the basis of the proposal of Hauenstein et al. (1988) and the evaluation scale of González (2000), which consider phytogeographic origin (the ratio between native and introduced species), and life forms (Raunkiaer's biological forms) as measurements of this form of disturbance. The result was an inventory of the flora in the study area (table 2), containing all the above elements.

The vegetation units were determined using the maps produced by CONAF & CONAMA (1999) and specific cartography provided by CONAF (National Forest Corporation), Valparaiso Region. The definition and nomenclature of the vegetation units followed the proposal of Gajardo (1995). The vegetation units identified in the map for study were subsequently verified on the ground.

2.3 Results

Flora. The vascular flora recorded (table 2) included 420 species distributed as: 12 Pteridophytes (3.0%), 3 Gymnosperms (1.0%), 290 Dicotyledons (69.0%) and 115 Monocotyledons (27.0%) (figure 2). The phytogeographic origin (figure 3) indicates that 49.0% of the species (207 sp.) are native, 28.0% (118 sp.) endemic and 23.0% (95 sp.) introduced. The biological spectrum (life-forms; figure 4) was represented by 122 hemicryptophytes (29.0%), 112 therophytes (26.4%) corresponding to annual and biannual plants, 72 cryptophytes (17.1%) including hydrophytes and geophytes, 52 nanophanerophytes (shrubs) (12.4%), 38 phanerophytes (trees) (9.0%), 14 vines (3.3%), 5 chamaephytes (low shrubs) (1.4%), 4 parasites (1.0%) and only one epiphyte (0.2%).

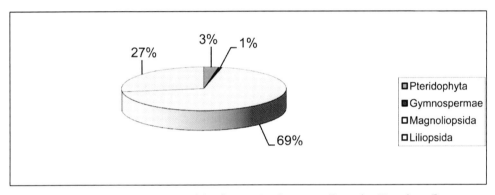

Fig. 2. Taxonomic distribution (%) of the flora in La Campana-Peñuelas Biosphere Reserve, Chile.

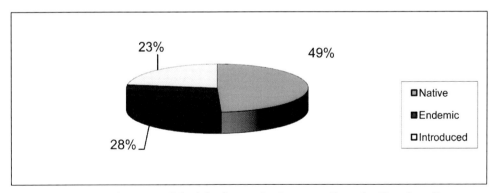

Fig. 3. Phytogeographic origin (%) of the flora in La Campana-Peñuelas Biosphere Reserve, Chile.

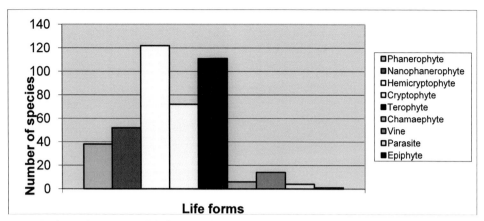

Fig. 4. Biological spectrum (life forms) of the flora in La Campana-Peñuelas Biosphere Reserve, Chile.

The state of conservation of the species (table 1), considering the total mentioned for the area, shows that the majority present no problems, with 18 species (4.5%) being identified as suffering from conservation problems. Of these, two are in the "Endangered" category, *Beilschmiedia miersii* (belloto del norte) and *Calydorea xiphioides* (tahay); 14 in the "Vulnerable" category, mainly species of bulbs (geophytes) such as *Herbertia lahue*, *Leucocoryne ixioides*, *L. violacescens* and *Phycella bicolor*; two in the "Rare" category, *Alstroemeria garaventae* and *A. zoellneri*; and just one species in the "Insufficiently known" category, *Blechnum cordatum*, of which a single specimen was recorded in the El Abrevadero sector.

Vegetation. The vegetation in the Reserve includes the following woody communities: sclerophyllous forest, hygrophilous forest, deciduous forest, thorny scrub and mixed sclerophyllous scrub. The herbaceous communities include: dry grassland, wet grassland, reed beds, sedge beds and the aquatic plant community. These communities are characterised as:

1. Peumo and boldo sclerophyllous forest. This is a woody formation typical of the region. This community corresponds to *Peumo-Cryptocaryetum albae* (Oberdorfer, 1960). Variants can be found with sclerophyllous scrub species (*Retanilla trinervia, Gochnatia foliolosa*), and one with Chilean palm tree (*Jubaea chilensis*). The average cover of the arboreal stratum is 70%, shrubs 35% and herbaceous 25%. The principal arboreal species are: *Cryptocarya alba, Peumus boldus, Persea lingue* and *Dasyphyllum excelsum*; shrubs are dominated by: *Retanilla trinervia, Schinus latifolia* and *Chusquea cumingii*,; creepers and lianas included: *Proustia pyrifolia, Lardizabala funaria* and *Cissus striata*; herbaceous vegetation comprised mainly: *Blechnum hastatum, Alonsoa meridionalis, Conyza floribunda* and *Anagallis arvensis*. The altitude ranges of this community is from 350 to 1,000 masl (Elórtegui & Moreira, 2002; Hauenstein et al., 2009).
2. Quillay and litre sclerophyllous forest. A more strongly xerophytic community than the first category described above, with cover between 40 and 90%. The principal arboreal species are: *Cryptocarya alba, Quillaja saponaria* and *Lithrea caustica*; shrubs: *Retanilla trinervia, Schinus latifolia, Colliguaja odorifera, Podanthus mitiqui* and *Escallonia pulverulenta*; and herbaceous species include: *Adiantum chilense, Solenomelus*

pedunculatus, Vulpia myuros, Alonsoa meridionalis, Conyza floribunda and *Anagallis arvensis*. The altitude range is from 500 to 1,050 masl. There is also a variant with *Jubaea chilensis*, in which *Aristeguietia salvia* and *Acacia caven* are also present (Elórtegui & Moreira, 2002; Hauenstein et al., 2009).

Species	EC	Bibliographical source
Alstroemeria garaventae	R	Ravenna et al. (1998)
Alstroemeria hookeri	V	Ravenna et al. (1998)
Alstroemeria zoellneri	R	Ravenna et al. (1998)
Beilschmiedia miersii	P	Benoit (1989)
Blechnum cordatum	IC	Baeza et al. (1998)
Blepharocalyx cruckshanksii	V	Benoit (1989)
Calydorea xyphioides	P	Ravenna et al. (1998)
Chloraea disoides	V	Ravenna et al. (1998), Novoa et al. (2006)
Chloraea heteroglossa	V	Ravenna et al. (1998), Novoa et al. (2006)
Conanthera trimaculata	V	Benoit (1989)
Eriosyce curvispina	V	Benoit (1989), Belmonte et al. (1998)
Herbertia lahue	V	Benoit (1989)
Jubaea chilensis	V	Benoit (1989)
Leucocoryne ixioides	V	Benoit (1989)
Leucocoryne violacescens	V	Benoit (1989)
Persea lingue	V	Benoit (1989),
Phycella bicolor	V	Benoit (1989)
Puya chilensis	V	Benoit (1989)

Table 1. Vascular plants with conservation problems of the La Campana-Peñuelas Biosphere Reserve (EC= condition of conservation, P= threatened, R= rare, V= vulnerable, IC= insufficiently known).

3. Temo and pitra hydrophilous laurifolious forest. Includes small remnants of forest associated with permanent watercourses and springs, present in the Vega del Alamo sector of RN Peñuelas. These are composed of hygrophilous species such as *Drimys winteri, Myrceugenia obtusa, Blepharocalyx cruckshanksii* and *Luma chequen*, and some creepers like *Lardizabala biternata* and *Cissus striata*, and correspond to the *Blepharocalyo-Myrceugenietum exsuccae* association (hualve or swamp forest) of myrtaceous plants in the central southern zone of Chile (Ramírez *et al*. 1996). These forests are included in the *Wintero-Nothofagetea* phytosociological class defined by Oberdorfer (1960). Other woody species present in this community were: *Maytenus boaria* and *Escallonia revoluta*, together with some creepers such as *Cissus striata* and *Lardizabala biternata*. The herbaceous stratum is scarce, probably due to the closed canopy, which hinders the penetration of

light; where present it is represented by *Uncinia trichocarpa* and ferns like *Blechnum hastatum* and *B. cordatum*. This community is found at 270 masl (Hauenstein et al., 2009).

4. Belloto hygrophilous laurifolious forest. This community is characterised by an arboreal stratum dominated by *Beilschmiedia miersii, Cryptocarya alba* and *Dasyphyllum excelsum*; there is an abundant shrub stratum dominated by *Azara celastrina, Chusquea cumingii, Adenopeltis serrata* and an abundance of creepers such as *Proustia pyrifolia, Lardizabala funaria* and *Bomarea salsilla*. It is found in low-lying areas, up to 500 masl (Elórtegui & Moreira, 2002).

5. Canelo hygrophilous laurifolious forest. The arboreal stratum, dominated by *Drimys winteri*; an abundant shrub stratum, with *Otholobium glandulosum, Escallonia myrtoidea, Maytenus boaria* and occasionally *Salix humboldtiana*; and in the herbaceous stratum, *Equisetum bogotense*. In altered sectors by human activities, cover may be diminished to as little as 30%. It is found at between 500 and 1,300 masl (Elórtegui & Moreira, 2002).

6. Deciduous roble forest. This community is characterised by a well differentiated arboreal stratum with *Nothofagus macrocarpa* and *Lomatia hirsuta*; a shrub stratum with *Azara petiolaris, Ribes punctatum, Schinus montanus, Berberis actinacantha* and *Aristotelia chilensis*; and an herbaceous stratum with, *Adiantum sulphureum, Loasa tricolor, Oxalis laxa* and *Alstroemeria zoellneri*, respectively. The cover is from 70 to 100% and it is found between 1,100 and 1,500 masl. At altitudes above 1,500 masl, a clear diminution may be seen in the tree and shrub cover, constituting a variant with other species such as *Schizanthus hookeri, Valeriana lepidota* and *Senecio anthemidiphyllus* (Elórtegui & Moreira, 2002) (photo 4).

Photo 4. *Nothofagus macrocarpa* forest in the sector of the Hill the Oak.

7. Espino thorny scrub. The dominant species in the community are espino (*Acacia caven* = *Vachellia caven*) and maitén (*Maytenus boaria*), with average cover of 50% and 10% respectively. It is characterised by the presence of a rich herbaceous stratum, especially *Agrostis capillaris, Leontodon saxatilis, Avena barbata, Bromus hordeaceus, Briza minor, B. maxima* and *Rhodophiala advena*, with clear predominance of forage graminids. This community is found around Peñuelas lake (Hauenstein et al., 2009) (photo 5).

8. Trevo thorny scrub. A little developed shrub community dominated by *Retanilla trinervia*, accompanied by shrub forms of *Lithrea caustica* and *Quillaja saponaria*, and by *Cuscuta chilensis*, an abundant, parasitic, herbaceous species. Average cover is above 70% and the community is found between 400 and 1,000 masl. It forms a variant with abundant presence of *Jubaea chilensis*, but only in the Ocoa sector (Elórtegui & Moreira, 2002).

Photo 5. *Acacia caven* steppe (hawthorn) in sectors near to the Peñuelas lake.

9. Chagual and quisco thorny scrub. An open community of succulents, with a shrub stratum of less than 4 m height, very diverse, with *Puya berteroana, Echinopsis chiloensis, Adesmia arborea, Aristeguietia salvia, Podanthus mitiqui, Retanilla trinervia* and occasionally *Puya chilensis*; the herbaceous stratum is poor, with *Helenium aromaticum, Vulpia myuros* and *Madia sativa*. Cover is very variable, between 10 and 80%, but generally not more than 40%. It is found at between 450 and 1,100 masl. It can also form a variant with abundant presence of *Jubaea chilensis*, but only in the Ocoa sector, with other species *Baccharis paniculata* and *Tristerix corymbosus* present (Elórtegui & Moreira, 2002) (photo 6).

Photo 6. *Jubaea chilensis* in the sector of Ocoa's Palms.

10. Chagualillo thorny scrub. A low, open shrub community, with two well differentiated strata: higher, with shrubs and succulents of 1-3 m in height, and lower, < 1 m. The higher stratum is dominated by *Puya coerulea, Eryngium paniculatum, Colliguaya odorifera, Retanilla ephedra* and *Calceolaria polifolia*; the lower by *Eriosyce curvispina, Tweedia birostrata, Gamochaeta americana, Senecio farinifer* and *Chorizanthe virgata*. Cover varies between 20 and 60%, and the altitude is above 1,100 masl. A variant may also be formed with presence of *Jubaea chilensis*, with very low cover, the individuals of Chilean palm tree being small and widely dispersed. These communities are found in certain high sectors of Cajón Grande, El Roble and Ocoa, and have cover ± 40% (Elórtegui & Moreira, 2002).
11. Mira and maicillo thorny scrub. A community with sparse distribution representing situations of sclerophyllous forest with severe disturbance in difficult environments for

plant life. The dominant species in the shrub stratum are *Gochnatia foliolosa, Baccharis rhomboidalis, B. linearis, Satureja gilliesii, Escallonia pulverulenta, Ageratina glechonophylla* and *Haplopappus velutinus*; in the herbaceous stratum, *Alstroemeria angustifolia, Solenomelus pedunculatus, Triptilion spinosum, Acaena pinnatifida* and *Azorella spinosa*. The community presents covers of 30 and 50%, at altitudes between 1,000 and 1,300 masl. It can also form a variant with *Jubaea chilensis*, with very low cover, between 1,000 and 1,200 masl. The differential character of this unit is given by the presence of *Vulpia myuros* and *Baccharis paniculata*, indicating a high level of degradation (Elórtegui & Moreira, 2002).

12. Mixed sclerophyllous scrub. This community is a mixture of species from the acacia steppe with sclerophyllous forest elements, such as *Quillaja saponaria, Schinus latifolia,* and *Cryptocarya alba*, together with shrub species, such as *Schinus polygama, Baccharis linearis, Berberis actinacantha, Eryngium paniculatum* and *Satureja gilliesii*. It may be characterised as open scrub of espino and romerillo (*Baccharis linearis*). The dominant woody species are *Acacia caven* with nearly 20% cover and *B. linearis* with 10%; these were followed by *Maytenus boaria, Quillaja saponaria* and *Schinus latifolia*; lower canopy comprises *Peumus boldus, Lithrea caustica, Escallonia pulverulenta* and the creeper *Muehlenbeckia hastulata*. The herbaceous stratum, with average cover 90%, was represented by the species *Agrostis capillaris, Avena barbata, Poa annua, Festuca sp., Bromus hordeaceus, Briza minor* and *Leontodon saxatilis* (Hauenstein et al., 2009).

13. Low altitude neneo scrub. A low, open community of dwarf shrubs (chamaephytes), principally *Mulinum spinosum, Chuquiraga oppositifolia, Viviania marifolia, Haplopappus ochagavianus, Ephedra chilensis* and a herbaceous stratum formed by *Phacelia secunda* and *Calceolaria campanae*. The cover is 10 to 40%; it is found on rocky substrates with variable exposure and gradient, above 1,750 masl, present only at the summits of La Campana and El Roble (Elórtegui & Moreira, 2002).

14. Dry grassland. The principal floral components of these grasslands were *Phyla canescens, Bromus hordeaceus, Agrostis capillaris, Hypochaeris radicata, Gamochaeta coarctata, Leontodon saxatilis, Rumex acetosella* and *Avena barbata*. They correspond to the *Bromo-Lolietum* community (Oberdorfer, 1960), and are found in the north-east sector of Peñuelas lake, at around 360 masl (Hauenstein et al., 2009).

15-16. Wet grasslands. There are two types. Apart from some of the species of the previous community, the first is distinguished by *Mentha pulegium* and reeds (*Juncus acutus, J. cyperoides, J. pallescens*), and corresponds to reedy wet grassland (*Juncetum acuti* ass. nov.). The second has marked presence of *Ludwigia peploides, Cotula coronopifolia, Distichlis spicata* and *Paspalum dilatatum*, corresponding to *Polygono-Ludwigietum peploidis* (Steubing et al. 1980). Both are found on the northern and southern plains close to Peñuelas lake, although *Polygono-Ludwigietum* occupies the strip closest to the water (Hauenstein et al., 2009).

17-18. Marsh communities. Two communities of marsh plants were identified, preferably located near the channel which flows into the eastern end of Peñuelas lake. The first, located closer to the water, corresponds to *Scirpetum californiae* (Ramírez & Añazco, 1982), the principal component being totora reed (*Schoenoplectus californicus*), accompanied by *Ludwigia peploides* and *Polygonum hydropiperoides*. The second community, found in a continuous strip a little further away from the water, corresponds to *Loto-Cyperetum eragrostidae* (San Martín et al., 1993), the principal species being *Cyperus eragrostis* and *Carex excelsa* (saw-sedges), accompanied by *Juncus pallescens* and *J. acutus*; it is a perennial, marshy association, typically found in depressions and on the banks of watercourses, associated with totora reeds or the border of swamp forest with temo (*Blepharocalyx cruckschanksii*) and pitra (*Myrceugenia exsucca*) (Hauenstein et al., 2002; San Martín et al., 2002). In both communities (reed beds and sedge beds), due to the periodic flooding of the site, hydrophytic life forms dominate absolutely, such as typical helophytes (marsh or swamp plants) and hydrophytes (Hauenstein et al., 2009).

19. Aquatic communities. These communities consist principally of hydrophytic species, particularly *Azolla filiculoides*, a free-floating pteridophyte, constituting the *Lemo-Azolletum filiculoidis* association (Roussine & Negre, 1952); moreover the high frequency of *Hydrocotyle ranunculoides*, *H. modesta* and *Eleocharis exigua* allow the existence of *Hydrocotyletum* to be inferred (San Martín et al., 1993), however more censuses would be required to confirm this inference. For example, in the least overgrown part of the spring in the El Abrevadero sector of RN Lago Peñuelas, a small body of water is formed which is colonised by aquatic and marsh species such as: *Azolla filiculoides*, *Gunnera tinctoria*, *Juncus cyperoides*, *Carex excelsa*, *Eleocharis exigua*, *Polygonum hydropiperoides*, *Hydrocotyle modesta* and *H. ranunculoides* (Hauenstein et al., 2009).

To resume, without considering the variants, 19 plant associations have been described for the Biosphere Reserve.

2.4 Discussion

Flora. When the general floral wealth of the area (420 species) is compared with other surveys done in protected wilderness areas in central Chile, we find for the Maule Region: RN Los Bellotos del Melado – 297 species (Arroyo et al., 2000); RN Los Ruiles – 139 species; RN Los Queules – 104 species (Arroyo et al., 2005). For the Metropolitan Region: Monumento Natural El Morado – 300 species (Teillier, 2003); RN Río Clarillo – approximately 600 species (Teillier et al., 2005). With the exception of the last, all others consist of considerably fewer species than recorded in the La Campana-Peñuelas Biosphere Reserve, demonstrating the floral wealth of this area.

The biological spectrum shows the predominance of therophytes, hemicryptophytes and cryptophytes over other life forms, which is consistent with the xeric conditions of the site, especially during summer, since therophytes (herbaceous plants with short life cycles, annuals or biannuals) and cryptophytes (geophytes or plants with enduring subterranean organs) represent this type of climate very well and are good environmental indicators. Meanwhile the abundance of hemicryptophytes is indicative of human intervention, since this life form accompanies man and corresponds to plants capable of surviving being trodden on and browsed by domestic animals (Cabrera & Willink, 1973; Ramírez, 1988; Grigera et al., 1996).

The general phytogeographic origin of the plants in the Reserve indicates that 23% are introduced species, 49% native and 28% endemic. This confirms the high value of this Biosphere Reserve as an area for the conservation of Chile's native and endemic flora. At the same time, when the relatively high percentage of allochthonous species is considered, and compared to the majority of the studies mentioned above where the values for allochthonous species do not exceed 20%, this would indicate a high level of human intervention in the study area. According to Hauenstein et al. (1988) and González (2000), a percentage distribution in which allochthonous plants reach values of between 20 and 30%

corresponds to the category of "moderate intervention". This high level of intervention is explained by the large flow of visitors and the presence of domestic animals at certain times of year, leading to the arrival of therophytes and hemicryptophytes which, as mentioned previously, correspond for the most part to fast-growing and strongly invasive plants. To this must be added the different soil structure in some sectors of RN Peñuelas under plantations of exotic species (*Pinus radiata, Eucalyptus globulus*) and the aggressive colonization by Australian acacias (*Acacia dealbata, A. melanoxylon*) (ICSA, 1980; CONAF, 1994).

Arroyo et al. (1995) indicate that the high percentage of endemic species is remarkable in all the protected wilderness areas of Chile's central zone, with values over 40%, which rise to 70% when native plants are included; introduced species thus do not exceed 30%. This is one of the characteristics which highlight the value of these areas as sites where our flora and vegetation units are preserved and protected (Luebert & Becerra, 1998; Hauenstein et al., 2009).

The high number of monocotyledons in the sector should also be noted, especially the abundance of bulb geophytes, which include the majority of species with conservation problems. It should be born in mind that these bear beautiful, brightly coloured flowers, and are especially important in the Bajo Los Lirios sector (CONAF, 1994, 2008; Hauenstein et al., 2009), where there are very brightly coloured and varied Orchidaceae, the Iridaceae, with lilies of the genera *Sisyrinchium* and *Olsynium*, and the tahay (*Calydorea xiphioides*) an "endangered" species which is very scarce throughout the central zone. There are other more frequent but no less beautiful species, like the lahue (*Herbertia lahue*), and flowers of the genera *Alstroemeria, Phycella, Rhodophiala* and *Calceolaria*, which are of aesthetic importance for recreation and use in gardening (Riedemann & Aldunate, 2001; Muñoz & Moreira, 2002).

Vegetation. The total phytosociology of the site consists of 19 plant associations, 6 of which are herbaceous, 7 scrub and 6 forest. With respect to the open espino scrub, also known as "espino or acacia steppe" (Pisano, 1956; CONAF, 1994), it should be made clear that strictly speaking it is not steppe but rather a savannah, as stated by Grau (1992), since the term steppe covers isolated vegetation with denuded surrounding soil and represents cold environments. Savannah on the other hand has isolated trees or thorny shrubs and a rich herbaceous stratum (Cabrera & Willink, 1973). This herbaceous stratum rich in forage species allows use of this type of community, at certain times of year, for grazing by domestic animals in RN Lago Peñuelas; this activity needs to be reviewed urgently by CONAF, since its effects on the biodiversity of the site are unknown.

The predominance of therophytes and geophytes in the biological spectrum of this plant formation is consistent with the climatic conditions and levels of precipitation in the area, since these life forms present morphological and physiological adaptations to the environmental conditions with extended periods of drought, and they are also important elements as food for wild fauna. The lower percentage of phanerophytes is explained by the low levels of precipitation in the area, and probably also the felling of native woody vegetation in the past. The scarcity of chamaephytes is due to the fact that this life form is adapted to conditions of low temperatures and greater altitude, such as are found only on El Roble and La Campana mountains (Cabrera & Willink, 1973; Ramírez, 1988; Grigera et al., 1996).

In the mixed sclerophyllous scrub, the important forms are the sclerophyllous phanerophytes and the therophytes; this also indicates the suitability of this type of plant formation to the climatic conditions of the area. The strong presence of hemicryptophytes indicates human alteration (Hauenstein et al., 1988). Numerous authors have carried out ecophysiological studies on mediterranean scrub species, explaining their ability to adapt to this type of environment, including Mooney & Kummerow (1971) on the drought response of *Flourensia thurifera, Kageneckia oblonga, Lithrea caustica* and *Proustia cinerea*; Montenegro et al. (1979) on the growth dynamic of *Colliguaja odorifera, Lithrea caustica, Satureja gilliesii* and *Retanilla trinervia*; Araya & Avila (1981) on the production of new shoots in scrub species affected by fire; and Avila et al. (1981) on the behaviour of herbaceous stratum species in scrub after a fire.

With respect to sclerophyllous forest, the predominance by phanerophytes and the importance of therophytes and hemicryptophytes are also consistent with the mediterranean climate of the area, since the species present adaptations which enable them to survive intense water stress, e.g. the presence of sclerophyllous leaves (Mooney & Kummerow, 1971), and the fires which are frequent in the area. For protection against fire, many develop a thick peridermis or a lignotuber, structures which enable them produce new shoots after a disaster (Araya & Avila, 1981). This community was classified by Oberdorfer (1960) in the class *Lithraeo causticae-Cryptocaryetea albae* and extends from 31° South to the limit of temperate territory (Aguilella & Amigo, 2001). There is no doubt that areas with this type of vegetation have been most affected by human action (Balduzzi et al., 1981, 1982).

At the same time, the remnants of hygrophilous forest, which requires conditions of greater humidity and precipitation, present absolute predominance of phanerophytes and nanophanerophytes (shrubs) over other life forms, indicating that the area would present high levels of rainfall, which is not the case. The explanation of the presence of this forest with hygrophilous characteristics at the site is based on the fact that it grows above a spring located in the Vega del Álamo sector, also known as the "abrevadero de caballos" [horse water-hole], which generates abundant water all year round, maintaining this community with azonal characteristics (Ramírez et al., 1996; Hauenstein et al., 2009).

The dry grassland is characterised by the predominance of hemicryptophytes, indicating strong human action, since these life forms are adapted to survive being trodden on and browsed by animals introduced by man; and of therophytes, indicating drought conditions (Hauenstein et al., 1988; Ramírez, 1988). In the wet grassland on the other hand, cryptophytes and hemicryptophytes predominate.

Reed beds are the most abundant and variable marsh associations in central-southern Chile, colonising marshes and banks of shallow lotic and lentic bodies of water (Ramírez & Añazco, 1982; Ramírez et al., 1987; San Martín et al., 1993). The same is true of sedge beds, which habitually accompany reed beds, forming a characteristic drier fringe. The principal species, *Cyperus eragrostis* and *Carex excelsa*, have sharp-edged leaves, from which their common name of "saw-sedges" derives (Hauenstein et al., 2002, 2005a).

The hydrophyte community corresponds to *Lemno-Azolletum*, characteristic of eutrophied aquatic environments (Palma et al., 1978); it is mentioned by Ramírez et al. (1987) as very abundant in lakes in Chile's central zone. Likewise, species such as *Hydrocotyle modesta* and

H. ranunculoides are indicators of organic contamination with nitrogen (Hauenstein et al., 2005b), which would corroborate the high trophic levels of Peñuelas lake; at the same time, this body of water and its associated flora form the habitat for a rich avifauna (Strang, 1983).

In this respect, Arroyo et al. (2005) propose, among other measures, the urgent need to complete inventories of flora of all the protected areas of central Chile; likewise they suggest analysing these areas at the scale of small polygons to evaluate their biological importance and vulnerability, elements which should form the basis for an integrated conservation strategy. Likewise, the findings of Armesto et al. (2002) and Simonetti (2004) are important, regarding the need to increase the number of protected areas, both public and private, to increase the size of existing areas, and to make use of the environment which surrounds them and which has suffered intervention, in order to create interconnectivity zones between them to serve as biological corridors. These units of conservation should consider representation of the different types of vegetation (Luebert & Becerra, 1998). Another interesting proposal is that of Elórtegui & Moreira (2002) for La Campana National Park, to zone areas for different types of use. In this context, Simonetti (1995) proposes the need for a model for planning the organization of these protected areas, linking the need to conserve natural resources with their sustainable use, which would be a basic support tool for defining which, where and how possible activities could be carried out, in such a way as would be compatible with conserving biodiversity.

3. Conclusions

The study reported here of the La Campana-Peñuelas Biosphere Reserve registered 420 species of vascular plants, of which 49% are of native origin, 28% correspond to endemic species and 23% to introduced species, respectively. Taxonomically, they are divided into 12 Pteridophytes, 3 Gymnosperms, 290 Dicotyledons (Magnoliopsida) and 115 Monocotyledons (Liliopsida), representing a high floral wealth. However, the percentage of allochthonous plants indicates a moderate degree of human disturbance. Eighteen species present conservation problems (2 Endangered, 13 Vulnerable, 2 Rare, 1 Insufficiently known). The vegetation at the site includes the presence of 19 plant communities, six of which are herbaceous, seven shrub and six arboreal. Important sectors exist of standing out in the Reserve such as: Los Lirios sector, which contains a high percentage of bulb species and numerous threatened species; the Vega del Álamo sector as a valuable remnant of hygrophilous forest; and the Ocoa sector due to the important presence of the endemic Chilean palm tree (*Jubaea chilensis*).

4. Appendix

Classification / Scientific name	Common name	Family	FV	OF	EC
PTERIDOPHYTA					
Adiantum chilense Kaulf. var. *hirsutum* Hook. et Grev.	Palito negro	Adiantaceae	Hc	N	
Adiantum excisum Kunze	Palito negro	Adiantaceae	Hc	N	
Adiantum scabrum Kaulf.	Palito negro	Adiantaceae	Hc	N	
Adiantum sulphureum Kaulf.	Palito negro	Adiantaceae	Hc	N	
Adiantum thalictroides Schlecht.	Palito negro	Adiantaceae	Hc	N	

Classification / Scientific name	Common name	Family	FV	OF	EC
Azolla filiculoides Lam.	Flor del pato	Azollaceae	Cr	N	
Blechnum cordatum (Desv.) Hieron	Costilla de vaca	Blechnaceae	Cr	N	IC
Blechnum hastatum Kaulf.	Palmilla	Blechnaceae	Hc	N	
Cheilanthes glauca (Cav.) Mett.	s.n.	Adiantaceae	Hc	N	
Cheilanthes hypoleuca (Kunze) Mett.	Doradilla	Adiantaceae	Hc	N	
Equisetum bogotense Kunth	Hierba del platero	Equisetaceae	Hc	N	
Notholaena tomentosa Desv.	s.n.	Adiantaceae	Hc	N	
GYMNOSPERMAE					
Cupressus macrocarpa Hartw.	Ciprés de Monterrey	Cupressaceae	F	I	
Ephedra chilensis C.Presl.	Pingo-pingo	Ephedraceae	Nf	N	
Pinus radiata D.Don	Pino insigne	Pinaceae	F	I	
MAGNOLIOPHYTA (ANGIOSPERMAE)					
MAGNOLIOPSIDA (Dicotyledoneae)					
Acacia caven (Molina) Molina	Espino	Mimosaceae	F	N	
Acacia dealbata Link	Aromo	Mimosaceae	F	I	
Acacia melanoxylon R.Br.	Aromo australiano	Mimosaceae	F	I	
Acaena pinnatifida Ruiz et Pav.	cadillo, pimpinela	Rosaceae	Hc	N	
Acrisione denticulata (Hook. et Arn.) B.Nord.	Palo de yegua	Asteraceae	Nf	E	
Adenopeltis colliguaya Bert.	Colliguay	Euphorbiaceae	Nf	N	
Adenopeltis serrata (W.T.Aiton) G.L.Webster	Colliguay macho	Euphorbiaceae	Nf	E	
Adesmia exilis Clos	Paramela	Fabaceae	Hc	E	
Adesmia tenella Hook. et Arn.	Trebillo	Fabaceae	Te	E	
Ageratina glechonophylla (Less.) R.M.King et H.Rob.	Barbón	Asteraceae	Nf	N	
Alonsoa meridionalis (L.f.) Kuntze	Ajicillo	Scrophulariaceae	Te	N	
Amsinckia calycina (Moris) Chater	Ortiguilla	Boraginaceae	Te	N	
Anagallis alternifolia Cav.	s.n.	Primulaceae	Hc	N	
Anagallis arvensis L.	Pimpinela azul	Primulaceae	Te	I	
Andeimalva chilensis (Gay) J.A.Tate	s.n.	Malvaceae	Nf	E	
Anemone decapetala Ard.	Centella	Ranunculaceae	Hc	N	
Anthemis cotula L.	Manzanillón	Asteraceae	Te	I	
Aristeguietia salvia (Colla) R.M.King et H.Rob.	Salvia macho	Asteraceae	Nf	E	
Aristolochia chilensis Bridges ex Lindl.	Oreja de zorro	Aristolochiaceae	Te	E	
Aristotelia chilensis (Molina) Stuntz	Maqui	Elaeocarpaceae	Nf	N	
Armeria maritima (Mill.) Willd.	s.n.	Plumbaginaceae	Hc	N	
Asteriscium chilense Cham. et Schltdl.	Anicillo	Apiaceae	Hc	N	
Astragalus berteroanus (Moris) Reiche	Yerba loca	Fabaceae	Te	N	
Azara celastrina D.Don	Lilén, corcolén	Flacourtiaceae	F	E	
Azara dentata Ruiz et Pav.	Corcolén	Flacourtiaceae	Nf	N	

Classification / Scientific name	Common name	Family	FV	OF	EC
Azara petiolaris (D.Don) I.M.Johnst.	Lilén, Corcolén	Flacourtiaceae	F	E	
Azara serrata Ruiz et Pav.	Lilén, Corcolén	Flacourtiaceae	F	E	
Azorella spinosa (Ruiz et Pav.) Pers.	Yerba santa	Apiaceae	Hc	N	
Baccharis linearis (Ruiz et Pav.) Pers.	Romerillo	Asteraceae	Nf	N	
Baccharis macraei Hook. et Arn.	Vautro	Asteraceae	Nf	N	
Baccharis paniculata DC.	Chilca	Asteraceae	Nf	N	
Baccharis rhomboidalis Remy	Vautro	Asteraceae	Nf	N	
Baccharis salicifolia (Ruiz et Pav.) Pers.	Chilca	Asteraceae	Nf	N	
Bahia ambrosioides Lag.	Manzanilla	Asteraceae	Nf	N	
Bartsia trixago L.	Belardia	Scrophulariaceae	Te	I	
Beilschmiedia miersii (Gay) Kosterm.	Belloto del norte	Lauraceae	F	E	P
Berberis actinacantha Mart.	Michay	Berberidaceae	Nf	E	
Blepharocalyx cruckshanksii (Hook. et Arn.) Nied.	Temo, Pitra, Picha	Myrtaceae	F	N	V
Bowlesia uncinata Colla	Barba de gato	Apiaceae	Te	N	
Brassica rapa L. ssp. *campestris* (L.) A.R.Clapham	Yuyo	Brassicaceae	Te	I	
Calandrinia compressa Schrad. ex DC.	s.n.	Portulacaceae	Te	N	
Calandrinia nitida (Ruiz et Pav.) DC.	s.n.	Portulacaceae	Te	N	
Calceolaria ascendens Lindl. ssp. *glandulifera*	Topa topa	Scrophulariaceae	Hc	E	
Calceolaria campanae Phil.	Topa topa	Scrophulariaceae	Hc	E	
Calceolaria corymbosa Ruiz et Pav.	Arguenita del cerro	Scrophulariaceae	Hc	E	
Calceolaria dentata Ruiz et Pav.	Capachito, topa-top	Scrophulariaceae	Hc	E	
Calceolaria glandulosa Poepp. ex Benth.	Capachito	Scrophulariaceae	Hc	E	
Calceolaria meyeniana Phil.	Capachito	Scrophulariaceae	Hc	E	
Calceolaria morisii Walp.	Capachito	Scrophulariaceae	Hc	E	
Calceolaria oreas Phil.	Capachito	Scrophulariaceae	Hc	E	
Calceolaria petioalaris Cav.	Capachito, topa-top	Scrophulariaceae	Hc	E	
Calceolaria polyfolia Hook.	Capachito	Scrophulariaceae	Nf	E	
Calceolaria purpurea Graham	Topa-topa	Scrophulariaceae	Hc	E	
Calceolaria thyrsiflora Graham	Hierba dulce	Scrophulariaceae	Hc	E	
Callitriche palustris L.	Huenchecó	Callitrichaceae	Cr	N	
Camissonia dentata (Cav.) Reiche	s.n.	Onagraceae	Te	N	
Capsella bursa-pastoris (L.) Medik.	Bolsa de pastor	Brassicaceae	Te	I	
Cardamine bonariensis Pers.	s.n.	Brassicaceae	Hc	N	
Cardionema ramosissimum (Weinm.) A.Nelson et J.F.Macbr.	Dicha	Caryophyllaceae	Hc	N	
Carduus pycnocephalus L.	Cardilla	Asteraceae	Te	I	
Centaurea melitensis L.	Cizaña, abrepuño	Asteraceae	Te	I	
Centaurea solstitialis L.	Cardo amarillo	Asteraceae	Te	I	

Classification / Scientific name	Common name	Family	FV	OF	EC
Cerastium fontanum Baumg.	Cerastio	Caryophyllaceae	Te	I	
Cestrum parqui L'Hér.	Palqui	Solanaceae	Nf	N	
Chaetanthera linearis Poepp. ex Less.	s.n.	Asteraceae	Te	E	
Chaetanthera microphylla (Cass.) Hook. et Arn.	s.n.	Asteraceae	Te	E	
Chenopodium album L.	Quinguilla	Chenopodiaceae	Te	I	
Chorizante virgata Benth.	Sanguinaria	Polygonaceae	Hc	E	
Chuquiraga oppositifolia D.Don	Hierba blanca	Compositae	Nf	N	
Cirsium vulgare (Savi) Ten.	Cardo negro	Asteraceae	Hc	I	
Cissus striata Ruiz et Pav.	Pilpilvoqui	Vitaceae	Ft	N	
Cistanthe grandiflora Lindl.	Pata de guanaco	Portulacaceae	Hc	E	
Citronella mucronata (Ruiz et Pav.) D.Don	Naranjillo	Icacinaceae	F	E	
Clarkia tenella (Cav.) F.H.Lewis et M.E.Lewis	Huasita	Onagraceae	Te	N	
Colletia hystrix Clos	Crucero	Rhamnaceae	Nf	N	
Colliguaja odorifera Molina	Colliguay	Euphorbiaceae	Nf	E	
Conium maculatum L.	Cicuta	Apiaceae	Te	I	
Convolvulus arvensis L.	Correhuela	Convolvulaceae	Hc	I	
Convolvulus chilensis Pers.	Correhuela rosada	Convolvulaceae	Hc	E	
Conyza sumatrensis (Retz.) E.Walker	s.n.	Asteraceae	Te	N	
Corrigiola squamosa Hook. et Arn.	Hierba del niño	Caryophyllaceae	Hc	N	
Cotula australis (Sieber ex Spreng.) Hook.f.	s.n.	Asteraceae	Te	I	
Cotula coronopifolia L.	Botón de oro	Asteraceae	Hc	I	
Crassula closiana (Gay) Reiche	Flor de piedra	Crassulaceae	Te	N	
Crinodendron patagua Molina	Patagua	Elaeocarpaceae	F	E	
Cryptantha aprica (Phil.) Reiche	s.n.	Boraginaceae	Te	E	
Cryptocarya alba (Molina) Looser	Peumo	Lauraceae	F	E	
Cuscuta chilensis Ker Gawl.	Cabello de ángel	Convolvulaceae	Fp	N	
Cyclospermum laciniatum (DC.) Constance	Capuchilla	Apiaceae	Te	N	
Cynara cardunculus L.	Cardo penquero	Asteraceae	Hc	I	
Dasyphyllum excelsum (DC.) Cabrera	Palo santo, Tayú	Compositae	F	E	
Dichondra sericea Sw.	Oreja de ratón	Convolvulaceae	Hc	N	
Diplolepis menziesii J.H.Schult.	Voquicillo	Asclepiadaceae	Ft	E	
Dipsacus sativus (L.) Honck.	Carda	Dipsacaceae	Te	I	
Drimys winteri J.R.Forst. et G.Forst.	Canelo	Winteraceae	F	N	
Echinopsis chiloensis (Colla) Friedrich et G.D.Rowley	Quisco	Cactaceae	Nf	E	
Eccremocarpus scaber Ruiz et Pav.	Chupa-chupa	Bignoniaceae	Ft	N	
Erigeron fasciculatus Colla	s.n.	Asteraceae	Hc	N	
Eriosyce curvispina (Bertero ex Colla) Katt.	Quisquito	Cactaceae	Nf	E	V
Erodium botrys (Cav.) Bertol.	Alfilerillo	Geraniaceae	Te	I	
Erodium cicutarium (L.) L'Hér. ex Aiton	Alfilerillo	Geraniaceae	Te	I	

Classification / Scientific name	Common name	Family	FV	OF	EC
Erodium malacoides (L.) L'Hér. ex Aiton	Alfilerillo	Geraniaceae	Te	I	
Erodium moschatum (L.) L'Hér. ex Aiton	Alfilerillo	Geraniaceae	Te	I	
Eryngium paniculatum Cav. et Dombey ex Delaroche	Cardoncillo	Apiaceae	Hc	N	
Escallonia myrtoidea Bertero ex DC.	Lun	Escalloniaceae	F	E	
Escallonia pulverulenta (Ruiz et Pav.) Pers.	Corontillo	Escalloniaceae	F	E	
Escallonia revoluta (Ruiz et Pav.) Pers.	Lun	Escalloniaceae	Nf	N	
Escallonia rubra (Ruiz et Pav.) Pers.	Siete camisas	Escalloniaceae	Nf	N	
Eschscholzia californica Cham.	Dedal de oro	Papaveraceae	Hc	I	
Eucalyptus globulus Labill.	Eucalipto	Myrtaceae	F	I	
Euphorbia klotzschii Oudejans	Pichoguilla	Euphorbiaceae	Te	N	
Euphorbia peplus L.	Pichoga	Euphorbiaceae	Te	I	
Fabiana imbricata Ruiz et Pav.	Pichi	Solanaceae	Nf	N	
Facelis retusa (Lam.) Sch. Bip.	s.n.	Asteraceae	Te	N	
Flourensia thurifera (Molina) DC.	Maravilla de campo	Asteraceae	Nf	E	
Fuchsia magellanica Lam.	Chilco	Onagraceae	Nf	N	
Fumaria capreolata L.	Flor de la culebra	Fumariaceae	Te	I	
Fumaria densiflora DC.	Flor de la culebra	Fumariaceae	Te	I	
Galium aparine L.	Lengua de gato	Rubiaceae	Te	I	
Galium hypocarpium (L.) Endl. ex Griseb.	Relbún	Rubiaceae	Hc	N	
Galium trichocarpum DC.	s.n.	Rubiaceae	Hc	E	
Gamochaeta coarctata (Willd.) Kerguélen	Nafalium	Asteraceae	Te	N	
Geranium berteroanum Colla	Core-core	Geraniaceae	Hc	N	
Geranium core-core Steud.	Core-core	Geraniaceae	Hc	N	
Glandularia laciniata (L.) Schnack et Covas	Hierba del incornio	Verbenaceae	Hc	N	
Gnaphalium cheiranthifolium Lam.	Vira-vira	Asteraceae	Hc	N	
Gochnatia foliolosa (D.Don) D.Don ex Hook. et Arn.	Mira-mira	Asteraceae	Hc	E	
Gunnera tinctoria (Molina) Mirb.	Nalca	Gunneraceae	Cr	N	
Haplopappus linifolius (Phil.) Reiche	s.n.	Asteraceae	C	N	
Haplopappus ochagavianus Phil.	s.n.	Asteraceae	C	N	
Haplopappus poeppigianus (Hook. et Arn.) A.Gray	s.n.	Asteraceae	C	N	
Haplopappus velutinus Remy ssp. *illinitus*	Buchú	Asteraceae	C	N	
Helenium aromaticum (Hook.) L.H.Bailey	Manzanilla	Asteraceae	Te	N	
Hydrocotyle modesta Cham. et Schlecht.	Sombrero de agua	Apiaceae	Cr	N	
Hydrocotyle ranunculoides L.f.	Sombrero de agua	Apiaceae	Cr	N	
Hypochaeris glabra L.	Hierba del chancho	Asteraceae	Te	I	
Hypochaeris radicata L.	Hierba del chancho	Asteraceae	Hc	I	
Hypochaeris scorzonerae (DC.) F. Muell.	Escorzonera	Asteraceae	Hc	E	

Classification / Scientific name	Common name	Family	FV	OF	EC
Kageneckia oblonga Ruiz et Pav.	Bollén	Rosaceae	F	E	
Lactuca serriola L.	Lechuguilla	Asteraceae	Te	I	
Lamiun amplexicaule L.	Gallito	Lamiaceae	Te	I	
Lardizabala biternata Ruiz et Pav.	Cóguil	Lardizabalaceae	Ft	N	
Lardizabala funaria (Molina) Looser	Cóguil	Lardizabalaceae	Ft	E	
Lathyrus berteroanus Colla	Clarincillo, arvejilla	Fabaceae	Hc	N	
Lathyrus magellanicus Lam.	Clarincillo, arvejilla	Fabaceae	Hc	N	
Leontodon saxatilis Lam.	Chinilla	Asteraceae	Hc	I	
Lepidium spicatum Desv.	s.n.	Brassicaceae	Hc	N	
Leucheria cerberoana Remy	s.n.	Asteraceae	Te	N	
Linum chamissonis Schied.	Nanco	Linaceae	Hc	N	
Linum macraei Benth.	Nanco	Linaceae	Hc	E	
Linum usitatissimum L.	Lino, linaza	Linaceae	Te	I	
Lithrea caustica (Molina) Hook. et Arn.	Litre	Anacardiaceae	F	E	
Loasa tricolor Ker Gawl.	Ortiga caballuna	Loasaceae	Te	N	
Loasa triloba Dombey ex Juss.	Ortiga blanca	Loasaceae	Te	N	
Lobelia excelsa Bonpl.	Tabaco del diablo	Campanulaceae	Nf	E	
Lobelia polyphylla Hook. et Arn.	Tupa	Campanulaceae	Nf	E	
Logfia gallica (L.) Coss. et Germ.	s.n.	Asteraceae	Te	I	
Lomatia hirsuta (Lam.) Diels ex J.F.Macbr.	Radal	Proteaceae	F	N	
Lotus subpinnatus Lag.	Porotillo	Fabaceae	Te	N	
Ludwigia peploides (Kunth) P.H.Raven ssp. *montevidensis*	Pasto de la rana	Onagraceae	Cr	N	
Luma chequen F.Phil.	Chequén, luma	Myrtaceae	F	E	
Lupinus microcarpus Sims	Arvejilla	Fabaceae	Te	N	
Lycium chilense Miers ex Bertero	Coralillo	Solanaceae	Nf	N	
Lythrum hyssopifolia L.	Hierba del toro	Lythraceae	Te	I	
Madia sativa Molina	Melosa, pegajosa	Asteraceae	Te	N	
Malesherbia linearifolia (Cav.) Pers.	s.n.	Malesherbiaceae	Hc	E	
Malesherbia sp.	s.n.	Malesherbiaceae	Hc	E	
Malva nicaensis All.	Malva	Malvaceae	Te	I	
Margyricarpus pinnatus (Lam.) Kuntze	Perlilla	Rosaceae	C	N	
Marrubium vulgare L.	Toronjil cuyano	Lamiaceae	Hc	I	
Matricaria chamomilla L.	Manzanilla	Asteraceae	Te	I	
Maytenus boaria Molina	Maitén	Celastraceae	F	N	
Medicago arabica (L.) Huds.	Hualputra	Fabaceae	Te	I	
Medicago polymorpha L.	Hualputra	Fabaceae	Te	I	
Melilotus indicus (L.) All.	Trevillo	Fabaceae	Te	I	
Mentha pulegium L.	Poleo	Lamiaceae	Hc	I	
Microseris pygmaea D.Don	s.n.	Asteraceae	Te	N	
Misodendrum linearifolium DC.	Injerto	Misodendraceae	Fp	N	
Monnina sp.	s.n.	Polygalaceae	Nf	N	

Classification / Scientific name	Common name	Family	FV	OF	EC
Muehlenbeckia hastulata (Sm.) I.M.Johnst.	Voqui negro	Polygonaceae	F	N	
Mulinum spinosum (Cav.) Pers.	Neneo	Umbelliferae	Nf	N	
Mutisia acerosa Poepp. ex Less.	Romerillo	Asteraceae	Nf	N	
Mutisia ilicifolia Cav.	Clavel del aire	Asteraceae	Ft	E	
Mutisia latifolia D.Don	Clavel del campo	Asteraceae	Ft	E	
Mutisia rosea Poepp. ex Less.	Clavel del campo	Asteraceae	Ft	E	
Mutisia subulata Ruiz et Pav.	Hierba del jote	Asteraceae	Ft	N	
Myrceugenia exsucca (DC.) O.Berg	Pitra, petra	Myrtaceae	F	N	
Myrceugenia obtusa (DC.) O.Berg	Arrayán, rarán	Myrtaceae	F	E	
Myriophyllum aquaticum (Vell.) Verdc.	Pinito de agua	Haloragaceae	Cr	N	
Nasturtium officinale R.Br.	Berro	Brassicaceae	Hc	I	
Nicotiana acuminata (Graham) Hook.	Tabaco del campo	Solanaceae	Te	N	
Nothofagus macrocarpa(A.DC.) F.M.Vázquez et R.A.Rodr.	Roble	Nothofagaceae	F	E	
Oenothera acaulis Cav.	Diego de la noche	Onagraceae	Hc	N	
Oenothera affinis Spach	Diego de la noche	Onagraceae	Te	N	
Otholobium glandulosum (L.) J.W.Grimes	Culén	Fabaceae	Nf	N	
Oxalis articulata Savign.	Culle	Oxalidaceae	Cr	N	
Oxalis corniculata L.	Vinagrillo	Oxalidaceae	Hc	I	
Oxalis laxa Hook. et Arn.	Culle	Oxalidaceae	Te	N	
Oxalis megalorrhiza Jacq.	Vinagrillo	Oxalidaceae	Hc	N	
Oxalis micrantha Bertero ex Colla	Vinagrillo	Oxalidaceae	Hc	N	
Oxalis perdicaria (Molina) Bertero	Flor de la perdiz	Oxalidaceae	Cr	N	
Oxalis rosea Jacq.	Culle, vinagrillo	Oxalidaceae	Te	E	
Persea lingue Nees.	Lingue	Lauraceae	F	N	V
Peumus boldus Molina	Boldo	Monimiaceae	F	E	
Phacelia circinata Jacq.	Té de burro	Hydrophyllaceae	Hc	N	
Phacelia secunda J.F.Gmel.	Té de burro	Hydrophyllaceae	Hc	N	
Phyla canescens (Kunth) Greene	Hierba de la Virgen	Verbenaceae	Hc	N	
Plagiobothrys procumbens A.Gray	s.n.	Boraginaceae	Te	N	
Plantago hispidula Ruiz et Pav.	s.n.	Plantaginaceae	Te	N	
Plantago lanceolata L.	Siete venas	Plantaginaceae	Hc	I	
Podanthus mitiqui Lindl.	Mitique	Asteraceae	Nf	E	
Polygonum hydropiperoides Michx.	Duraznillo	Polygonaceae	Hc	N	
Polygonum persicaria L.	Duraznillo	Polygonaceae	Te	I	
Populus deltoides Bartram ex Marshall	Alamo americano	Salicaceae	F	I	
Populus nigra L.	Alamo negro	Salicaceae	F	I	
Porliera chilensis Johnst.	Guayacán	Zygophyllaceae	Nf	E	
Proustia pyrifolia DC.	Voqui blanco	Asteraceae	Ft	E	
Quillaja saponaria Molina	Quillay	Rosaceae	F	E	
Quinchamalium chilense Molina	Quinchamalí	Santalaceae	Hc	N	
Ranunculus chilensis DC.	Ranúnculo	Ranunculaceae	Te	N	
Ranunculus muricatus L.	Centella, huante	Ranunculaceae	Hc	I	

Classification / Scientific name	Common name	Family	FV	OF	EC
Rapistrum rugosum (L.) All.	Falso yuyo	Brassicaceae	Te	I	
Retanilla ephedra(Vent.) Brongn.	Retamilla	Rhamnaceae	Nf	E	
Retanilla trinervia (Guillies et Hook.) Hook. et Arn.	Trevo, tebo	Rhamnaceae	Nf	E	
Rhaphithamnus spinosus (Juss.) Moldenke	Arrayán macho	Verbenaceae	F	N	
Ribes punctatum Ruiz et Pav.	Zarzaparrilla	Grossulariaceae	Nf	N	
Robinia pseudoacacia L.	Falsa acacia	Fabaceae	F	I	
Rosa moschata Herrm.	Mosqueta, coral	Rosaceae	Nf	I	
Rubus ulmifolius Schott	Zarzamora, mora	Rosaceae	Nf	I	
Rumex acetosella L.	Vinagrillo	Polygonaceae	Hc	I	
Rumex conglomeratus Murray	Romaza	Polygonaceae	Hc	I	
Rumex crispus L.	Romaza	Polygonaceae	Hc	I	
Rumex pulcher L.	Romaza	Polygonaceae	Hc	I	
Sagina apetala Ard.	s.n.	Caryophyllaceae	Te	I	
Salix babylonica L.	Sauce llorón	Salicaceae	F	I	
Salix humboldtiana Willd.	Sauce amargo	Salicaceae	F	N	
Sanicula crassicaulis Poepp. ex DC.	Pata de leon	Apiaceae	Hc	N	
Satureja gilliesii (Graham) Briq.	Menta de árbol	Lamiaceae	Nf	E	
Schinus latifolia (Gillies ex Lindl.) Engler	Molle	Anacardiaceae	F	E	
Schinus montana (Phil.) Engler	Litrecillo	Anacardiaceae	F	E	
Schinus polygama (Cav.) Cabrera	Huingán, borocoi	Anacardiaceae	Nf	N	
Schizanthus hookeri Gillies ex Graham	Mariposita	Solanaceae	Hc	N	
Schizanthus litoralis Phil.	Pajarito, mariposita	Solanaceae	Te	E	
Schizanthus pinnatus Ruiz et Pav.	Mariposita	Solanaceae	Te	E	
Schizanthus tricolor Grau et E.Gronbach	Mariposita	Solanaceae	Te	N	
Senecio adenotrichius DC.	s.n.	Asteraceae	Hc	N	
Senecio anthemidiphyllus Remy	Senecio	Asteraceae	Nf	E	
Senecio farinifer Hook. et Arn.	Senecio	Asteraceae	Nf	E	
Senecio aff. macratus Kunze	Senecio	Asteraceae	Nf	N	
Senna candolleana (Vogel) H.S.Irwin et Barneby	Quebracho	Caesalpiniaceae	F	N	
Silene gallica L.	Calabacillo	Caryophyllaceae	Te	I	
Silybum marianum (L.) Gaertn.	Cardo blanco	Asteraceae	Te	I	
Solanum furcatum Dunal ex Poir.	Yerba mora	Solanaceae	Te	N	
Solanum tomatillo (J.Remy) F.Phil.	Tomatillo	Solanaceae	Te	N	
Soliva sessilis Ruiz et Pav.	Dicha	Asteraceae	Te	N	
Sonchus asper (L.) Hill.	Ñilhue	Asteraceae	Te	I	
Sonchus oleraceus L.	Ñilhue	Asteraceae	Te	I	
Sophora macrocarpa Sm.	Mayu	Fabaceae	F	E	
Sphacele salviae (Lindl.) Briq.	Salvia	Lamiaceae	Nf	E	
Spergularia media (L.) C.Presl ex Griseb.	Tiqui-tiqui	Caryophyllaceae	Te	I	
Stachys albicaulis Lindl.	Hierba Santa Rosa	Lamiaceae	Hc	E	

Classification / Scientific name	Common name	Family	FV	OF	EC
Stachys grandidentata Lindl.	Toronjilcillo	Lamiaceae	Hc	E	
Stellaria chilensis Pedersen	Quilloi-quilloi	Caryophyllaceae	Hc	N	
Stellaria media (L.) Cirillo	Quilloi-quilloi	Caryophyllaceae	Te	I	
Stenandrium dulce (Cav.) Nees	Hierba de la piñada	Acanthaceae	Hc	N	
Teucrium bicolor Sm.	Oreganillo	Lamiaceae	Nf	E	
Torilis nodosa (L.) Gaertn.	s.n.	Apiaceae	Te	I	
Trifolium dubium Sibth.	Trebillo	Fabaceae	Te	I	
Trifolium polymorphum Poir.	Trébol	Fabaceae	Hc	N	
Trifolium subterraneum L.	Trébol subterráneo	Fabaceae	Cr	I	
Triptilion spinosum Ruiz et Pav.	Siempre viva	Asteraceae	Hc	N	
Tristerix aphyllus (Miers ex DC.) Tiegh. ex Barlow et Wiens	Quintral del quisco	Loranthaceae	Fp	E	
Tristerix corymbosus (L.) Kuijt	Quintral del álamo	Loranthaceae	Fp	N	
Tropaeolum azureum Miers.	Violeta del campo	Tropaeolaceae	Ft	E	
Tropaeolum ciliatum Ruiz et Pav.	Pajarito	Tropaeolaceae	Ft	E	
Tropaeolum tricolor Sweet	Soldadito, relicario	Tropaeolaceae	Ft	E	
Tweedia birostrata (Hook. et Arn.) Hook. et Arn.	Zahumerio	Asclepiadaceae	Ft	E	
Urtica urens L.	Ortiga	Urticaceae	Te	I	
Valeriana lepidota Clos	Valeriana	Valerianaceae	Cr	E	
Valeriana verticillata Clos	Valeriana	Valerianaceae	Hc	N	
Verbascum virgatum Stokes	Raspa la choica	Scrophulariaceae	Te	I	
Verbena bonariensis L.	Verbena	Verbenaceae	Hc	N	
Veronica anagallis-aquatica L.	No me olvides	Scrophulariaceae	Cr	I	
Veronica arvensis L.	Verónica	Scrophulariaceae	Te	I	
Vicia magnifolia Clos	Arvejilla	Fabaceae	Te	N	
Vicia mucronata Clos	Arvejilla	Fabaceae	Te	N	
Vicia nigricans Hook. et Arn.	Arvejilla	Fabaceae	Te	N	
Vicia sativa L.	Arvejilla, clarincillo	Fabaceae	Te	I	
Viola sp.	Violeta del campo	Violaceae	Te	N	
Viviania marifolia Cav.	Te de burro	Vivianiaceae	Nf	N	
LILIOPSIDA (Monocotyledoneae)					
Agrostis capillaris L.	Chépica	Poaceae	Hc	I	
Aira caryophyllea L.	s.n.	Poaceae	Te	I	
Alisma plantago-aquatica L.	Llantén de agua	Alismataceae	Cr	I	
Alstroemeria angustifolia Herb.	Lirio del campo	Alstroemeriaceae	Cr	E	
Alstroemeria garaventae Ehr.Bayer	Lirio del campo	Alstroemeriaceae	Cr	E	R
Alstroemeria hookeri Lodd. ssp. *recumbens*	Lirio del campo	Alstroemeriaceae	Cr	N	V
Alstroemeria ligtu L. ssp. *simsii*	Lirio del campo	Alstroemeriaceae	Cr	E	
Alstroemeria pulchra Sims ssp. *pulchra*	Lirio del campo	Alstroemeriaceae	Cr	E	

Classification / Scientific name	Common name	Family	FV	OF	EC
Alstroemeria revoluta Ruiz et Pav.	Lirio del campo	Alstroemeriaceae	Cr	E	
Alstroemeria zoellneri Ehr.Bayer	Lirio del campo	Alstroemeriaceae	Cr	E	R
Avena barbata Pott ex Link	Teatina	Poaceae	Te	I	
Bipinnula mystacina Lindl.	Orquídea	Orchidaceae	Cr	E	
Bipinnula plumosa Lindl.	Orquídea	Orchidaceae	Cr	E	
Bomarea salsilla Mirb.	Bomarea	Alstroemeriaceae	Ft	N	
Brachystele unilateris (Poir.) Schltr.	Orquídea	Orchidaceae	Cr	E	
Briza maxima L.	Tembladera	Poaceae	Te	I	
Briza minor L.	Pasto perdiz	Poaceae	Te	I	
Bromus catharticus Vahl	Pasto lanco	Poaceae	Hc	N	
Bromus hordeaceus L.	Bromo, cebadilla	Poaceae	Te	I	
Calydorea xiphioides (Poepp.) Espinosa	Tahay, violeta	Iridaceae	Cr	E	P
Carex excelsa Poepp. ex Kunth	Cortadera	Cyperaceae	Cr	N	
Carex setifolia Kunze ex Kunth	Cortadera	Cyperaceae	Cr	N	
Carex sp.	s.n.	Cyperaceae	Cr	N	
Chascolytrum subaristatum (Lam.) Desv.	Tembladera	Poaceae	Te	N	
Chloraea barbata Lindl.	Orquídea	Orchidaceae	Cr	E	
Chloraea bletioides Lindl.	Lengua de loro	Orchidaceae	Cr	E	
Chloraea chrysantha Poepp.	Orquídea	Orchidaceae	Cr	E	
Chloraea cylindrostachya Poepp.	Orquídea	Orchidaceae	Cr	N	
Chloraea disoides Lindl. var. *picta*	Orquídea	Orchidaceae	Cr	N	V
Chloraea galeata Lindl.	Orquídea	Orchidaceae	Cr	E	
Chloraea heteroglossa Rchb. f.	Orquídea	Orchidaceae	Cr	E	V
Chloraea incisa Poepp.	Orquídea	Orchidaceae	Cr	E	
Chloraea multiflora Lindl.	Orquídea	Orchidaceae	Cr	E	
Chloraea virescens (Willd.) Lindl.	Orquídea	Orchidaceae	Cr	N	
Chusquea cumingii Nees	Coligüe, quila	Poaceae	Nf	E	
Conanthera bifolia Ruiz et Pav.	Pajarito del campo	Tecophilaeaceae	Hc	E	
Conanthera campanulata (D.Don) Lindl.	Flor de la viuda	Tecophilaeaceae	Hc	E	
Conanthera trimaculata (D.Don) F.Meigen	Flor de la viuda	Tecophilaeaceae	Hc	E	V
Cynodon dactylon (L.) Pers.	Pasto bermuda	Poaceae	Hc	I	
Cynosurus echinatus L.	Cola de zorro	Poaceae	Te	I	
Cyperus eragrostis Lam. var. *compactus*	Cortadera	Cyperaceae	Cr	N	
Cyperus eragrostis Lam. var. *eragrostis*	Cortadera	Cyperaceae	Cr	N	
Dioscorea parviflora Phil.	Papa cimarrona	Dioscoreaceae	Cr	N	
Dioscorea saxatilis Poepp.	Papa cimarrona	Dioscoreaceae	Cr	N	
Distichlis spicata (L.) Greene	Pasto salado	Poaceae	Cr	E	
Echinochloa colonum (L.) Link	Hualcacho	Poaceae	Te	N	
Eleocharis pachycarpa E.Desv.	Rume	Cyperaceae	Hc	N	
Eleocharis radicans (Poir.) Kunth	Rume	Cyperaceae	Hc	N	
Elodea potamogeton (Bertero) Espinosa	Luchecillo	Hydrocharitaceae	Cr	N	
Eragrostis virescens J.Presl.	s.n	Poaceae	Te	N	
Festuca sp.	Coirón	Poaceae	Hc	N	

Classification / Scientific name	Common name	Family	FV	OF	EC
Gastridium phleoides (Nees et Meyen) C.E.Hubb.	s.n	Poaceae	Te	I	
Gavilea venosa (Lam.) Garay et Ormd.	Orquídea	Orchidaceae	Cr	N	
Gavilea leucantha Poepp. et Endl.	Orquídea	Orchidaceae	Cr	N	
Gavilea longibracteata (Lindl.) Sparre ex L.E.Navas	Orquídea	Orchidaceae	Cr	E	
Gilliesia graminea Lindl.	Junquillo	Alliaceae	Cr	N	
Herbertia lahue (Molina) Goldblat	Lahue	Iridaceae	Cr	N	V
Hordeum chilense Roem. et Schult.	Cebadilla	Poaceae	Hc	N	
Hordeum murinum L.	Cebadilla	Poaceae	Te	I	
Isolepis cernua (Vahl) Roem. et Schult.	s.n	Cyperaceae	Cr	N	
Jubaea chilensis Baill.	Palma de coquitos	Arecaceae	F	E	V
Juncus acutus L.	Junquillo	Juncaceae	Hc	N	
Juncus bufonius L.	Junquillo	Juncaceae	Te	I	
Juncus cyperoides Laharpe	Junco	Juncaceae	Hc	N	
Juncus imbricatus Laharpe	Junquillo	Juncaceae	Cr	N	
Juncus pallescens Lam.	Junco	Juncaceae	Hc	N	
Juncus stipulatus Nees et Meyen	Junco	Juncaceae	Hc	N	
Lemna valdiviana Phil.	Lenteja de agua	Lemnaceae	Cr	N	
Leucocoryne ixioides (Sims) Lindl.	Huilli	Alliaceae	Cr	E	V
Leucocoryne violacescens Phil.	Huilli	Alliaceae	Cr	E	V
Lolium multiflorum Lam.	Ballica italiana	Poaceae	Te	I	
Lolium perenne L.	Ballica inglesa	Poaceae	Hc	I	
Luzula racemosa Desv.	Lúzula	Juncaceae	Cr	N	
Melica longiflora Steud.	s.n	Poaceae	Hc	E	
Melica violacea Cav.	s.n	Poaceae	Te	N	
Nassella chilensis (Trin.) E.Desv.	Coironcillo	Poaceae	Hc	E	
Nassella gigantea (Steud.) M.Muñoz	s.n	Poaceae	Hc	E	
Nassella manicata (Desv.) Barkworth	s.n	Poaceae	Hc	N	
Nothoscordum gramineum (Sims) Beauverd	Huilli de perro	Alliaceae	Cr	N	
Olsynium junceum (E.Mey. ex C.Presl) Goldblatt	Huilmo, ñuño	Iridaceae	Cr	N	
Oziroe biflora (Ruiz et Pav.) Speta	Cebolleta	Hyacinthaceae	Cr	N	
Pasithea caerulea (Ruiz et Pav.) D.Don	Azulillo	Anthericaceae	Cr	N	
Paspalum dilatatum Poir.	Chépica gigante	Poaceae	Cr	N	
Paspalum vaginatum Sw.	Chépica	Poaceae	Cr	N	
Phleum pratense L.	s.n	Poaceae	Hc	I	
Phycella bicolor Herb.	Añañuca	Amaryllidaceae	Cr	E	V
Phycella ignea (Lindl.) Lindl.	Añañuca de fuego	Amaryllidaceae	Cr	E	
Piptochaetium montevidense (Spreng.) Parodi	s.n	Poaceae	Hc	N	
Piptochaetium panicoides (Lam.) Desv.	s.n	Poaceae	Hc	N	
Piptochaetium stipoides (Trin. et Rupr.) Hackel	s.n	Poaceae	Te	N	

Classification / Scientific name	Common name	Family	FV	OF	EC
Placea ornata Miers	Macaya	Amaryllidaceae	Cr	E	
Poa annua L.	Pasto piojillo	Poaceae	Te	I	
Potamogeton pusillus L.	Huiro	Potamogetonaceae	Cr	N	
Puya berteroana Mez.	Chagual	Bromeliaceae	Hc	N	
Puya chilensis Molina	Chagual, cardón	Bromeliaceae	Hc	N	V
Puya coerulea Lindl.	Chagual chico	Bromeliaceae	Hc	E	
Rhodophiala advena (Ker Gawl.) Traub.	Añañuca	Amaryllidaceae	Cr	E	
Rostraria cristata (L.) Tzvelev	s.n	Poaceae	Te	I	
Schoenoplectus californicus (C.A.Mey.) Soják	Totora, estoquilla	Cyperaceae	Cr	N	
Scirpus asper J.Presl et C.Presl	s.n	Cyperaceae	Cr	N	
Setaria parviflora (Poir.) Kerguélen	s.n	Poaceae	Hc	N	
Sisyrinchium arenarium Poepp.	Huilmo, ñuño	Iridaceae	Hc	E	
Sisyrinchium chilense Hook.	Huilmo, ñuño	Iridaceae	Hc	E	
Sisyrinchium cuspidatum Poepp.	Huilmo, ñuño	Iridaceae	Hc	E	
Solenomelus pedunculatus (Guillies ex Hook.) Hochr.	Maicillo	Iridaceae	Hc	N	
Solenomelus segethii (Phil.) Kunze	Clavelillo	Iridaceae	Hc	N	
Stipa sp.	Flechilla	Poaceae	Hc	N	
Tecophilaea violiflora Bertero ex Colla	Violeta hojas largas	Tecophilaeaceae	Hc	N	
Tillandsia usneoides L.	Barba de viejo	Bromeliaceae	Fe	N	
Trichopetalum plumosum (Ruiz et Pav.) J.F.Macbr.	Flor de la plumilla	Anthericaceae	Cr	N	
Tristagma bivalve (Lindl.) Traub	s.n	Alliaceae	Cr	N	
Typha angustifolia L.	Vatro	Typhaceae	Cr	N	
Uncinia trichocarpa C.A.Mey.	Clin clin	Cyperaceae	Hc	N	
Vulpia bromoides (L.) Gray	Pasto sedilla	Poaceae	Te	I	
Vulpia myuros (L.) C.C.Gmel.	Pasto sedilla	Poaceae	Te	I	

Table 2. Checklist of the flora of La Campana-Peñuelas Biosphere Reserve (FV = life form, OF = geographic origin, EC = condition of conservation, F = phanerophyte, Nf = nanophanerophyte, Hc = hemicryptophyte, Cr = cryptophyte, C = chamaephyte, Te= terophyte, Fe= epiphyte, Ft= vine, Fp= parasite, N = native, I = introduced, P = threatened, R= rare, V = vulnerable, IC = insufficiently known).

5. Acknowledgements

Thanks to the staff and park wardens of CONAF, Valparaiso Region, for their support in field activities and for facilitating the photographies, bibliographic and cartographic information.

6. References

Aguilella, A.; Amigo, J. (2001). Vegetation transects in central-southern Chile. In: Gómez-Mercado, T. & Mota-Poveda, J. eds. *Vegetación y cambios climáticos*. Spain. University of Almería. pp. 87-101.

Araya, S. & Avila, G. (1981). New shoot production in shrubs affected by fire in "Chilean scrub". *Anales del Museo de Historia Natural de Valparaíso* 14:99-105.

Armesto, J.; Papic, C. & Pliscoff, P. (2002). Importance of small wilderness areas for the conservation of biodiversity in native forest. *Ambiente y Desarrollo* 18:44-50.

Arroyo, MTK.; Cavieres, L.; Marticorena, C. & Muñoz, M. (1995). Convergence in the mediterranean floras of central Chile and California: Insights from comparative biogeography. In: Arroyo, MTK.; Fox, M. & Zedler, P. eds. *Ecology and biogeography of mediterranean ecosystems in Chile, California, and Australia*. New York. Springer-Verlag. pp. 43-88.

Arroyo, MTK.; Matthei, O.; Marticorena, C.; Muñoz, M.; Pérez, F. & Humaña, AM. (2000). The vascular plant flora of the Bellotos del Melado National Reserve, VII Region, Chile: a documented checklist. *Gayana Botánica* 57:117-139.

Arroyo, MTK.; Matthei, O.; Muñoz-Schick, M.; Armesto, JJ.; Pliscoff, P.; Pérez, F. & Marticorena, C. (2005). Flora of four National Reserves in the Coastal Range of the VII Region (35°-36° S), Chile, and its role in the protection of regional biodiversity. In: Smith-Ramírez, C.; Armesto, JJ. & Valdovinos C. eds. *Historia, biodiversidad y ecología de los bosques costeros de Chile*. Santiago, Chile. Editorial Universitaria. pp. 225-244.

Avila, G.; Aljaro, ME. & Silva, B. (1981). Observations in the herbaceous stratum of scrub after fire. *Anales Museo de Historia Natural de Valparaíso* 14:107-113.

Baeza, M.; Barrera, E.; Flores, J.; Ramírez, C. & R. Rodríguez. 1998. Conservation categories of native Chilean Pteridophyta. *Boletín Museo Nacional de Historia Natural, Chile* 47: 23-46.

Balduzzi, A.; Serey, I.; Tomaselli, R. & Villaseñor, R. (1981). New phytosociological observations on the mediterranean type of climax vegetation of central Chile. *Atti Istituto Botanico Laboratori Crittogamico di Pavia*, serie 6, 14:93-112.

Balduzzi A.; Tomaselli, R.; Serey, I. & Villaseñor, R. (1982). Degradation of the mediterranean type of vegetation in central Chile. *Ecología Mediterránea* 8:223-240.

Belmonte, E.; Faúndez, L.; Flores, J.; Hoffmann, A.; Muñoz, M. & Teillier, S. (1998). Conservation categories of native Chilean cactaceae. *Boletín Museo Nacional de Historia Natural, Chile* 47:69-89.

Benoit, IL. ed. 1989. *Red book of Chilean terrstrial flora*. Santiago, Chile. CONAF. 157 pp.

Braun-Blanquet, J. (1964). *Plant sociology – features of phytosociology*. Wien. Springer Verlag. 865 pp.

Braun-Blanquet, J. 1979. *Phytosociology. Basis for the study of plant communities*. Madrid. Blume. 686 pp.

Cabrera, AL. & Willink, A. (1973). *Biogeography of Latin America*. Washington D.C. Biology Series, Monograph N° 13, Regional Program for Scientific and Technological Development, Departament of Scientific Affairs, General Secretariat of the Organization of American States. 120 pp.

Cavieres, LA.; Mihoc, M.; Marticorena, A.; Marticorena, C.; Matthei, O. & Squeo, FA. (2001). Determination of priority areas for conservation: Parsimony Analysis of Endemicity (PAE) in the flora of the IV (Coquimbo) Region. In: Squeo FA, G Arancio, JR Gutiérrez eds. *Libro rojo de la flora nativa y de los sitios prioritarios para su conservación: Región de Coquimbo*. La Serena, Chile. University of La Serena. pp. 159-170.

CONAF (Corporación Nacional Forestal). (1986). *Management plan, Lago Peñuelas Forest Reserve*. Santiago, Chile. National Forest Corporation. Working Document N° 77. 20 pp.

CONAF (Corporación Nacional Forestal). (1992). *National Parks and Natural Monuments of Chile*. Santiago, Chile. National Forest Corporation. 2 pp.

CONAF (Corporación Nacional Forestal). (1994). *Report on the floral wealth of the Bajo los Lirios sector. Lago Peñuelas National Reserve*. Santiago, Chile. National Forest Corporation. Technical Bulletin u/n. 15 pp.

CONAF (Corporación Nacional Forestal). (2008). *Biosphere Reserve La Campana-Peñuelas (Offer of Extension)*. Document bases MAB Program - UNESCO. National Forest Corporation, Region of Valparaiso, Chile. 188 pp.

CONAF & CONAMA (Corporación Nacional Forestal, CL - Comisión Nacional del Medio Ambiente, CL). (1999). *Register and evaluation of Chilean native vegetation resources*. Regional Report, Fifth Region. Santiago, Chile. Project CONAF-CONAMA-BIRF. 141 pp.

DI Castri, F. & Hajek, E. (1976). *Bioclimatology of Chile*. Santiago, Chile. Catholic University of Chile. 128 pp.

Ellenberg, H. & Mueller-Dombois, D. (1966). A key to Raunkiaer plant life forms with revised subdivisions. *Ber. Geob. Inst. ETH Stiftung Rubel, Zurich* 37:56-73.

Elórtegui, S. & Moreira-Muñoz, A. eds. (2002). *La Campana National Park: Origin of a Biosphere Reserve in central Chile*. Santiago, Chile. Taller La Era. 176 pp.

Gajardo, R. (1995). *Chile's natural vegetation. Classification and distribution*. Santiago, Chile. Editorial Universitaria. 165 pp.

González, A. (2000). *Evaluation of vegetation resources in the Budi river basin, current situation and proposals for management*. Degree Thesis in Natural Resources. Temuco, Chile. Faculty of Sciences, Catholic University of Temuco. 110 pp.

Grau, J. (1992). The central zone of Chile. In: Grau, J. & Zizka, G. eds. Chilean wild flora. *Palms, special volume* 19: 1-154.

Grigera, D.; Brion, C.; Chiapella, JO. & Pillado, MS. (1996). Plant life forms as indicators of environmental factors. *Medio Ambiente* 13:11-29.

Hauenstein, E.; Ramírez, C.; Latsague, M. & Contreras, D. (1988). Phytogeographic origin and biological spectrum as a means for measuring the degree of human intervention in plant communities. *Medio Ambiente* 9:140-142.

Hauenstein, E.; González, M.; Peña, F. & Muñoz, A. (2002). Classification and characterization of the flora and vegetation coastal wetlands of Toltén (IX Region, Chile). *Gayana Botánica* 59:87-100.

Hauenstein, E.; González, M.; Peña-Cortés, F. & Muñoz-Pedreros, A. (2005a). Plant diversity in coastal wetlands of the Araucanía Region. In: Smith-Ramírez, C.; Armesto, JJ. & Valdovinos, C. eds. *Historia, biodiversidad y ecología de los bosques costeros de Chile*. Santiago, Chile. Editorial Universitaria. pp. 197-205.

Hauenstein, E.; González, M.; Peña-Cortés, F. & Falcón, L. (2005b). Indicator plants for eutrophisation in lakes of southern Chile. In: Vila, I. & Pizarro, J. eds. *Eutrofización de lagos y embalses*. Iberoamerican Programme of Science and Technology for Development Desarrollo (CYTED – University of Chile). Santiago, Chile. Patagonia Impresores. pp. 119-133.

Hauenstein, E.; Muñoz-Pedreros, A.; Yáñez, J.; Sánchez, P.; Möller, P.; Guiñez, B. & Gil, C. (2009). Flora and vegetation of the Peñuelas Lake National Reserve. Biosphere Reserve, Region of Valparaiso, Chile. *BOSQUE* 30(3): 159-179.

Hoffmann, A. (1978). *Wild flora of Chile, central zone*. 1st ed. Santiago, Chile. Claudio Gay Foundation. 255 pp.

Hoffmann, A. (1991). *Wild flora of Chile, Araucana zone*. 2nd ed. Santiago, Chile. Claudio Gay Foundation. 257 pp.

Hoffmann, A.; Arroyo, MK.; Liberona, F.; Muñoz, M. & Watson, J. (1998). *High Andean plants in the wild flora of Chile*. Santiago, Chile. Claudio Gay Foundation. 281 pp.

ICSA (Ingenieros Consultores Asociados S.A.). (1980). *Study of proposals for integrated development potential of the Lago Peñuelas Forest Reserve*. Santiago, Chile. ICSA-CONAF. 187 pp.

IPNI (International Plant Names Index). (2011). The International Plant Names Index. Consulted 10 Jul 2011. Available at http://www.ipni.org/

Koeppen, W. (1931). The earth's climate. Berlin. Outline of climatology. 182 pp.

Koeppen, W. (1948). *Climatology*. Mexico. Economic Culture Fund. 350 pp.

Looser, G. (1927). *Nothofagus, Cyttaria* and *Myzodendron* on El Roble mountain. *Revista Chilena de Historia Natural* 31: 288-290.

Looser, G. (1944). Phytosociological notes on the Quintero region. *Revista Universitaria* 29: 27-33.

Luebert, F. & Becerra, P. (1998). Plant representativeness of the National System for State-Protected Wilderness Areas (SNASPE) in Chile. *Ambiente y Desarrollo* 14(2):62-69.

Luebert, F. & Pliscoff, P. (2006). *Bioclimate and vegetation synopsis of Chile*. Santiago, Chile. Editorial Universitaria. 307 pp.

Marticorena, C. & Quezada, M. (1985). Catalogue of the vascular flora of Chile. *Gayana Botánica* 42:1-155.

Marticorena, C & Rodríguez, R. eds. (1995). *Flora of Chile. Vol. 1. Pteridophyta- Gymnospermae*. Concepción, Chile. University of Concepción. 351 pp.

Marticorena, C. & Rodríguez, R. eds. (2001). *Flora of Chile. Vol. 2(1) .Winteraceae-Ranunculaceae*. Concepción, Chile. University of Concepción. 99 pp.

Marticorena, C & Rodríguez, R. eds. (2003). *Flora of Chile. Vol. 2(2). Berberidaceae-Betulaceae*. Concepción, Chile. University of Concepción. 93 pp.

Marticorena, C. & Rodríguez, R. eds. (2005). *Flora of Chile.Vol. 2(3). Plumbaginaceae - Malvaceae*. Concepción, Chile. University of Concepción. 128 pp.

Marticorena, C.; Von Bohlen, C.; Muñoz, M. & Arroyo, MK. (1995). Dicotyledons. In: Simonetti, JA. & Arroyo, MK.; Spotorno, AE. & Lozada, E. eds. *Biological diversity of Chile*. Santiago, Chile. Artegrama. pp. 77-89.

Matthei, O. (1995). *Manual of the weeds which grow in Chile*. Santiago, Chile. Alfabeta Impresores. 545 pp.

Mittermeier, RA.; Myers, N.; Thomsen, JB.; Da Fonseca, GA. & Olivieri, S. (1998). Biodiversity hotspots and major tropical wilderness area: approaches to setting conservation priorities. *Conservation Biology* 12:516-520.

Montenegro, G.; Aljaro, ME. & Kummerow, J. (1979). Growth dynamics of Chilean matorral shrubs. *Botanical Gazette* 140:114-119.

Mooney, HA. & Kummerow, J. (1971). The comparative water economy of representative evergreen sclerophyll and drought deciduous shrubs of Chile. *Botanical Gazette* 132:246-252.

Muñoz, C. (1966). *Synopsis of Chilean flora*. 2nd ed. Santiago, Chile. University of Chile. 500 pp.

Muñoz, M. & Moreira, A. (2003). *Alstroemerias de Chile: diversidad, distribución y conservación*. Santiago, Chile. Taller La Era. 139 pp.

Myers, N.; Mittermeier, RA.; Mittermeier, CG.; Da Fonseca, GA. & Kent, J. (2000). Biodiversity hotspots for conservation priorities. *Nature* 403: 853-858.

Navas, ME. (1973). *Flora of the Santiago basin, Chile. Vol I*. Santiago, Chile. University of Chile. 299 pp.

Navas, ME. (1976). *Flora of the Santiago basin, Chile. Vol II*. Santiago, Chile. University of Chile. 559 pp.

Navas, ME. (1979). *Flora of the Santiago basin, Chile. Vol III*. Santiago, Chile. University of Chile. 509 pp.

Novoa, P.; Espejo, J.; Cisternas, M.; Rubio, M. & Dominguez, E. (2006). *Field guide to Chilean orchids*. Concepción, Chile. Corporación Chilena de la Madera. 120 pp.

Oberdorfer, E. (1960). *Plant sociology studies in Chile – A comparison with Europe*. Weinheim. Cramer. 208 pp.

Oltremari, J. (2002). *Protected areas and the conservation of biological diversity*. Santiago, Chile. Catholic University of Chile. 11 pp.

Palma, B.; San Martín, C.; Rosales, M.; Zúñiga, L. & Ramírez, C. (1978). Spatial distribution of aquatic and marsh flora and vegetation of the Marga-Marga river in central Chile. *Anales del Instituto de Ciencias del Mar y Limnología Universidad Nacional Autónoma de México* 14(2):125-132.

Pisano, E. (1956). Scheme for the classification of plant communities in Chile. *Agronomía* 2:30-33.

Ramírez, C. (1988). Life forms, phytoclimates and plant formations. *El Arbol Nuestro Amigo* 4:33-37.

Ramírez, C. & Añazco, N. (1982). Seasonal variations in the development of *Scirpus californicus, Typha angustifolia* and *Phragmites communis* in Valdivian swamps, Chile. *Agro Sur* 10:111-123.

Ramírez, C. & Westermeier, R. (1976). Study of the spontaneous vegetation of the Botanic Garden of the Austral University of Chile (Valdivia), as an example of phytosociological tabulation. *Agro Sur* 4:93-105.

Ramírez, C.; San Martín, C. & San Martín, J. (1996). Floral structure of the swamp forests of central Chile. In: Armesto, JJ.; Arroyo, MK. & Villagrán, C. eds. *Ecología de los bosques nativos de Chile*. Santiago, Chile. Editorial Universitaria. pp. 215-234.

Ramírez, C.; San Martín, J.; San Martín, C. & Contreras, D. (1987). Floral and vegetation study of El Peral lake, Fifth Region, Chile. *Revista Geográfica de Valparaíso* 18:105-120.

Ravenna, P.; Teillier, S.; Macaya, J.; Rodríguez, R. & Zöllner, O. (1998). Conservation categories of Chilean native bulb plants. *Boletín Museo Nacional de Historia Natural, Chile* 47:47-68.

Riedemann, P. & Aldunate, G. (2001). *Native flora of ornamental value, identification and propagation. Chile, central zone*. Santiago, Chile. Andrés Bello. 566 p.

Rodríguez, A. (1979). Plant formations of the Campanita Range. *Boletín Informativo Instituto Geográfico Militar Chile*, IV Trimestre 1979: 11-31.

Rodríguez, A. (1982). Principal plant formations of La Campana National Park. *Boletín Informativo Instituto Geográfico Militar Chile*, I Trimestre 1982: 11-31.

Rodríguez, A. & Calderón, F. (1982). Determination of plant structures on La Campana mountain. *Boletín Informativo Instituto Geográfico Militar Chile*, II Trimestre 1982: 29-47.

Roussine, N. & Negre, R. (1952). *Plant groupings of mediterranean France*. Montpellier. CNRS. 297 pp.

Rundel, P. & Weisser, P. (1975). La Campana, a new National Park in Central Chile. *Biological Conservation* 8: 35-46.

San Martín, C.; Medina, R.; Ojeda, P. & Ramírez, C. (1993). Plant biodiversity in the Río Cruces Nature Sanctuary (Valdivia, Chile). *Acta Botanica Malacitana* 18:259-279.

San Martín, C.; Ramírez, C. & Rubilar, H. (2002). Ecophysiology of saw-sedge marshes in Valdivia, Chile. *Ciencia e Investigación Agraria* 29:171-179.

Simonetti, JA. (1995). Biological diversity: something more than names, something more than numbers. In: Simonetti, JA.; Arroyo, MK.; Spotorno, AE. & Lozada, E. eds. *Diversidad biológica de Chile*. Santiago, Chile. Artegrama. pp. 1-4.

Simonetti, JA. (2004). Connect to conserve. *Ambiente y Desarrollo* 20:2-4.

Strang, C. (1983). The birds of Peñuelas lake. *Chile Forestal* 95:20-22.

Steubing, L.; Ramírez, C. & Alberdi, M. (1980). Energy content of water and bog plant associations in the region of Valdivia (Chile). *Vegetatio* 43:153-161.

Steubing, L.; Godoy, R. & Alberdi, M. (2002). *Plant ecology methods*. Valdivia, Chile. University texts collection, Austral University of Chile. 345 pp.

Teillier, S. (2003). Flora of the El Morado Natural Monument: *Addenda et Corrigenda*. *Gayana Botánica* 60:94-101.

Teillier, S.; Aldunate, G.; Riedemann, P. & Niemeyer, H. (2005). *Flora of the Río Clarillo National Reserve. Guide to species identification*. Santiago, Chile. Impresos Socías. 367 pp.

Villaseñor, R. (1980). Physionomic and floral units of the La Campana National Park. *Anales del Museo de Historia Natural de Valparaíso* 13:65-70.

Villaseñor, R. (1986). *Guide to the recognition of the most frequent tree and shrub species in La Campana National Park*. Valparaíso, Chile. CONAF – University of Playa Ancha. 190 pp.

Villaseñor, R. & Serey, I. (1980-1981). Phytosociological study of the vegetation of La Campana mountain (La Campana National Park) in central Chile. *Atti Ist. Bot. Lab. Critt. Univ. Pavia* 6:69-91.

Weber, C. (1986). Conservation and rational use of nature in protected areas. *Ambiente y Desarrollo* 2:165-181.

WWF. & IUCN. (1997). *Centres of plant diversity: a guide and strategy for their conservation. The Americas*. IUCN Publications Units, Cambridge, United Kingdom Vol. 3.

Zöllner, O.; Olivares, M. & Varas, ME. (1995). The genus *Fumaria* L. (Fumariaceae) in the central zone of Chile. *Anales del Museo de Historia Natural de Valparaíso* 23: 21-31.

12

Investigation of Reproductive Birds in Hara Biosphere Reserve, Threats and Management Strategies

Elnaz Neinavaz, Elham Khalilzadeh Shirazi,
Besat Emami and Yasaman Dilmaghani
*Department of Environment and Energy, Islamic Azad University,
Science and Research Branch, Tehran
Iran*

1. Introduction

Mangroves usually occur in the intertidal zone (Naidoo, 2009) within tropical and subtropical coastal rivers, estuaries and bays of the world (Zhou et al., 2010) , where they may receive organic materials from estuarine or oceanic ecosystems (Ellison & Farnsworth, 2000). In addition, mangroves have been defined by Hamilton and Snedaker (1984) as salt tolerant ecosystems of the intertidal regions along coastlines. Mangroves generally grow in loose, wet soils, saltwater and are periodically submerged by tidal flows along sheltered coastal, estuarine and riverine areas in tropical and subtropical latitudes (Kasawaniet al., 2007). In order to fulfill their important socio-economic and environmental services, issues concerning mangrove functions and their biodiversity have grown in importance in conservation biology all over the world (Badola&Hussain, 2005; Jennerjohn&Ittekkot, 2002; Radhika, 2006; Simardet al., 2006). Furthermore, mangrove roots become habitat for terrestrial species as well as marine plants, algae, invertebrates and vertebrates. Mangroves provide a habitat for a wide variety of species with high populations. Mangroves are also important to humans for a variety of reasons, but especially in aquaculture, agriculture, forestry and protection against shoreline erosion, and as a source of firewood, building material, and other local subsistence use (Walters et al, 2008).

Mangrove forests were first recorded in the Persian Gulf and Oman Sea by Eratosthenes (194 to 276 BC), a geographer from Alexandria (Safiari, 2002). Nowadays, mangroves are found along the Iranian coasts of the Persian Gulf and Oman Sea, as well as around Bahrain, Qatar, Saudi Arabia and the United Arab Emirates (Danehkar, 1996). Iran has the largest acreage of natural mangrove forests and its area ranks 43rd in the world and 10th in Asia (FAO, 2007). Mangrove forests on the southern coast of Iran, are given the common title "The Hara forests", and they cover several locations between the 25°11' and 27°52' parallels longitude (Safiari, 2002). Mangrove growth in Iran consists of only two species of trees *Avicennia marina* and *Rhizophora mucronata*, with *Avicennia marina* scrub being the most prolific, covering over 90% of the Oman Sea and Persian Gulf mangrove habitats. *Rhizophora mucronata* growth is limited and is usually restricted to creeks of the Syric region (included Gas and Hara River Delta). Other sites, however, are dominated by *Avicennia marina*, known locally as the "Hara". Hara Biosphere Reserve is located on Tor'ehyeKhoran-e Bostanu strait between Khamir port and Qeshm island in the environs of Hormozgan Province within 40 km of Qeshm city as well as northwest wetland and swamp margin of Qeshm island between latitudes 26°40_ to 37° N and longitudes 55°21_ to 55°22_ E.

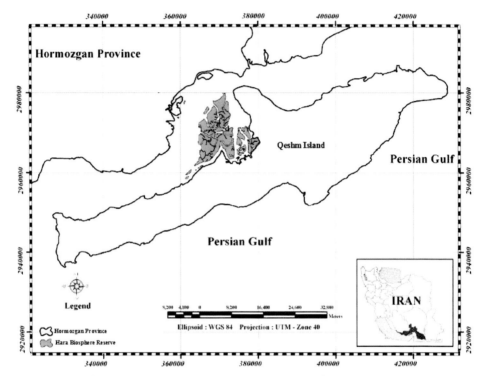

Fig. 1. Location of the study area in southern Iran, Persian gulf.

This area was recognized by the Man and the Biosphere program (MAB) of UNESCO in 1977. On the other hand, this region is one of the protected areas in Iran under the authority of the Department of Environment. The whole region was declared as a Wetland of international importance under the category: Wetlands of International Importance as the Habitat of Aquatic Birds under the Ramsar Convention. In addition, this region was introduced as one of the important bird areas By the Birdlife international.

The land cover of this area comprises two main life forms. The first includes the halophyte plant communities of coastal areas, extending up to the high tide zone. These plant communities are dominated by members of the Chenopodiaceae family which are at times accompanied by tamarisk bushes. The second growth form is the mangrove which constitutes relatively dense forests on tidal islands. In some parts of the region, no other plants accompany mangrove. More than 100 species of sea and shoreline birds, including nine classes and 35 families, have been recognized in the Hara Biosphere Reserve. Moreover, two threatened species and 32 species of Persian Gulf and Oman Sea fishes have been recorded. In addition, 6 and 10 species of prawn and crabs, respectively, have been identified in this area. Furthermore, five species of sea snake, two species of sea turtle and only one kind of marine mammal have been reported in this region. In addition, *Echiscarinatus,* which lives on land, has been observed along the north cost of this area.

2. Diversity and distribution of reproductive birds in Hara biosphere reserve

Since the species of birds which inhabit and reproduce in the Hara Biosphere Reserve are considered wetland-dependent they occupy and important niche in the food chain and their status serves as an indicator for understanding biological changes, impacts of pollutions, and other human activities on Hara Biosphere Reserve. Investigation of the reproductive behaviours of birds is also important for management planning in order to maintain the health of the mangrove ecosystem and control factors threatening the sensitive biodiversity of area (Neinavaz et al, 2010).

In the recent years, studies revealed that only six species of birds have successfully bred the Hara Biosphere Reserve. They are four: the Great Egret (*Ardea alba*), Western-reef Heron (*Egretta gularis*), Indian Pond Heron (*Ardeola grayii*), Eurasian Spoonbill (*Platalea leucorodia*), Great Stone Plover (*Esacus recurvirostris*) and Kentish Plover (*Charadrius alexandrinus*) Reproduction of Great Egret, Western-reef Heron, Indian Pond Heron and Great Stone Plover had been reported in the 1970's. However, there are no reports dating back to the same decades regarding the reproduction of the Eurasian Spoonbill and the Kentish Plover.

Great Egret, Western-reef Heron, Indian Pond Heron and Eurasian Spoonbill build their nests on mangrove trees. On the other hand, the Kentish Plover and the Great Stone Plover build their nests on the ground (sandbar). Today, the Great Egret and the Western-reef Heron have the largest reproductive population in Hara Biosphere Reserve and the Kentish Plover and the Great Stone Plover have the (smallest) least one (Etezadifar et al, 2010; Scott 2007).

In the past years, the breeding range of the Great Egret was in southern provinces of Iran (Fars and Hormozgan); due to the recent multi-year drought, reduction of precipitation and habitat destruction, it appears that the breeding places of the Great Egret are now limited to the Hara Biosphere reserve in the Hormozgan Province. It is considered (Neinavaz et al, 2011) as the largest breeding colony of the Great Egret in the Persian Gulf. In addition, a recent study by Etezadifar (2010) shows that the biggest breeding population of the Western-reef Heron in Iran occurs in the Hara Biosphere Reserve too.

3. The threats

3.1 Natural threats

3.1.1 Invasive species threats

One of the main factors which has a severe impact on breeding success of herons is the status of invasive species. Black Rat (*Rattus rattus*) is the only rodent species which exists in the mangrove forests of the Hara Biosphere Reserve. This has been reported as an invasive species with a significant impact on the reproduction of forest birds and breeding seabirds.

Field observations have revealed that Black Rats (*Rattus rattus*) feed on the eggs and chicks of the Western-reef Heron, Great Egret and the Indian Pond Heron in this region. For instance, Black Rat (*Rattus rattus*) was responsible for about 50% of nest failures of the Western-reef Heron in the Hara Biosphere Reserve (Etezadifar et al, 2010).

Considering the impact of the Black Rats on the breeding success of the Great Egret and the Western-reef Heron, nest site selection by the birds and food preferences of the Black Rat are important factors. In the Hara Biosphere Reserve, Black Rats build their nests on mangrove trees that are greater than the average height. Incidentally, herons use similar trees for nesting as well. Ghadirian (2007) has shown that Black Rats and the Great Egret both prefer the densest areas of the mangrove forest to build their nests. Such an overlap of niches of the rats and the herons has increased the rate of egg/chick predation of the herons by Black Rats.

On the other hand, abandoned cats by local people have a significant impact on the breeding success of the Kentish Plover and the Great Stone Plover which nest on the ground.

3.1.2 Climate change as a threat to mangroves

Climate change components that affect mangroves include changes in sea-level, high water events, storm surges, precipitation, temperature, atmospheric CO_2, ocean circulation patterns, health of functionally linked neighboring ecosystems, as well as human responses to climate change. Of all the outcomes from changes in the atmosphere's composition and alterations to land surfaces, relative sea-level rise may be the greatest threat (Field, 1995; Lovelock and Ellison, 2007). Although to date it has been a smaller threat than anthropogenic activities such as conversion for aquaculture and filling (IUCN, 1989; Primavera, 1997; Valiela et al., 2001;Alongi, 2002; Duke et al., 2007) relative sea-level rise is a substantial cause of recent and predicted future reductions in the area and health of mangroves and other tidal wetlands (IUCN, 1989; Ellison and Stoddart, 1991; Nichols et al., 1999;Ellison, 2000; Cahoon and Hensel, 2006; McLeod and Salm,2006; Gilman et al., 2006, 2007a,b). Mangroves perform valued regional and site-specific functions (e.g., Lewis, 1992; Ewel et al., 1998; Walters et al., 2008). Reduced mangrove area and health will increase the threat to human safety and consequently its species, particularly birds; shoreline development that can impact risks arising from coastal hazards such as erosion, flooding, storm waves and surges, and tsunami, as most recently observed following the 2004 Indian Ocean tsunami (Danielsen et al., 2005; Kathiresan and Rajendran, 2005; Dahdouh-Guebas et al., 2005a,b, 2006) can also adversely impact mangrove integrity. Mangrove loss will also reduce coastal water quality, reduce biodiversity, eliminate fish and crustacean nursery habitats, adversely affect adjacent coastal habitats, and eliminate a major resource for human communities that rely on mangroves for numerous products and services (Ewel et al., 1998; Mumbyet al., 2004; Nagelkerken et al., 2008; Walters et al., 2008). Mangrove destruction can also release large quantities of stored carbon and exacerbate global warming and other climate change impacts.

To date, relative sea-level rise has perhaps been a smaller threat to mangroves than non-climate related anthropogenic stressors, which have mostly accounted for the global average annual rate of mangrove loss, estimated to be 1–2%, with losses during the last quarter century ranging between 35 and 86%(Valiela et al., 2001; FAO, 2003; Duke et al., 2007) .

However, relative sea-level rise may constitute a substantial proportion of predicted future losses: Studies of mangrove vulnerability to change in relative sea-level, primarily from the western Pacific and Wider Caribbean regions, have been documented and the results indicate that the majority of mangrove sites have not been keeping pace with current rates of relative sea-level rise (Cahoon et al., 2006; Cahoon and Hensel, 2006; Gilman et al., 2007b; McKee et al., 2007). Based on this limited information, relative sea-level rise could be a substantial cause of future reductions in regional mangrove area, contributing about 10–20% of total estimated losses of mangroves. Rising sea-level will have the greatest impact on mangroves experiencing net lowering in sediment elevation, that are in a physiographic setting that provides limited area for landward migration due to obstacles or steep gradients. Direct climate change impacts on mangrove ecosystems are likely to be less significant than the effects of associated sea-level rise. Rise in temperature and the direct effects of increased CO_2 levels are likely to increase mangrove productivity, change the timing of flowering and fruiting, and expand the ranges of mangrove species into higher latitudes. Changes in precipitation and subsequent changes in aridity may affect the distribution of mangroves. The combined effect of just three stresses, climate change and other anthropogenic and natural stressors on the mangrove could result in an accelerated rate of rise in sea-level relative to the mangrove sediment surface; decreased mangrove productivity may compromise that ecosystem's resistance and resilience to stresses from climate change and other sources. (Gilman et al., 2007)

3.2 Anthropogenic threats

Anthropogenic pollution in many coastal regions due to rapid population growth, urbanization and industrialization has received considerable public concern, as it may reduce biodiversity and productivity of marine ecosystems, thus posing direct or indirect potential threats to marine species and human health . (Lin and Mendelssohn, 1996; Duke et al., 1997)

3.2.1 Local people

Lack of awareness about reproductive ecology and behaviour of birds in Hara biosphere reserve is the other problem among local communities. Therefore some risks and damages might be caused by them. For instance, in some cases, local people picked up the eggs of Great Egret, Western-reef Heron and Eurasian Spoonbill and used them.

3.2.2 Industrial pollutants

Worldwide, mangrove forests are under threat due to the accumulation of pollutants, which may be imported into mangrove ecosystems through the waters from rivers and streams. The distribution, behaviour, and accumulation of these imported chemicals in the ecosystem are largely defined by the hydrology of the mangroves, the geochemical properties of sediments, and the class of pollutants (e.g. heavy metals, organotins, organochlorine pesticides (OCPs), polychlorinated biphenyls (PCBs)). The properties of the mangrove sediments provide good binding opportunities for a number of these pollutants: hydrophobic organic pollutants adsorb to the extensive surfaces that are provided by the fine particulate sediments of estuaries and mangroves. Metals are trapped in mangrove sediments through the formation of complexes with sulphides (Lacerda et al., 1991), particulate organic carbon, or iron oxyhydroxides (Chapman et al., 1998). As a consequence, anthropogenic pollutants are filtered from the water layer and accumulate in the sediments of estuaries and mangroves (Bayen et al., 2005; Bhattacharya et al., 2003;Tam and Wong, 1995; Tam and Yao, 2002). Depending on the speciation of chemicals, the pollutants can accumulate in the tissues of biota. Some examples of pollutants are heavy metals, considered as natural elements used in high concentrations in many industrial processes. PCBs are included as key representatives of industrial persistent organic pollutants (POPs). Hexachlorobenzene (HCB) was included since it is known to be a by-product of a wide variety of industrial processes and is also present in some pesticides (Bailey, 2001), while dichlorodiphenyl-trichloroethane (DDT) and hexachlorohexanes (HCHs) are representatives of agrochemical POPs.

These pollutants have adverse effects on mangroves, in particular, their species which choose these ecosystems as their habitats. Birds are sensitive ones to industrial pollution like oil contamination, heavy metals, sewage and solid residues. Unfortunately these contaminants put the bird species of Iran mangrove under pressure. **Factors threatening or adversely impacting mangroves in Iran include:**

1. Overexploitation of mangrove leaves and branches as fodder for domestic animals by rural populations, especially in Qeshm Island and Khamir (Danehkar 1996; Khosravi 1992).
2. Severe waves and extreme tidal action can uproot mangroves, decreasing growth and development in some regions (Zahed 2002).
3. The alternation of rivers and delta directions to sea reduce mangrove forests area (Khosravi 1992).
4. Climatic change resulting in drying of land and retardation of inundation periods contribute to mangrove destruction (Zahed 2002).
5. Mangrove trees, resistant to most plant diseases, are susceptible to insects of the Cynipidae family, genus Neuroterus (specie unknown, possibly Neuroterus lenticuloris or Neuroterus lanuginosus) these insects damage *A. marina* leaves in Khamir estuaries and Qeshm Island. Although, current damage is not extensive; it may increase in the future (Khosravi 1992).
6. Mammals, such as *Rattus rattus* and *Gerbillus nanus*, eat mangrove seeds, fruits and leaves, causing damage to these forests (Khosravi 1992).
7. Road construction within the Nayband forests created a seawater connection in the part of these regions, destroying part of the mangrove forests (Danehkar 1996).
8. The lack of forest conservation program as well as control and preservation plan in order to prevent destruction by villagers and other people (Mjnounian 1998; Khosravi 1992).
9. Even though some mangrove species show adaptability to salt stress (Yan and Guizhu 2007); freshwater extraction may cause excessively high salt level and reduced mangrove forests in Persian Gulf (Zahed 2002).
10. Urban and industrial wastewater and sewage discharge into forest areas generate microbial pollution. Marine Pollution, especially oil pollution in the Nayband Bay, as a result of development of the southern Pars oil field in Assaluyeh region (Zahed 2002) has caused damage.

Cutting seawater connections as a result of road construction and overexploitation as forage for domestic animals are the main threats for Iranian mangroves. Moreover, chronic oil pollution in Persian Gulf exerts continual stress and pressure on these mangrove ecosystems.

3.2.2.1 Oil pollution and its impacts on birds

Oil pollution is a common pollution type that would cause a devastating impact on coastal marshes, including mangrove wetlands and salt marshes (Lin and Mendelssohn, 1996; Duke et al., 1997). They are of high ecological and conservation value, which serve as important nurseries for fish ,crustaceans and birds; while on the other hand, this low-energy environment makes the mangrove habitat especially vulnerable to oil contamination (Jackson et al., 1989). The environmental impact of oil contamination in marine environments is potentially serious and increasing attention has been focused on the fate of oil in the environment and the weathering mechanisms (Garrett et al., 1998). Polycyclic aromatic hydrocarbons (PAHs), among the major components in crude oil, are defined as a group of organic compounds with two, or more than two fused benzene rings. PAHs have been targeted as priority pollutants by the United States Environmental Protection Agency (US EPA) for study in oil-contaminated sites due to their carcinogenic, mutagenic and toxic properties. Although contamination of PAHs can result from natural and anthropogenic processes, inputs of PAHs from human activities such as oil spill, offshore production, transportation and combustion are very significant and pose serious threats to coastal habitats such as mangroves (Corredor et al., 1990). Mangrove ecosystems are exposed to oil pollution by accidental or chronic oil spills (Getter et al., 1984). The unique features of mangrove systems such as high primary productivity, abundant detritus, rich organic carbon, anaerobic and reduced conditions favor the retention and accumulation of PAHs (Bernard et al., 1996). Oil spills that enter estuarine and marine intertidal systems can produce a number of deleterious effects .When spilled oil enters mangrove systems, oil soaks into sediments and/or coats exposed trunks, prop routes and pneumatophores sometimes causing extensive mortality of mangroves, declines in productivity or growth irregularities (Lewis 1983; Getter et al., 1985; Duke 1991; Devlin and Priffitt, in press). However the effects of oil go well beyond simple problems for survival and growth. Long-term chronic oiling or oil spills that result in elevated concentrations of polynuclear aromatic hydrocarbons in the sediment can produce chlorophyll-deficient mutations in Rhizophora mangle (Klekowski et al., 1994).

Unfortunately, one of the main destruction factors of Iran mangroves which at the same time have fatal impact on birds' species is chronic oil pollution in the Persian Gulf. It exerts continual stress and pressure on these mangrove ecosystems and their species.

Some of the important health effects of oil pollution on mangrove birds are:

- Hypothermia in birds by reducing or destroying the insulation and waterproofing properties of their feathers;
- Birds become easy prey, as their feathers being matted by oil, making them less able to fly away;
- Birds sink or drown because oiled feathers weigh more and their sticky feathers cannot trap enough air between them to keep them buoyant;
- Birds lose body weight as their metabolism tries to combat low body temperature;
- Birds become dehydrated and can starve as they give up or reduce drinking, diving and swimming to look for food;

- Damage to internal anatomy; for example ulcers or bleeding in their stomachs if they ingest the oil by accident

Non-sticky oils such as refined petroleum products are not likely to stick to a bird, but they are much more poisonous than crude oil or bunker fuel. While some of the following effects on sea birds can be caused by crude oil or bunker fuel, they are more commonly caused by refined oil products .Oil in the environment or oil that is ingested can cause:

- poisoning of species which are higher up in the food chain if they eat large amounts of other organisms that have taken oil into their tissues;
- interference with breeding by making the bird too ill to breed, interfering with nesting behaviour, or by reducing the number of eggs a bird will lay;
- irritation or ulceration of skin, mouth or nasal cavities;
- damage to red blood cells;
- organ damage and failure;
- damage to a bird's adrenal tissue which interferes with a bird's ability to maintain blood pressure, and concentration of fluid in its body;
- decrease in the thickness of egg shells;
- stress;
- damage to estuaries, coral reefs, sea grass and mangrove habitats which are the breeding areas;

3.2.2.2 Impacts of Heavy metals on mangrove ecosystem

Heavy metals are common pollutants in urban aquatic ecosystems and in contrast to most pollutants, are not biodegradable and are thus persistent in the environment. Many metals, a number of which are non-essential to plant and animal metabolism, are often toxic in low concentrations (Baker,1981). Metal inputs arise from industrial effluents and wastes, urban runoff, sewage treatment plants, boating activities, agricultural fungicide runoff, domestic garbage dumps and mining operations. The metals of greatest concern are copper (Cu), lead (Pb), and zinc (Zn), both because of potential threat they pose to aquatic organisms and to human consumption (Luoma, 1990; MacFarlane; 2000). Mangrove systems have the capacity to act as a sink or buffer and remove or immobilize heavy metals before they reach nearby aquatic ecosystems. Because mangrove sediments have a large, heavy proportion of fine particles, high organic content and low PH, they effectively trap metals, often by immobilizing them in the anaerobic sediments either by adsorption on ion exchange sites on the surface of sediment particles, or precipitation as insoluble sulfides (Hrbison, 1986).

3.2.2.3 Sewage and solid residues impacts on mangroves

Coastal areas of the developing world are often extensively populated and in many tropical regions peri-urban population concentrations also coincide with the existence of mangrove ecosystems. Consequently, many fringing urban communities depend heavily on mangroves for both subsistence and commercial harvesting of products (MA, 2005; Ro¨nnback et al., 2007). The biofiltering function of natural mangroves limits coastal sewage pollution to some extent. However, sewage effluents are also likely to affect other ecosystem services. Increased nutrients will enhance tree growth but pathogens and heavy metals are a potential health hazard for people exposed through use of mangrove resources or consumption of mangrove associated marine products. The filtering service of mangroves have nonetheless been put forth as one sewage management option whereby mangroves are strategically reforested or conserved for biofiltration. It is well known that mangroves have a high capacity for filtering suspended and particulate matter (Hemminga et al., 1994) and that mangrove sediments make efficient 'sinks' of nutrients (Alongi, 1990, 1991, 1996; Boto et al., 1989; Hemminga et al., 1994; Holmboe et al., 2001; Rivera-Monroy et al., 1995), but it is uncertain how this capacity will translate to efficiently filter sewage. In addition, understanding of the capacity of mangroves to filter pollutants and pathogens is based on very limited work which has been paid to domestic sewage (Clark, 1998). Effects of sewage and sewage related pathogens on human health has been looked at in terms of infectious disease spread (e.g. Louis et al., 2003; Olago et al., 2007; Rogers, 1996; Singh et al., 2004) as well as use of sewage sludge and water for irrigation of crops (e.g. Rogers, 1996; Singh et al., 2004), but to our knowledge societal and cultural impacts on communities affected by sewage effluent in mangroves have not been previously studied. The deep cultural roots of human interactions with mangroves can be observed in the rituals and cults of many civilizations as well as in the techniques used to extract their biological resources (Kathiresan and Bingham, 2001). The wide traditional use of mangroves includes its exploitation as a source of food (hunting, fishing and harvesting mollusks) and wood (fire wood, extraction of tannins, house construction, the manufacture of utilitarian items, etc.). With growing human occupation of mangrove areas, many of the economic activities centered on the use of estuary resources have grown at scales far superior to the carrying capacity of these environments, contributing to their degradation. The advance of urban areas towards the coast precipitates additional environmental problems derived from the inability of the government to control the occupation of these areas and to provide basic sanitary needs (Field, 1999; Rosso and Cirilo, 2002).Only part of the residues produced by human activities can be assimilated by the ecosystem, depending on the river volume, the tidal regime, the types and frequencies of the discharges, and the over-all productivity of the estuary (Allongi, 2002)[but most of the time solid wastes are not degraded and accumulate in these natural areas. The presence of solid anthropogenic residues in natural areas has always been a problem, but they have become increasingly evident as coastal areas previously known for their scenic beauty become degraded by pollution – at larger scales near urban centers, but visible even in remote islands – creating problems of global dimensions that are accelerated by the rapid and uncontrolled growth of human populations. In addition to their visual impact, solid residues generate costs to the ecosystem and the economy, resulting in the direct mortality of animal species by strangulation or ingestion, habitat suppression, the avoidance by tourists of low quality beaches and damage to boats (Ribic et al., 1992; Ress and Pond, 1995; Frost and Cullen, 1997; Walker et al., 1997; Macauley et al., 1999; Klein et al., 2004; Araujo and Costa, 2006

4. Management strategies

Mangrove biodiversity and conservation of different species especially birds have received considerable attention in recent years since research has increased the understanding of the values of biodiversity and functions and attributes of mangrove ecosystems. At the same time, coastal habitats across the world are under heavy population and development pressures. The scale of human impact on mangroves has increased dramatically over the past three decades or so, with many countries showing losses of 50-80% or more, compared to the mangrove forest cover that existed as recently as in the 1960s. Recognition of the environmental, social and economic impacts associated with the decline and degradation of mangroves are now being addressed through legislative, management, conservation and rehabilitation efforts aimed at mitigating the negative impacts of development on mangrove ecosystems. These include the introduction of new legislation and new governing bodies with clearer administrative or advisory roles on environmental issues; stronger conservation status for some mangrove areas of outstanding value (e.g. as World Heritage sites in Sunderbans of India and Bangladesh, respectively); and more emphasis is being given to raising public awareness and promoting education about mangrove biodiversity. An understanding of the many aspects of human population influences on biological diversity and their underlying driving forces is of crucial importance for setting priorities and directing conservation and sustainable use measures (UNEP, 1995).

The adaptations of mangrove birds to live in the intertidal environment needs conserving . Mangroves support a high diversity of fauna, micro- and macroscopic, terrestrial and aquatic (marine and freshwater), temporary and residential. In this case, the many bird species that have been recorded from mangrove forests may only spend part of their time in mangroves. They migrate seasonally, commute daily or at different states of the tide and may use it as a feeding, nesting or refuge area.

The importance of conserving bird species is now widely appreciated amongst the scientific community, international agencies, governments and NGOs. For this reason, one of the fastest growing sectors of international trade and a powerful tool for national development is ecotourism or nature-based tourism which reduces the negative social and environmental impacts of tourist visits to an area and the money which is gained from that can be used for the conservation of birds. But for best results the government should immediately institute an extensive program that develops awareness among the different sectors/stakeholders (in particular local decision makers) of the value of the mangrove ecosystem resources, functions and services. Also promote and encourage understanding and the importance of mangrove ecosystems through the media and education programs. In many countries the greatest strength of NGOs is public awareness. NGOs should review the information activities of others and prepare programs that fill important gaps. Work with schools and colleges, and with the general public through campaigns; media events are likely to be particularly effective, and will draw upon the unique qualities of many NGOs (Dugan, 1990). NGOs can generate public awareness on policy issues, and site-specific conservation problems and either support government action or press when it is lacking (Dugan, 1990). There are now several examples of successful partnerships (e.g. community-NGO, government- university, and researchers-public media) and stewardship schemes helping to achieve sustainable management of mangrove birds. A key aspect of their success lies in importing a sense of ownership over the mangroves to the local communities concerned.

Other useful strategies that may be considered for conserving birds in mangrove forests:

1. Preparation of environmental management plans for industries and factories located on fringes of the area
2. Environmental monitoring of human user units located along the circumference of the area

3. Preparation of mitigation plan with non-indigenous species (control, prevention, replacement and so on)
4. Preparation of training packages and public participation programs to support and protection of mangrove ecosystems and change people's behavior towards protected species like birds
5. Managing of solid waste disposal in the protected area in partnership with local groups to promote the culture of proper waste disposal.
6. Providing harvest program in the framework of production capacity, harvesting period, site selection and development of mangrove forests located outside the protected area
7. Environmental monitoring programs in partnership with local community
8. Infrastructure development plans related to healthcare and water in marginal habitats within protected area
9. Codification of participatory management plans using local communities as well as financial support from international organizations with the aims of the conservation and sustainable use of bio-regions
10. Use of a local environmental activist groups as a local guide in tourism projects
11. Surveillance on environmental privacy observations in the region for any future development activities (what does this mean?– not clear)

To support research projects in universities to further understanding of structure, process and performance of mangrove forests as well as wildlife protection and conservation in the broader region (Dilmaghani, 2011).

5. Conclusion

Mangrove forests on the southern coast of Iran, "The Hara forests", are considered as one of the most important and unique ecosystems in the world, due to their potential to be an appropriate habitat for birds. Accordingly, this region has been introduced as one of the important bird areas By the Birdlife international. Subsequently, its conservation have become of greatest importance in terms of biodiversity in recent years. Unfortunately, these forests are susceptible to a few natural and anthropogenic threats such as invasive species, climate change, local use of biological resources, oil pollution, sewage and solid residues and heavy metals. These main threats have adverse effects on Hara birds, in particular, their productivity; the birds choose this region for feeding, nesting and resting during their seasonal migration. For comprehensive conservation, management strategies are necessary. In the first place, public awareness should be raised to protect these forests and their species from overexploitation; for example through the participation of local people in environmental management plans .Environmental management and monitoring plans should be prepared to reduce the negative impacts of sewage and waste disposal in marine ecosystems which put mangrove forests under pressure.

6. Acknowledgements

The work of several colleagues has been utilized significantly in the preparation of this inventory; in particular Farzaneh Etezadifar and Taher Ghadirian deserve special mention. The author acknowledges the continued support of Dr Afshin Danehkar, Prof. Mahmoud Karami, Dr esmaeil Kahrom and Dr Reza Marandi and Dr Ali Jozi.

7. References

Alongi, D.M., (1990). Effect of mangrove detrital outwelling on nutrient regeneration and oxygen fluxes in coastal sediments of the Central Great Barrier Reef Lagoon. Estuarine Coastal and Shelf Science 31, 581–598.

Alongi, D.M., (1991). The role of intertidal mudbanks in the diagenesis and export of dissolved and particulate materials from the Fly Delta, Papua New Guinea.Journal of Experimental Marine Biology and Ecology 149, 81–107.

Alongi, D.M., (1996). The dynamics of benthic nutrient pools and fluxes in tropical mangrove forests. Journal of Marine Resources 54, 123–148.

Alongi, D.M., (2002). Present state and future of the world's mangrove forests. Environmental Conservation 29 (3), 331–349.

Araujo, M.C.B., Costa, M.F., 2006. Municipal services on tourist beaches: costs and benefits of solid waste collection. Journal of Coastal Research 22 (5), 1070–1075.

Badola, R., &Hussain, S. A. (2005).Valuing Ecosystem Functions: an Empirical Study on the Storm Protection Functions of Bhitarkanika Mangrove Ecosystem, India.*Environmental Conservation, 32*(1), 85-92.

Bailey, R.E., (2001). Global hexachlorobenzene emissions. Chemosphere 43, 167–182.

Baker,A.J. (1981). Accumulators and excluders: strategies in the response of plants to heavy metals .Journal of plant nutrition 3(4), 643-654.

Bayen, S., Wurl, O., Karuppiah, S., Sivasothi, N., Lee, H.K., Obbard, J.P. (2005). Persistent organic pollutants in mangrove food webs in Singapore. Chemosphere 61, 303–313.

Bernard, D., Pascaline, H., Jeremie, J.J., (1996). Distribution and origin of hydrocarbons in sediments from lagoons with fringing mangrove communities. Marine Pollution Bulletin 32, 734–739.

Boto, K.G., Alongi, D.M., Nott, A.L.J., (1989). Dissolved organic carbon–bacteria interactions at sediment–water interface in a tropical mangrove system. Marine Ecology Progress Series 51, 243–251.

Cahoon, D.R., Hensel, P., (2006) . High-resolution global assessment of mangrove responses to sea-level rise: a review. In: Gilman, E. (Ed.), Proceedings of the Symposium on Mangrove Responses to Relative Sea Level Rise and Other Climate Change Effects, 13 July 2006, Catchments to Coast, Society Of Wetland Scientists 27th International Conference, 9–14 July 2006, Cairns Convention Centre, Cairns, Australia. Western Pacific Regional Fishery Management Council, Honolulu, HI, USA, ISBN: 1-934061-03-4 pp. 9–17.

Chapman, P.M., Wang, F.Y., Janssen, C., Persoone, G., Allen, H.E. (1998). Ecotoxicology of metals in aquatic sediments: binding and Release, bioavailability, risk assessment, and remediation. Canadian Journal of Fisheries and Aquatic Sciences 55, 2221–2243

Clark, M.W., (1998). Management of metal transfer pathways from a refuse tip to mangrove sediments. Science of the Total Environment 222, 17–34.

Cordeiro, C.A.M.M,.Costa, T.M., (2010). Evaluation of solid residues removed from a mangrove swamp in the Sao Vicente Estuary, SP, Brazil. *Marine pollution bulletin,*Vol. 60 , 1762-1767.

Corredor, J.E., Morell, J.M., Castillo, C.E.D., (1990). Persistence of spilled crude oil in a tropical intertidal environment. Marine Pollution Bulletin 21, 385–388.

Danehkar, A. (1996). Iranian Mangroves Forests.*The Environment Scientific Quarterly Journal 8*, 8-22. [*In Persian*]

Danielsen, F., Soerensen, M., Olwig, M., Selvam, V., Parish, F., Burgess, N., Hiraishi, T., Karunagaran, V., Rasmussen, M., Hansen, L., Quarto, A., Nyoman, S.,(2005). The Asian tsunami: a protective role for coastal vegetation. Science 310, 643

Devlin, D. J. & Proffitt, C. E. (In press). Experimental analysis of the effects of oil on mangrove seedlings and samples. In *Proceedings: Gulf of Mexico and Caribbean Oil Spills in Coastal Ecosystems* (C. E. Proffitt & P.Roscigno, eds). US Minerals Management Service Publ., GOM Program New Orleans, LA.

Dilmaghani, Y.,. et al. (2011), Codification of Mangrove forests management strategies: case study of Hara Protected Area-Iran, *Journal of Food, Agriculture & Environment* Vol.9 (1).

Donald J. Macintosh and Elizabeth C. Ashton, 2002, A Review of Mangrove Biodiversity Conservation and Management, *Centre for Tropical Ecosystems Research (cenTER Aarhus)*

Dugan, P.J., 1990. Wetland Conservation: A review of current issues and required action. IUCN, Gland, Switzerland, 96 pp.

Duke, N. (1991). Mangrove forests. In *Long-term Assessment of the Oil Spill at Bahia Las Minas, Pananma. Interim Report. Vol. II: Technical Report* (B. Keller & J. B. C. Jackson, eds), pp. 153-178. US Dept. of the Interior, Minerals Management Serv. OCS study MMS 90-0031.

Duke, N.C., Pinzon, Z.S., Prada, M.C., (1997). Large-scale damage to mangrove forests following two large oil spills in Panama. Biotropica 29, 2–14.

Ellison, A. M., & Farnsworth, E. J. (2000).Mangrove Communities. In: Bertness, M. D., S.D. Gaines, and M. E. Hay (Eds.). *Marine Community Ecology*, 423-442.

Ellison, J., Stoddart, D., (1991). Mangrove ecosystem collapse during predicted sea level rise: Holocene analogues and implications. J. Coast. Res. 7, 151–165.

Etezadifar, F., Barati.A, Karami.M, &Danehkar, A., &Khalighizadeh, A. 2010.Breeding Success of Western Reef Heron in Hara Biosphere Reserve, Persian Gulf.Waterbirds 33(4): 527- 533

Ewel, K.C., Twilley, R.R., Ong, J.E., (1998). Different kinds of mangrove forests provide different goods and services. Global Ecol. Biogeogr. 7, 83–94.

Field, C., (1995). Impacts of expected climate change on mangroves. Hydrobiologia, 295, 75–81.

Field, C.D., (1999). Mangrove rehabilitation: choice and necessity. Hydrobiologia 413, 47–52.

Frost, A., Cullen, M., (1997). Marine debris on Northern New South Wales Beaches (Australia): sources and the role of beach usage. Marine Pollution Bulletin 34 (5), 348–352.

Garrett, R.M., Pickering, I.J., Haith, C.E., Prince, R.C., (1998). Photooxidation of crude oils. Environmental Science and Technology 32, 3719–3723

Getter, C. D., Ballou, T. G. & Koons, C. B. (1985). Effects of dispersed oil on mangroves. Synthesis of a seven-year study. *Mar. Pollut. Bull.* 16, 318-324.

Getter, C.D., Cintron, G., Dicks, B., Lewis III, R.R., Seneca, E.D., (1984). The recovery and restoration of salt marshes and mangroves following an oil spill. In: Cairns Jr., J.Buikema Jr., A. (Eds.), Restoration of Habitats Impacted by Oil Spills. Butterworth Publishers, Boston, pp. 65–113.

Ghadirian, T. (2007).Habitat and Population Density and Abundance of Black Rat (*Rattusrattus*) in Hara Biosphere Reserve, Hormozgan Province, Iran.Msc. Dissertation, Islamic Azad University, Science and Research Branch. Tehran, Iran. [*In Persian*]

Gilman, E., Ellison, J., Coleman, R., (2007a). Assessment of mangrove response to projected relative sea-level rise and recent historical reconstruction of shoreline position. Environ. Monit. Assess. 124, 112–134

Gilman, E., Ellison, J., Sauni Jr., I., Tuaumu, S. (2007b). Trends in surface elevations of American Samoa mangroves.Wetl. Ecol. Manag. 15, 391–404.Hamilton, L. S., &Snedaker, S. C. (1984).*Handbook of Mangrove Area Management.United Nations Environment Programme, and East West Center Environment and Policy Institute*.COE/IUCN, Gland-Switzerland.

Gilman, Eric L., Ellison, Juanna., Duke, Normann C., & Field Kolin . (2008). Threats to mangroves from climate change and adaptation options:A review. *Aquatic botany* , Vol. 89, 237-250.

Hemminga, M.A., Slim, F.J., Kazungu, J.M., Ganssen, G.M., Niewenhuize, J., Kruyt, N.M., (1994) . Carbon outwelling from a mangrove forest with adjacent seagrass beds and coral reefs (Gazi Bay Kenya). Marine Ecology Progress Series 106, 291–301.

Holmboe, N., Kristensen, E., Andersen, F.Ø., (2001) . Anoxic decompositions in sediment from a tropical mangrove forest and the temperate Waden Sea: implica-tions of N and P addition experiments. Estuarine Coastal and Shelf Science 53, 125–140.

Hrbison, P.(1986) Mangrove muds a sink or source for trace metals. Marine Pollution Bulletin 17, 246 250.

http://www.amsa.gov.au/marine_environment_protection/educational_resources_and_in formation/teachers/the_effects_of_oil_on_wildlife.asp

IUCN, (1989) . The impact of climatic change and sea level rise on ecosystems. Report for the Commonwealth Secretariat, London.

Jackson, J.B.C., Cubit, J.D., Keller, B.D., Batista, V., Burns, K., Caffey, H.M., Caldwell, R.L., Garrity, D.D., Getter, C.D., Gonzalez, C., Guzman, H.M., Kaufmann, K.W., Knap, A.H., Levings, S.C., Marshall, M.J., Steger, R., Jennerjohn, T. C., &Ittekkot, V. (2002).Relevance of Mangroves for the Production and Deposition of Organic Matter along Tropical Continental Margins.*Naturwissenschaften, 89*, 23–30.

Kasawani, I., Kamaruzaman, J., &Nurun-Nadhirah, M. I. (2007).Biological Diversity Assessment of Tok Bali Mangrove Forest, Kelantan, Malaysia.*WSEAS Transactions on Environment and Development, 3*(2), 30-385.

Kathiresan, K., Bingham, B.L.,(2001). Biology of mangroves and mangrove ecosystems. Advances in Marine Biology 40, 81–251.

Ke, L., Wong, Teresa W.Y., Wong, Y.S., Tam, Nora F.Y., (2002) .Fate of polycyclic aromatic hydrocarbon contamination in a mangrove swamp in Hong Kong following an oil spill. *Marine pollution bulletin* ,Vol. 45 , 339-347.

Ke, Lin ., Zhang, Chunguang ., Wong, YukShan& Tam, Nora Fung Yee ., (2011) .Dose and accumulative effects of spent lubricating oil on four common mangrove plants in South China. *Ecotoxicology and Environmental Safety*, Vol. 74 , 55-66

Khosravi,M., (1992). Ecological studies plan of Iranian mangrove forests. Report of mangrove forests identification phase, Department of the Environment, Tehran, Iran.

Klein, Y.L., Osleeb, J.P., Viola, M.R., (2004) . Tourism-generated earnings in the Coastal Zone: a regional analysis. Journal of Coastal Research 20 (4), 1080–1088

Klekowski, E. J., Corredor, J. E., Morell, J. M. & Del Castillo, C. A. (1994). Petroleum pollution and mutation in mangroves, *Mar. Pollut. Bull.* 28, 166-169

Kruitwagen, G., Pratap, H.B., Covaci, A &WendelaarBonga, S.E ., (2008). Status of pollution in mangrove ecosystems along the coast of Tanzania. *Marine Pollution Bulletin* , Vol. 56 , 1022–1042

Lacerda, L.D., Rezende, C.E., Aragon, G.T., Ovalle, A.R., (1991). Iron and chromium transport and accumulation in a mangrove ecosystem. Water Air and Soil Pollution, 513–520.

Lewis III, R.R., (1992). Scientific perspectives on on-site/off-site, in-kind/out-of kind mitigation. In: Kusler, J.A., Lassonde, C. (Eds.), Effective Mitigation: Mitigation Banks and Joint Projects in the Context ofWetland Management Plans. Proceedings of the National Wetland Symposium, 24–27 June 1992, Palm Beach Gardens, FL, USA, pp. 101–106.

Lewis, R. R. (1983). Impact of oil spills on mangrove forests. In *Biology and Ecology of Mangroves. Tasks for Vegetation Science 8* (H. J. Teas, ed.), pp. 171-183. W. Junk, The Hague, The Netherlands

Lin, Q., Mendelssohn, I.A., (1996). A comparative investigation of the effects of South Louisiana crude oil on the vegetation of fresh, brackish and salt marshes. Mar. Pollut. Bull. 32, 202–209.

Liu, Y., Tam, N.F.Y., Yang, J.X., Pi b,N., Wongd, M.H., Ye, Z.H., (2009). Mixed heavy metals tolerance and radial oxygen loss in mangrove seedlings.*Marine Pollution Bulletin*, Vol. 58 ,1843–1849

Louis, V.R., Russek-Cohen, E., Choopun, N., Rivera, I.N.G., Gangle, B., Jiang, S.C., et al., (2003). Predictability of Vibrio cholerae in Chesapeake Bay. Applied Environmental Microbiology 69, 2773–2785.

Luoma, S.N. (1990). Processes affecting metal concentrations in estuarine and coastal marine sediments. In Heavy Metals In The Marine Environment, eds.R.W. Fumess and P.S. Rainbow, pp. 51-66. CRC Press, Boca Raton, FL.

MA (Millennium Ecosystem Assessment), (2005). Ecosystems and Human Wellbeing: Biodiversity Synthesis. Island Press, Washington, DC.

Macauley, J.M., Summers, J.K., Engle, V.D., (1999). Estimating the ecological condition of the estuaries of the Gulf of Mexico. Environmental Monitoring and Assessment 57, 59–83.

MacFarlane, G. R (2000) Mangrove and Pollution. In Mangroves: An Ecosystem Between Land and Sea, ed. U. Ganslober. Filander Press, Fruth, Germany.

Macfarlane, G.R., Burchett, M.d., (2001) .Photosynthetic Pigments and Peroxidase Activity as Indicators of Heavy metal Stress in the Grave Mangrove ,Avicennia marina (Forsk.) Vierh. *Marine pollution bulletin* ,Vol. 42, N. 3, 233-240.

Majnounian,H., Danehkar, A (1998) .(National parks and marine and coastal protected areas. Case study: Nayband Bay, The Environment Scientific Quarterly Journal 24 , 65–74.

McKee, K.L., Cahoon, D.R., Feller, I., (2007). Caribbean mangroves adjust to rising sea level through biotic controls on change in soil elevation. Global Ecol. Biogeogr. 16, 545–556.

Naidoo, G. (2009). Differential Effects of Nitrogen andPhosphorus Enrichment on Growth of dwarf*Avicennia marina* Mangroves. *AquaticBotany, 90*(2), 184-190.

Neinavaz, E. (2009). Study of Breeding Success of Great egret (*Egretta alba*) in the Mangrove Forest Located in Hara Protected Area-Hormozgan province, Iran.Msc. Dissertation, Islamic Azad University, Science and Research Branch. Tehran, Iran. [*In Persian*]

Neinavaz, E., Karami,M., Danehkar, A.(2011)Investigation of Great Egret (Casmerodius albus) breeding success in Hara Biosphere Reserve of Iran, *Environmental Monitoring and Assessment (Publisher: Springer)*, Vol. 179, No. 1-4, August 2011 , pp. 301-307(7)

Olago, D., Marshall, M., Wandiga, S.O., et al., (2007). Climatic, socio-economic, and health factors affecting human vulnerability to cholera in the Lake Victoria Basin, East Africa. Ambio 36 (4), 350–358.

Primavera, J., (1997). Socio-economic impacts of shrimp culture. Aquacult. Resour. 28, 815–827.

Proffitt , C. Edward., Devlin, Donna J & Lindsey Mark., (1995) . Effect of Oil on Mangrove Seedlings Grown Under Different Environmental Conditions .*Marine pollution bulletine*, Vol . 30, No. 12 , 788-793.

Ress, G., Pond, K., (1995). Marine litter monitoring programmes – a review of methods with special reference to national surveys. Marine Pollution Bulletin 30 (2), 103–108.

Ribic, C.A., Dixon, T.R., Vining, I., (1992). Marine Debris Survey Manual. NOAA Technical Report. NMFS 108, US Department of Commerce, 94p.

Rivera-Monroy, V.H., Twilley, R., Boustany, R.R.G., Day, J.W., Veraherrera, F., Ramirez, M.D., (1995). Direct denitrification in mangrove sediments in Terminos Lagoon Mexico. Marine Ecology Progress Series 126, 97–109.

Ro¨nnback, P., Crona, B., Ingwall, L., (2007). The return of ecosystem goods and services in replanted mangrove forests: perspectives from local communities in Kenya. Environmental Conservation 34 (4), 313–324.

Rogers, H.R., (1996). Sources, behavior and fate of organic contaminants during sewage treatment and in sewage sludges. Science of the Total Environment 185, 3–26.

Rosso, T.C.A., Cirilo, J.A., (2002). Water Resources Management and Coastal Ecosystems: Overview of the Current Situation in Brazil. Littoral 2002. The Changing Coast. EUROCOAST/EUCC, Porto, Portugal, pp. 221–229.

Safiari, S. (2002). Mangrove Forests, *Vol. 2:* Mangrove forests in Iran: *Research Instituteof Forests and Rangelands* (RIFR) press, Tehran.

Scott, D. A. (2007). A Review of the Status of the Breeding Waterbirds in Iran in the 1970s. Podoces 2: 1-21.

Simard, M., Zhang, K., Rivera-Monroy, V. H., Ross,M. S., Ruiz, P. L., &Castañeda-Moya, E. (2006). Mapping Height and Biomass of Mangrove Forests in Everglades NationalPark with SRTM Elevation Data.*Photogrammetric Engineering and RemoteSensing*, 209-311.

Singh, K.P., Mohan, D.S.S., Dalwani, R., (2004). Impact assessment of treated/untreated wastewater toxicants discharged by sewage treatment plants on health, agricultural, and environmental quality in the wastewater disposal area. Chemosphere 55, 227–255.

Thompson, R.C., Weil, E., (1989). Ecological effects of a major oil spill on Panamanian coastal marline communities. Science 243, 37–44.

United Nations Environment Program, 1995. *Global Biodiversity Assessment. Cambridge University Press.*

Valiela, I., Bowen, J., York, J., (2001). Mangrove forests: one of the world's threatened major tropical environments. Bioscience 51, 807–815.

Walker, T.R., Reid, K., Arnould, J.P.Y., Croxall, J.P., (1997). Marine debris surveys at Bird Island, South Georgia 1990–1995. Marine Pollution Bulletin 34, 61–65.

Walters, B. B., Rönnbäck, P., Kovacs, J. M., Crona, B., Hussain, S. A., &Badola, R.(2008). Ethnobiology, Socio-economic and Management of Mangrove Forests: A review. [doi: DOI: 10.1016/j.aquabot.2008.02.009]. *Aquatic Botany, 89*(2), 220-236.

Yan, L., Guizhu, C. (2007). Physiological adaptability of three mangrove species to salt stress, Acta Ecologica Sinica 27 (6) , 2208–2214.

Zahed, Mohammad Ali., Rouhani, Fatemeh., Mohajeri, Soraya, Bateni, Farshid&Mohajeri, Leila. (2010) . An overview of Iranian mangrove ecosystems, northern part of the Persian Gulf and Oman Sea .*ActaEcologicaSinica* .Vol. 30, 240–244.

Zahed,M.A, (2002). Effect of pollution on Persian Gulf mangroves. Ministry of jihad – eagriculture final report. Tehran, Iran.

Zhou, Y., Zhao, B., Peng, Y., & Chen, G. (2010).Influence of Mangrove Reforestation on Heavy Metal Accumulation and Speciation in Intertidal Sediments.*Marine Pollution Bulletin.*

13

Livelihoods and Local Ecological Knowledge in Cat Tien Biosphere Reserve, Vietnam: Opportunities and Challenges for Biodiversity Conservation

Dinh Thanh Sang, Hyakumura Kimihiko and Ogata Kazuo
Kyushu University
Japan

1. Introduction

Many studies have found that protected areas (PAs) have been conventionally designed based on top-down approaches by government agencies without consulting the locals (Batisse, 1997; Kelsey at al., 1995; Wells & Brandon, 1992). Consequently, experience has shown that top-down approaches to PA management are often ineffective in reaching conservation objectives (Brown, 2002). Beyond conventional PAs, biosphere reserves are found to take into account both biodiversity conservation and socioeconomic development of local populations. Local people are considered as a key element for sustainable management of biosphere reserves (Sang, 2006). They are major stakeholders who play direct and indirect roles in resource conservation. For generations, they have not only relied on forest resources but also accumulated their local ecological knowledge (LEK) of forest resource use. To conserve biodiversity as well as securing local livelihoods, sustainable use of resources in biosphere reserves is considered crucial. The sustainable use depends on ecological knowledge, flexible policy implementation and adaptive management in terms of local realities (Olsson & Folke, 2001; Ostrom, 1990, 2005).

With regard to the case of Cat Tien Biosphere Reserve (CTBR), both local livelihoods and local ecological knowledge have brought not only opportunities but also challenges for biodiversity conservation. The forest resources in the reserve have continued to be degraded owing to a lot of pressures and several species living there are threatened. About 60-80% of the populations depend mainly on forest resources for their living (CTBR, 2006). Most of the local people within and around CTBR have depended much on forest resources for generations and have caused the major loss of the reserve's natural features (Sang et al., 2009). They harvest forest resources for both subsistence and cash income. Biodiversity in CTBR is endangered mainly by encroachment of forest land, illegal logging and poaching. This results in biological invasions, endangering endemic fauna and flora.

Natural growth and immigration lead to an increasing population in the areas within and around the reserve, which increases the threats to CTBR. One of the biggest problems for the authorities of CTBR is that the areas where hamlets and lands for cultivation occur now were previously in the core zones (CZs) before CTBR was established. The most important conservation issue of the reserve is the large human population living within its boundaries, which is an indirect source for a lot of threats. Consequently, with a growing human population and economy, the demand for agricultural land, food, forest resources for subsistence and cash income increases. For instance, most of the hills surrounding the reserve have been stripped of trees and converted into agricultural land by the local communities. As reported in Tuoi Tre Newspaper (June 9th, 2006), from January to June of 2006 a 60 ha area of forest in the CZ and a 10 ha area in the buffer zone (BZ) of the reserve were destroyed by local residents, mostly ethnic minorities.

Despite being one of the largest protected areas in Vietnam, Cat Tien continues to be stalked by illegal poachers. Poaching and wildlife trade pose perhaps the greatest threat. According to Vietnam News on July 29th, 2004, forest guards said they discover over 2,000 animal traps in the reserve every month. On April 29th, 2010 a Javan Rhino was found dead by poachers in Cat Loc Sector, a CZ of CTBR belonging to Lam Dong Province, and its horn had been removed (CTBR, 2010). While mainly birds, monkeys, wild pigs and deer are hunted for consumption, larger mammals such as the gaurs and bantengs are also poached. Furthermore, the reserve is too small for the larger species found inside it; this leads to either erosion of conservation values or to conflicts with local people as these animals move beyond the confines of the reserve (CTBR, 2010). This problem is particularly intense for the elephant population of the reserve, which is prone to wandering and is considered too small to be self sustainable (CTBR, 2010).

The aims of this paper are to identify the feature of local livelihood strategies and LEK in CTBR of Vietnam and to clarify the opportunities as well as challenges for biodiversity conservation. Recommendations are made for biodiversity conservation and socioeconomic development in the area.

2. Description of Cat Tien Biosphere Reserve

The study was conducted in CTBR where Vietnam Central Highlands give way to Southern Delta, at E107°09'05"-107°35'20", N11°20'50"-11°50'20". It covers an area of approximately 71,350 ha and consists of three sectors: South Cat Tien (39,627 ha), West Cat Tien (4,193 ha) and Cat Loc (27,539 ha) (CTBR, 2011). Its topography comprises a mosaic of landforms of both Truong Son Mountain Range and of lowland rivers, semi-plains, medium hills, relatively flat land and scattered lakes, ponds and wetland of eastern parts of the Southern Delta (Cox et al., 1995). Most of the area of CTBR lies on a bedrock of basalt and five soil types can be recognized: ferralitic soils developed on basalt stone, sandstone, old alluvium, shale and conglomerate.

The reserve lies in the tropical monsoon region with two main seasons: the rainy season from May to November, and the dry season from December to April. February and March are the driest. The mean annual rainfall is 2,435 mm, the mean annual temperature is 25.5º C and the mean relative humidity is 80%. Rain falls on an average of 150-190 days per year, mainly from July to October. During the rainy season, rainfall is about 300 mm per month. In Bao Loc City (Lam Dong), it is up to 245 mm of rain per month; while in Phuoc Long Town (Binh Phuoc) 167 mm of rainfall has been recorded in one day (Cox et al., 1995).

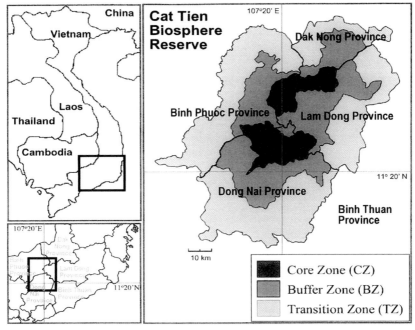

Fig. 1. Location and the map of CTBR (Source: Sang et al., 2010)

CTBR has been shown to have a high diversity of fauna and flora, especially for a relatively small area with a limited altitudinal range. The high biodiversity of the reserve stems from its location in an area between the biogeographically distinct Da Lat Plateau and eastern parts of the Southern Delta. The habitat types include primary evergreen forest (0.9%), secondary evergreen forest (23.1%), semi- evergreen forest (7%), mixed forest (bamboos and other plants, 19.3%), bamboo forest (40.1%), and wetland (2.2%). There are about 103 species of mammals, 348 species of birds, 79 species of reptiles, 159 species of fish, 49 species of amphibians, and 439 species of butterflies (table 1). The only Javan Rhino population outside of Java, Indonesia, occurs in the Cat Loc Sector. Globally endangered species found in CTBR include: *Arborophila davidi, Bos javanicus, Bubalus arenee, Cairina scutulata, Capricornis sumatraenis, Elephas maximus, Nomascus gabriellae, Panthera tigris corbetti, Pseudibis davisoni, Pygathrix nigripes*. So far 1,610 species of flora have been recorded on six types of habitats and belong to 724 genera, 162 families and 75 orders; 34 species are listed in the Vietnam Red Data Book; 31 are listed as rare and 23 are endemic (CTBR, 2006). The large number of plant species in CTBR has conservation as well as economic values: 38 species in 13 families for protection of genetic resources, 23 species are of indigenous endemic value, 511 species are wood trees (of which 176 species are precious timber), 550 species are medicinal plants; and hundreds of species have essential oil and special use (CTBR, 2006).

According to decision 360 / TTg of the Prime Minister on July 7th, 1978, Cat Tien National Park, the first name of Cat Tien Biosphere Reserve, was protected initially as two sectors, South Cat Tien in Dong Nai Province and West Cat Tien in Binh Phuoc Province. Then based on decision 8 / CT of the Chairman of the Council of Ministers on January 13th, 1992,

South Cat Tien was approved as a national park. Gazetted as a Rhino Reserve in 1992 upon the discovery of a Javan Rhino population, Cat Loc received the protected status from Lam Dong Province in the same year. Subsequently, decision 38-QD of February 16th, 1998 approved the integration of South Cat Tien, West Cat Tien and Cat Loc into Cat Tien National Park. On November 10th, 2001 the park was recognised by United Nations Educational, Scientific and Cultural Organisation (UNESCO) as a biosphere reserve of the world.

Taxa	All species		VN Red List 2000 species		IUCN Red List 2000
	Total	% of all species in Vietnam	Total	% of all species in Vietnam	
Amphibians	23 (49)	22	1 (2)	9 (18)	0
Birds	283 (348)	34	18 (21)	22 (26)	15 (16)
Butterflies	435 (439)	43	n.a.	n.a.	n.a.
Fresh water fish	99 (159)	21	6 (9)	18 (27)	1
Mammals	60 (103)	25	15 (32)	20 (39)	16 (26)
Reptiles	46 (60)	18	14 (17)	33 (40)	5 (8)

Note: Non-bracketed figures are confirmed records, figures in ( ) include possible records

Table 1. Fauna species recorded in Cat Tien Biosphere Reserve (Source: CTBR, 2010)

CTBR is divided into three zones. Core zone is strictly protected; some activities and sustainable resource uses can be acceptable if they are in accordance with its conservation goals. There are two core zones in CTBR; South Cat Tien (Dong Nai Province) and West Cat Tien (Binh Phuoc Province) constitute the first CZ; and the Cat Loc Sector (Lam Dong Province) is the second CZ located in the north of the reserve. Buffer zone may provide a variety of sustainable uses which ensure the protection and conservation of the reserve, and improve the local socioeconomic conditions. Transition zone (TZ) is for sustainable socioeconomic development to reduce pressure on the reserve. At present, the board of CTBR under Vietnam Ministry of Agriculture and Rural Development has overall responsibility over the CZs. However, the socioeconomic development and management in the BZ is in the hands of many other departments and local organizations under the governance of the three provinces.

CTBR is the home of about 2,000 people in the CZs and approximately 200,000 residents in the BZ. Local people in CTBR can be categorized into three main groups - Kinh people, indigenous ethnic minorities and migrant ethnic communities. The first group includes Kinh people who lived there before 1975, Kinh households from Mekong Delta or south - east provinces of Vietnam, and many Kinh families from northern provinces of Vietnam settled under the program of development of new economic zones in 1986; the indigenous ethnic minorities comprise Chau Ma, S'Tieng, Chau Ro, and Saray who have lived there for generations; the migrant ethnic communities consist of Tay, Nung, Dao, H'Mong, Hoa from the northern provinces and a few Kho Me from Mekong Delta. Each ethnic group has different histories, various traditions, and diverse material lives. About 60-80% of the population has depended primarily on forest resources for their livelihood (CTBR, 2006). Most of them, especially ethnic minorities within and around CTBR have depended much on forest resources for generations and have caused the major loss of biological resources (Sang et al., 2009). The locals have cultivated areas with high value industrial crops like cashew, cassava, coffee, pepper; rice and fruit such as jack-fruit, mandarin, pineapple, mango, banana, guava, papaya, custard-apple, and longan.

3. Methodologies

Field research was conducted to gain intensive understanding of people through discussions and interactions. The research reported here was carried out in 2005, 2006 and 2010. Primary data were gathered initially through household interviews based on questionnaires, Rapid Rural Appraisal (RRA), and the "walk-in-the-wood" method (Prance et al., 1987). Interviews were also carried out with community leaders, government officials, staff of CTBR and non-governmental organizations (NGOs). The data covered qualitative and quantitative information including socioeconomic status, forest resource use, LEK, crops, land holding, and cultural practices. Secondary data used in the study were mostly drawn from the author's previous studies as well as the reports of CTBR and the local government; books, and newspapers. The Statistical Package for Social Sciences (SPSS) was used to analyze the quantitative information obtained through the questionnaire survey.

CTBR was chosen as a case study because the forest resources in the reserve continue to be degraded owing to a lot of pressures and various species of plants and animals have been threatened by the local people. To reflect the livelihoods and local ecological knowledge, the research data were gathered in sites where there were natural forests in or near the hamlets, local residents were dependent on forest resources, and the hamlets were accessible.

4. Opportunities and challenges in terms of local livelihood strategies and local ecological knowledge

The locals in CTBR have livelihood strategies combining farming, collection of non-timber forest products (NTPFs), logging, livestock raising, manufacture of handicraft, aquaculture, participation in forest management activities by government, and other employment. Many of them depend on upland cultivation, wetland rice cultivation, NTFP collection, and raising livestock. Differences in livelihood strategies reflected differences in the levels of poverty among three zones. For instance, the BZ had the highest rate of poor households (75% of the surveyed households in the zone) and the CZs the lowest (28%); the TZ had 40% (Sang et al., 2009). On an average, the poor households made up 58.3% of the total sample. Moreover, the mean family size was high in three zones (5.4 people per household). These people usually faced some months of food shortage and had low living standards. To meet the basic needs, they tended to exploit the forest resources and encroach on the forest land in the reserve. With regard to occupation, agriculture accounted for 47.7% of all population surveyed from ages six to sixty, and not including invalid or disabled persons; the distribution across the three zones was follows: the CZs 50.9%, BZ 48.2%, and TZ 44.6% (figure 2), respectively. Further more, the percentage of off-farm and non-farm jobs including wage and salary earnings, trade and handicraft was highest in the TZ (31.3%), but only 1.8% in the CZs and 6.4% in the BZ (figure 2), respectively. These showed that the closer to the biosphere reserve people were the less they were involved in off-farm and non-farm jobs. It is very easy to see at the chart that the ratio of off-farm and non-farm activities decreased significantly from the TZ to the CZs. Likewise, the threat on the forest resources in the TZ may be less than that in two other zones.

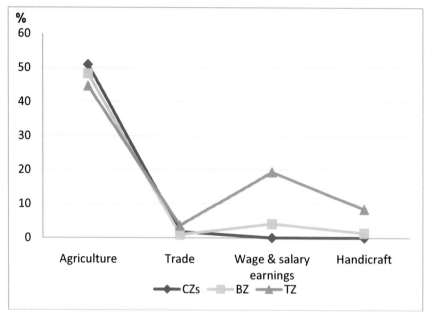

Fig. 2. Ratio of main occupation in different zones (Data appendix used for the compilation of the graph are available from the authors)

The area of cultivated land plays a very important role in the local people's livelihoods as it is the area in which productive activities take place. The bigger the land size is, the better opportunities for the people ensure food security. Table 2 gives an overview on land resources of households in the surveyed hamlets in the three zones. The average per capita area of cultivated land decreased significantly from the CZs to the TZ: CZs 3.8 ha (16 times larger than that in TZ), BZ 1.23 ha (5 times larger than that in TZ), TZ 0.24 ha (table 3). This may mean that the livelihood of the people in the CZs and the BZ depended much on land resource, but much less in the TZ.

The local people have got high income from upland cultivation; so many areas of natural forests in CTBR have been converted illegally into crop lands. Land conversion is the biggest threat to biodiversity conservation. The closer people lived to the core of the reserve, the more they depended on upland cultivation of cashew and cassava. For example, the residents in the CZs had the highest upland cultivation income (95.1%), and the TZ the lowest (11.1%); with the BZ with 60.7% falling in between (Sang et al., 2010). It indicates that upland cultivation in the CZs and BZ is the most important for the local livelihood strategy in terms of income generation. As a result of the practice of this livelihood strategy, forest land encroachment has been common. Certain areas in each zone, particularly both in the CZs and BZ have been converted illegally from forest into crop lands. For instance, in a 7 year period ratio of households involved in upland cultivation increased from 46.7% in the TZ through, 54.9% in the BZ to 88.9% in the CZs (Sang et al., 2009). Using Wilcoxon signed ranks test to compare the change of the upland cultivation area per local household during a 7 year period (1998-2005), it is evident (table 2) that the increase (129.5%) in the BZ was statistically significant. The increases in the CZs and the TZ were not significant. This indicates that the demand for cultivation land has been very high in the BZ where there were 59% of very poor households and 14% of poor households in 1998 (Sang et al., 2010). According to the field survey, the poor households in the BZ encroached 98.4% (36.4 ha) of the total area of the encroached uplands. So, it is obvious that the local poverty in the BZ is the reason for the encroachment into forest areas. To avoid getting caught, locals usually use special ways to encroach forest land. "Many households who have arable land near the boundary of the forest in CTBR have encroached very small areas of natural forest land gradually, making it difficult for local officials and reserve staff to detect the area encroached", said the officials in Phuoc Cat 2 and Bu Dang Commune. Another way is that both indigenous and migrant ethnic minorities have cleared areas deep inside the forest which the forest guards could not reach. According to Vietnam Forest Ranger (VFR), in 2010 the local people encroached 5.6 ha of natural forest in CTBR for crop upland, but in reality the lost area may be much higher. Forest conversion into agriculture land resulted in loss of habitat of many species, so conservation impact of this activity was very high. Furthermore, recently some outsiders have begun purchasing land from indigenous ethnic minorities illegally. Some local government officials believe that some outside individuals enticed indigenous ethnic people to sell their land at a low price; and the indigenous ethnic minorities then tended to give up land use rights and became wage earners. Consequently, further forest land has been encroached. In addition, a lot of indigenous ethnic minorities could not manage their livelihood strategy well; they had to sell their cashew nuts on trees for cheap prices before being ripe, and then they continued to be poor.

Zones	Z	Asym.sig. (2- Tailed)
CZ	-1.826	0.068
BZ	-5.053	0.000**
TZ	-1.604	0.109

Note: Statistic Test: Wilcoxon Signed Ranks Test (This Test is denoted by the test statistic T although the Z statistic can be reported instead, and Asym.sig. is the P value for the test)
*, ** denotes significance at 0.05 and 0.001 level

Table 2. Comparison of upland area per household in a 7 year period (Data appendix used for the compilation of the table are available from the authors, a 7 year period between 1998 and 2005)

Rice, a staple food of Vietnamese, is used mostly for consumption in CTBR. Most of the areas of paddy fields in the surveyed hamlets of the BZ were used to produce rice for two or three seasons a year, whereas in the TZ paddy was mostly produced for one season only. This was based on the irrigation system and characteristic of the land resource. For instance, in Ta Lai in the BZ there were three rice seasons per year in some areas thanks to Vam Ho Irrigation Dam. Paddy cultivation was an important component of on-farm operation in both the BZ (0.17 ha per capita) and the TZ (0.077 ha per capita), whilst the paddy rice area in the CZs was very low 0.0024 ha per capita (table 3). So the people in the CZs could not produce enough rice for their subsistence. Additionally, the areas of paddy cultivation in the surveyed hamlets in the CZs and the TZ area did not change during the 7 year period from

Land use types	CZs				BZ				TZ			
	A (ha)	RT (%)	AC (ha)	RH (%)	A (ha)	RT (%)	AC (ha)	RH (%)	A (ha)	RT (%)	AC (ha)	RH (%)
Agriculture												
Paddy land	0.2	0.07	0.0024	11.11	49.95	13.5	0.17	90.2	6.85	34.72	0.077	80
Vegetable land	0	0	0	0	0.4	0.11	0.001	1.96	0	0	0	0
Sub-total	0.2	0.07	0.0024	11.11	50.35	13.61	0.19	92.2	6.85	34.72	0.077	80
Home garden	0.19	0.06	0.002	83.3	5.47	1.48	0.02	29.4	5.03	37.91	0.057	66.7
Forestry land												
Upland	104.7	33.98	1.28	100	117.7	31.8	0.39	82.4	7	16.22	0.08	40
Protection forest (01) *	203	65.89	2.48	83.3	117	31.6	0.39	15.7	0	0	0	0
Production forest (02) **	0	0	0	0	76	20.5	0.25	25.5	0.5	2.53	0.006	6.7
Sub-total	307.7	99.87	3.75		310.7	83.97	1.03		7.5	18.75	0.085	
Other lands	0	0	0	0	3.9	1.05	0.013	49.02	1.4	8.62	0.016	26.7
Total	308	100	3.8		370	100	1.23		21	100	0.24	

Note: A: area (ha);
RT: ratio of the land type area to the total land (%);
AC: area per capita (ha);
RH: ratio of households having that land to the total surveyed households of each zone (%)
(The total households surveyed: 84);
Other lands including aquaculture and grassland;
* Forested land allocated to local households according to Decree 01/CP 1995 for forest protection, the land holders having right to gather a small amount of NTFPs on that land and to get subsidies for the forest protection incentive;
** Forestry land (with or without forest) allocated to households according to Decree 02/CP 1994 for forestry development, the land holders having rights to use the land for a long term but no finance paid for their management of that land.

Table 3. Distribution of the land holding size per capita by land use types (Data appendix used for the compilation of the table are available from the authors)

1998 to 2005; the reasons were that most areas in those hamlets in the CZs were upland and low lands for paddy in the BZ were in short supply; the same was true for the TZ (Sang et al., 2009). Conversely, in the 7 year period, the areas of paddy land of surveyed households in the BZ increased 14.2 ha (about 0.28 hectares per household); this increase was the result of the forest encroachment activity (Sang et al., 2009); the people there were very poor and short of food in some months, so they must encroach the forest land for agriculture cultivation so as to meet their basic needs. According to the field survey, all of the increase area for paddy land in this zone was encroached by most of the poor households, but not by any rich households. So, it is apparent that the high rate of poverty in the BZ was the reason for the encroached forest areas and the conversion of forest land to paddy fields. Of course, the impact on biodiversity conservation of this livelihood strategy was very high.

The wealth of each household was classified into one of the four levels: well-off households (group I), medium households (group II), poor households (group III) and very poor households (group IV) (Data appendix used for these wealth categories are available from the authors). Using Wilcoxon signed ranks test to compare the change of the total land area per surveyed household in the 7 year period, the increase of the total land area in the CZs and the BZ was significant, but not in the TZ (table 5). Besides, in 1998 the people's average gross incomes in the CZs and the BZ were below the poverty line (Sang et al., 2009). Likewise, the increase of the land area of the poor and very poor households in the CZs made up 74.8% of the total area increase, and 86.7% in the BZ (table 4). So it is clear that the low income led to the increase of the area of land use in the CZs and the BZ, and the increase of the land use area of the rich and medium households was for cash income. The increase of the total land area in the TZ was not significant (table 5).

Zone	Total land area increase of the surveyed households (ha)	Increase of the land area of the poor and very poor households Surveyed	
		Ha	% in the total area increase
CZs	214.3	160.3	74.8
BZ	224.2	194.4	86.7
TZ	5.8	5	86.2
Total	444.3	359.7	81

Table 4. Increase of the total land area of the poor and very poor households in the 7 year period (Data appendix used for the compilation of the table are available from the authors)

Zones	Z	Asym.sig (2- Tailed)
CZs	-3.628	0.000**
BZ	-5.152	0.000**
TZ	-2.366	0.18

Note: Statistic Test: Wilcoxon Signed Ranks Test
  *,** denotes significance at 0.05 and 0.001 level

Table 5. Comparison of the total land area per household in the 7 year period (Data appendix used for the compilation of the table are available from the authors)

With the development trend of cashew plantations in south-east provinces in Vietnam, since 1993 most of the households in Cat Loc including the surveyed hamlets in CZs and BZ have encroached on the forest land to plant cashew because the soil in this area is very good for the crop. The people in South Cat Tien comprising the hamlets in BZ and TZ cultivated rice on low land and cashew on hills. The people got high cash income from cashew and used rice mainly for self consumption.

Zone	Total encroached area (ha)	Area per household (ha)	Area per capita (ha)
CZs	11.42	0.63	0.14
BZ	51.2	1	0.17
TZ	5.31	0.34	0.06
Total	67.68	0.8	0.14

Table 6. Encroached forest areas by the local people in the 7 year period (Data appendix used for the compilation of the table are available from the authors)

In 1998 about 56.07% of the sample population in three zones was poor (Sang et al., 2009). Hence, many of them encroached on the forest land to cultivate crops. According to table 6, the total encroached forest area of sample households in three zones was 67.68 ha (about 0.8 ha per surveyed household). The people in the BZ had the largest encroached area per household (1 ha), the residents in the TZ the smallest (0.34 ha per household) (table 6). The ones in the CZs had 0.63 ha. The people in the BZ encroached on much forest land since the poor people made up 73% of the sample people in the hamlets in 1998 (Sang et al., 2009).

Zones	Z	Asym.sig (2- Tailed)
CZs	-1.483	0.138
BZ	-4.760	0.000**
TZ	-2.366	0.018*

Note: Statistic Test: Wilcoxon Signed Ranks Test for means of two related samples
 *,** denotes significance at 0.05 and 0.001 level
 The increase of cultivated areas came from encroached land.

Table 7. Comparison of cultivated land area per household in the 7 year period (Data appendix used for the compilation of the table are available from the authors)

The Wilcoxon signed ranks test was used to check the significance of the change of cultivated areas of households in each surveyed site among the three zones. The result in table 7 shows that the change was significant in the surveyed sites of the BZ and the TZ. According to the field survey, the sampled poor households in the BZ encroached 99.4% (50.9 ha) of the total encroached area, and 94.2% (5ha) in the TZ. Only a few well-off and medium households encroached on the forest land for cash income. This means that the low income in these two zones caused the encroachment. On the other hand, the locals in the CZs encroached large cultivated land areas from 1990 to 1998 (5.2 ha per surveyed household) (Sang et al., 2009); that is the reason why they were very poor in 1998 when their cashew was at young ages, but they have encroached on a small cultivated area since 1998. To sum up, the low income of the sample people in three zones of CTBR is the reason for the encroached forest areas.

Regarding NTFP usage, most of the locals, especially the EMs have harvested these resources for both subsistence and cash income (Sang et al., 2009). Many species are commonly exploited for a variety of uses such as fuelwood, foodstuffs, materials for handicraft and construction, agricultural tools, fodder, medicine and resin products. Nearly 100% of EM households collected the fuelwood which mainly came from natural forest; the demand for fuelwood for subsistence was very high; only one family of indigenous ethnic minorities had gas for cooking. About 58.8% of surveyed households of Kinh people gathered fuel wood mostly for their self consumption, only one of them for income generation (Sang, 2010). The demand for fuel wood reduced from the CZs to the TZ (Sang, 2006). The average amount of fuelwood for subsistence in the area was about 22 kg per household per day (Sang, 2006). Many Kinh households in BZ and TZ used gas for cooking because they were busy with on-farm, off-farm, and non-farm jobs. All of them collected branches or stems for firewood from natural forests, forest plantations and allocated forest land in CTBR, but primarily from natural forests. All EMs indicated to gather the firewood and carry it in a dosser (Gui) on one's back, but Kinh people used ox carts or even bicycles to carry fire wood. Each dosser of fuelwood weighs about 15-20 kg. From these results, it is clear that the EMs have depended much on fuelwood than Kinh people. The conservation impact of this utilization is likely to be moderate.

All of the EMs and more than 35% of Kinh families surveyed gathered and harvested some or many species of the edible forest plants for both subsistence and income generation, primarily for daily consumption. The edible forest plants were used as vegetable, fruit, traditional medicine, yeast, spice, water, agar, and fodder. The species were collected from natural forests, forest plantations and allocated forest land (table 3) in CTBR, but primarily from natural forests. Plants used as vegetables were the most important food for all of indigenous ethnic minorities. Particularly, leaves of *Gnetum gnemon* and shoots of some rattan species were used extensively as daily food by indigenous ethnic minorities; but only a small percentage of migrant minorities exploited these NTFPs and they seldom used these species for food (Sang & Ogata, 2011). Moreover, these species are also among 15 favorite main foods for Javan Rhinos in CTBR. With regard to forest plants used to make pickle, 5 out of 9 species are bamboos whose shoots were used for that product (Sang et al., 2011). The material for pickle also came from fruits of 2 wood species and petioles of 2 species belonging to Araceae. That product is very good for long term use, especially during the dry season. Edible forest fruits are commonly used in the area, they collected those from species in the natural forest of CTBR for subsistence, and half of them are wood species. Many households reported that their children were the primary collectors. Especially, one important species for high cash income is *Scaphium macropodium* whose nuts are collected on trees by males or on the forest floor; the harvest season stretches from March to May and it only flowers at an interval of 3 or 4 years. "Last season (2010) I harvested nuts of this species and earned about 17 million VND (USD 1 = 20,700 VND, approximate exchange rate in August of 2011), this income is really high for us", said a Chau Ma man in Brun Hamlet of Lam Dong Province. However, because the trees are high the people often cut down them to harvest nuts for convenience. And more than 10 respondents cut down trees for collecting the products. The conservation impact of this way of exploitation is likely to be high, so it should be strictly prohibited and changed to more sustainable ways. Additionally, the indigenous ethnic minorities in the area used the forest edibles as spice or for their water content. The majority of species used for spice were herbs (5 species); but each of the following life forms had only one species: tree, climber and shrub (Sang et al., 2011). They also drank water from 2 climber species in case of lacking in boiled water during the time in jungles. Noticeably, bark of *Cinnamomum iners* or leaf of *zingiber sp.* is the key material for yeast to make an indigenous traditional alcohol called "Ruou Can". These species have been retained in their traditional drink which appeared in their traditional festivals, parties of the lunar new years, weddings, or other events. Remarkably, *Canarium tramdenum* is a wood species providing fruit found in the Red Data Book of Vietnam (2007). This is vulnerable, possibly threatened to be extinct. *Mangifera dongnaiensis* that also provides fruits is an endemic flora species in CTBR (CTBR, 2010). However, poor harvesting practices and high intensity of collection were threatening their sustainability, the local uses and even the food sources for wildlife. Additionally, most of the gathering was officially illegal since it occurred in national protected forests.

Using some of the NTFPs for handicraft and other home appliances also brings both opportunities and challenges to the biodiversity conservation in CTBR. Particularly rattans and bamboos were very important for making handicraft for self consumption and cash income of the locals. Additionally, a majority of families who got income from the activity were poor or very poor, or had idle labors. About 9% of the total surveyed Kinh households in the TZ depended much on bamboo baskets and joss-sticks which brought daily main income to their family, 17.6% of interviewed households of EMs earned from indigenous indigo textile fabric and other kinds of handicraft (Sang & Ogata, 2010; Sang, 2006). This activity had a low conservation impact, the people needed only a certain amount of material from forest and they could add much more value to their final products thanks to their skill and time intensive at home. Regarding NTFPs for other home use products, there was a high subsistence demand in the area. 100% of the surveyed households used the forest resources for self consumption. They used many species of rattan and bamboo for weaving home appliances such as baskets, trays, boxes, dossers (Gui), fishing traps, and so on. Considerably, the most useful and traditional tool of the indigenous ethnic minorities is a dosser made from rattan species such as *Calamus dioicus, Calamus poilanei, Calamus sp, Calamus tetradactylus, Deamonorops pierreanus, Korthalsia laciniosa,* and *Plectocomiopsis geminiflorus*. 100% of indigenous respondents had at least one Gui in their families, but migrant ones and Kinh people did not use them. Canes of *Calamus poilanei, Calamus rhabdocladus, Deamonorops pierreanus,* and *Korthalsia laciniosa* have been used for furniture frames (Sang & Ogata, 2011). Particularly, mature stem logs of some bamboo species were used to cook their traditional bamboo-tube rice (Com Lam) and used as tubes to imbibe traditional "tube wine" (Ruou Can) from a jar. Moreover, many NTFPs were gathered for wrapping foods, and for making thatch roofs, bamboo walls, brooms, ladders, joss-sticks, chopsticks, fishing-rods, cages, ethnic musical instruments, looms, conduits, bamboo strings, hunting arrows, other tools.

Timber was a valuable forest product for which there was both high self consumption and high income generation in CTBR, especially those in the BZ and CZs. However, the conservation impact of timber logging is high because whole stems of trees were cut. The dry season is the favourite time for them to harvest timber as it is easier to harvest and transport timber out of the forest. Wood species logged are as follows: *Afzelia xylocarpa* (Go do), *Dipterocarpus intricatus* (Dau long), *Dipterocarpus alatus* (Dau rai)), *Dalbergia mammosa* (Cam lai), *Lagerstroemia calyculata* (Bang lang), *Sindora siamensis* (Go mat), *Tectona grandis* (Go tech) and so on. According to the field survey in 2010, 18.2 % of the sampled households involved in the illegal logging activity for cash income, while the figure was high in 1998 - around 27.4% (Sang, 2006, 2010). Although the ratio has been decreased, the activity is strictly prohibited in the CZs and natural forests in the other zones. One example, in 2010, 4 indigenous ethnic ones in the survey hamlet in Ta Lai and 3 outsiders who logged 12.039 m^3 of *Afzelia xylocarpa* were put in jail (CTBR, 2011). Recently, it has occurred that outsiders hired indigenous ethnic minorities for carrying or harvesting timber "illegally" because they had their local knowledge (Sang, 2010). This trend was common in the surveyed hamlets in Da Nhar (BZ) and Da Oai (TZ), but not in the surveyed areas of the CZs and Brun Hamlet (BZ) because the topography was full of obstacles and difficult of access. To carry timber out of the reserve; motorcycles, buffaloes, and means of river transportation were used. Their transportation income from forest to outside was reported about 150,000 – 300,000 VND per time (Field survey, 2010). And many of them were hired 2 times per day. Concerning timber for local subsistence demand, 100% of the surveyed households in the CZs used both converted timber and wood poles for the construction of their houses, pigsties, hen-coops, and stalls. For the indigenous ethnic minorities in the BZ, before the program of settlement and fixed cultivation the construction of their houses was also made of the above material; then thanks to the program the government built concrete houses for them.

Wildlife poaching has been carried out by both Kinh people and the EMs for self consumption, high cash income, and occasional barter. However, this activity makes a severe conflict between local residents and forest rangers in CTBR in term of biodiversity conservation. The percentage of local households involved in wildlife poaching reduced from the CZs to the TZ: the CZs 50%, the BZ 52.9%, and the TZ 20%, respectively (Sang et al., 2010). After 7 years, the ratio became lower with every passing day; its reduction was as follows: the CZs 5.6%, the BZ 13.8%, and the TZ 13.3%. Likewise, the annual income from wildlife poaching per capita decreased at the following rates: the CZs 8.1%, the BZ 10.4%, and the TZ 11.3% (Sang et al., 2010). According to our own survey in 2010, the reduction in income and involvement rate of wildlife poaching was mainly caused by further distance or longer time for the activity and more severe punishment. However, forest guards often turn a blind eye when seeing the locals fishing with traditional ways in the reserve. Recently, many households in the CZ in West Cat Tien and some in other parts of the reserve have used electric shock fishing that has killed not only fish but also all aquatic life in wetlands, rivers, and streams of CTBR. It can endanger one of the food chains of the ecosystem of the reserve.

In respect of purposes of forest resource uses, 80.6% of the surveyed households used the biological forest products for their subsistence; whereas 96.4% of them sold the products from upland cultivation to the market. Most of the products from the cultivation were cashew which gave high income to the local people; only 3.6% of households used the products from the upland cultivation for self consumption because some areas were cultivated to grow maize, green bean, cassava (table 8). They usually consumed most amounts of products from fishing, NTFP collection, hunting for daily foodstuffs; 17.5% of them sold products from the biological forest resources and 27.6% of households used timber harvested for cash income. In 2005, 60.8% of the total sampled households used the products from the biological forest resources and uphill cultivation for subsistence, 37% for cash income, and only 2.2% for storage (table 8).

Zones	Upland cultivation			Logging			NTFPs			Hunting & fishing			Forest resource uses		
	Subsistence	Cash income	Storage	Subsistence	Cash income	Storage	Subsistence	Cash income	Storage	Subsistence	Cash income	Storage	Subsistence	Cash income	Storage
CZs	0	100	0	50	50	0	90	5	5	88.9	0	11.1	57.2	38.8	4
BZ	5.9	94.1	0	84.3	15.7	0	76.5	19.6	3.9	92.2	7.8	0	64.7	34.4	0.9
TZ	0	100	0	50	50	0	62.5	37.5	0	100	0	0	53.1	46.9	0
Mean	3.6	96.4	0	72.6	27.6	0	76.2	20.2	3.6	92.9	4.8	2.3	60.8	37	2.2

Table 8. Purposes of forest resource uses (in %) (Data appendix used for the compilation of the table are available from the authors)

As for the forest protection program, the government has implemented this for poor households in CTBR since 2001. In local reality, a group of households in a hamlet took responsibility, on a rotational basis, for patrolling a large forest area; the group was managed by a head of a hamlet or a village, or a staff of local government who was responsible for signing of forest protection contract, patrol plans and benefit sharing. Under this program, the participants have got subsidies for the forest protection incentive (table 3). The annual payment was 50,000 VND per hectare, since 2010 the amount has been 100,000 VND. However, to compare the income with the total forest resource income per capita, the rate was very small: the CZs 0.97%, the BZ 0.49, the TZ 5.82% (Sang et al., 2010). This could not encourage the locals to participate in the program. For example, in 2009, the household group in the surveyed hamlet (CZ) in Bu Dang withdrew from the forest protection program.

Another program initiated by CTBR was ecotourism in South Cat Tien in 1995. Since then only a few indigenous ethnic minorities in Ta Lai (BZ) have had a chance to earn their living as guides or handicraft sellers. All of ecotourism guides were Chau Ma minorities in Ta Lai and their participation was at both active and passive levels. Handicraft products were sold at Ta Lai Cultural Centre for the Indigenous Ethnic Minorities by a S'Tieng staff and at the Headquarters of CTBR by a Chau Ma woman. Handicraft materials came from rattan, bamboo and some other NTFPs; thanks to the low amount of materials harvested, the conservation impact of harvesting was low. Before these people did not have any jobs and all of them ventured into CTBR illegally in search of wood, NTFPs and even encroached forest land for cultivation. Recently they have had alternative sources of income from ecotourism activities for their basic needs in the daily lives. As a result, they have gradually changed from high intensive of forest resource use to a more sustainable way. Since 2009, some further activities such as the programs of long houses and gong performances were implemented (gong is a traditional musical percussion instrument of the indigenous ethnic minorities in the Central Highlands of Vietnam which was recognized by UNESCO as a Masterpiece of the Intangible Heritage of Humanity on November 25th 2005). However, only 12.5% of indigenous participants had got dominant sources of incomes from guides or handicraft sellers (Sang & Ogata, 2010). In reference to local participants' satisfaction about the income derived from ecotourism, only 2.5% of them rated it as satisfactory. On the whole, the indigenous ethnic minorities in CTBR were employed in ecotourism activities on a small scale rather and not at the decision-making and management level (Sang & Ogata, 2010). And most of their involvement in ecotourism in CTBR was at the passive participation level (Sang & Ogata, 2010). Passive participation meant an elementary level participation in tourism development and its object was as much to prevent a too hurried residents' intervention into tourism development by providing residents with longer term sustainable participation opportunities (Tosun, 2000). Moreover, the distribution of the ecotourism program was in unbalanced among the various sectors of CTBR.

Change of forest resource income (FRI) in the 7 year period was dramatically in the CZ and the TZ, but slightly in the BZ (Sang et al., 2010). The rate of FRI to the total income per capita of surveyed households in the CZ increased significantly, from 66.8% to 97.2%. This rate in the BZ rose slightly, from 57.5% to 64.2%. In comparison, the rate decreased rapidly in the sample hamlet in the TZ, from 45.1% to 19.1% (Sang et al., 2010). Furthermore, the proportion of the annual increase of FRI to the increase of total annual income per capita in the CZ was very high (105.5%); this rate in the sample hamlets of the BZ was high (68.4%); but the sample one in the TZ had only 4.8% (Sang et al., 2010). This indicates that the income of the residents in the CZ and the BZ increased significantly thanks to the forest resources, but the contribution of forest resources to the rise of total income in the TZ was very small. According to table 9, the increase of FRI per capita was significant in the sampled hamlets in the CZ and the BZ, but not in the hamlet of the TZ.

Zones	Pairs	T	Df	Sig. (2-tailed)
CZ	Average gross FRI per capita in the 7 year period	-5.212	17	0.000**
BZ	Average gross FRI per capita in the 7 year period	-2.031	50	0.048*
TZ	Average gross FRI per capita in the 7 year period	-0.103	14	0.898

Note: *, ** denote significance at level of 0.05 and 0.001
Method: Paired Sample T Test for the income in 1998 and 2005.

Table 9. Comparison of average gross FRI per capita in the period of 7 years (Data appendix used for the compilation of the table are available from the authors)

The locals in CTBR have accumulated their LEK of forest resource use which includes both potentials and challenges for biodiversity conservation. From generations to generations, they have not only depended on forest resources but also summed up the experience in upland cultivation and NTFP harvest. With regard to the LEK of upland cultivation, they have long experience in land distinction and management of its fertility (table 10). The people could not distinguish different land types by color or texture, but they were able to identify fertile land from ground objects. In fact, to have good upland for cultivation, they selected a cultivated area on a less steep hill without stone surface, bamboo and species of dipterocarpaceae; their experience showed that the presence of these plant species may be a reliable indicator of poor land to cultivate; they also preferred to choose the arable land deriving from forest of plant diversity, especially near water sources. In addition, upland cultivation has been the traditional activity of the local residents, especially ethnic minorities. It was the result of the shifting cultivation in the past time. Some years ago, with unsustainable forest management and unawareness of forest protection, shifting cultivation systems were popular within and around the reserve. According to local elders, the fallow period of upland fields was up to 20 years, and the fallow period came after 1 or 2 years of cultivation. The period had reduced to 10 years, in which upland cultivation was conducted in 2-3 years, and then a fallow period lasted only 3-5 years. At present, due to the strict management of the reserve, it is more difficult to encroach on the forest land, so there are no fallow periods in upland cultivation. This leads to soil erosion and land degradation, which makes soil unable for cultivation.

Kind of LEK	Effects	Trend of change	Applicability for local residents, CTBR and local government
Land distinction for upland farming	Selecting forest upland for farming, then asserting land ownership and tenure in the community, but no official land titles	Unofficial ownership, so decrease in area	Participatory planning for upland use
Crop diversification	Multi crop products, plant diversity, higher output	Genetic degradation of native crop species	Conservation of native crop species
Soil fertility management	Periods for recovering fertility	Traditional upland farming with fallows, recently industrial crops and no more fallows	Recycling and composting organic materials, manure from livestock

Table 10. LEK of upland farming (Source: Sang, 2006, 2010; Sang & Diep, 2007)

As for the LEK of NTFP harvest, the locals can know when, where and how to harvest the NTFPs such as rattan species, nuts of *Scaphium macroporium* and resin from some wood species belonging to *Dipterocarpaceae* (Sang & Ogata, 2011; Sang & Diep, 2007). As for rattan use, they can know the mature age of rattan through its color of leaves or canes; for example, the rattan can be harvested when its stem becomes yellow, gray to almost white and relatively smooth with scars left by fallen leaves (table 11); after harvesting, rattan stumps need more than five years to provide the stem products; a near total of households (98.8%) identified that most rattan species were overharvested nearby and they had to go further in order to harvest young shoots or stems. Many species harvested for stem products in CTBR included *Calamus tetradactylus, Calamus scipionum, Calamus poilanei, Calamus rhabdocladus, Calamus dioicus, Calamus sp., Deamonorops pierreanus, Korthalsia laciniosa, Plectocomiopsis geminiflorus*. Regarding *Scaphium macroporium*, the residents have summed up the knowledge that this species produces fruits irregularly one in every three to four years

NTFPs	Signals of ripeness or mature	Places to harvest	Time	Harvesting tools	Use
Seed of *Scaphium macroporiu-m*	Nuts falling on ground, changing from green to light red	Evergreen rain forest	March to May, producing nuts only every 3-4 years	Collecting seed on forest floor or on trees, cutting down trees and harvesting	After being soaked and the seed kernel removed, the flesh is used with granulate-d sugar
Cane of rattan species	Stem becoming yellow, gray to almost white; relatively smooth with scars left by fallen leaves	Relatively high land with relative high canopy, along streams	Mainly dry season	Long knife, sickle	Handicrat, home appliance, ropes to combine parts in construct-on
Resin of some Dipterocar-p species	Tree trunk greater than 20 cm in diameter	Steep slope in deciduous forest with species of Dipterocar-p species	Mainly dry season	Collecting solid resin on forest floor or directly on trees by using a pole; liquid resin on trees by cutting a large notch on trunks near the ground, then burning, liquid oozing out	Used for making boats, baskets, ceilings, and bamboo walls watertigh; handicraf, lighting

Table 11. LEK of NTFP harvesting (Source: Sang, 2006, 2010)

(table 11); before the harvesting season, they marked on the trees to assert informal ownership in the community, then built a hut near the trees for laborers to stay many days there, take care and harvest nuts; they usually used all of their household labor, even children to gather this NTFP, during the harvesting season many children of the indigenous ethnic minorities played truant in order to participate in the activity for high cash income; this livelihood strategy is not sustainable, young generation continued to be forest resource dependents. On the subject of resin harvesting, resin of *Anisoptera costata*, *Shorea obtuse*, *Dipterocarpus alatus* and some others in CTBR used to be harvested a lot for high cash income thanks to high market demand, for example, a litter of *Dipterocarpus alatus* cost about

30,000 VND; but recently only some surveyed households have involved in this activity because the resource was overexploited some years ago. The people from the different ethnic groups had a lot of experience in harvesting the product (table 11); the species producing resin were abundant on steep slope with rocky soil; tree trunks which had many branches and were attacked by insects gave much resin; solid resin harvested on trees had better quality than that gathered on the ground. In addition, liquid resin was extracted from *Dipterocarpus alatus* by cutting a large notch on trunks near the ground, then firing, waiting for ooze of liquid, and collecting the liquid resin; this not only destroyed the species but also caused forest fire which damaged the habitat of wildlife and the ecosystem; so the conservation impact caused by these activities was high; sustainable extraction methods should be applied.

In the local reality, they must encroach upon forest lands for cultivation; poach wildlife; collect NTFPs for self consumption and income generation; and some of them must join illegal logging activities. Obviously, they have a variety of livelihood strategies, but it is very limited for them to select more sustainable ways.

5. Discussion and conclusions

It appears from the research that CTBR is naturally endowed with a great variety of forest resources on which the locals, especially the EMs in the CZs and the BZ depend (table 4 & 5). As mentioned above, the resources and their LEK have been contributing very much to the local livelihoods in the CZs and BZ of the reserve, especially richer households; the income of the residents in these two zones increased significantly thanks to the forest resources (table 9), but the contribution of forest resources to the rise of total income in the TZ was very small because many of them involved in wage and salary earnings, trade and handicraft (figure 2) (Sang et al., 2010). So it was said that low rate of off-farm and non-farm jobs in the area led to the high rate of the local people involved in forest resource use activities. Likewise some of the households who participated in the programs of forest protection and ecotourism initiated by CTBR with very low income withdrew from them; of course, they continued to involve in forest resource use illegally (Sang & Ogata, 2010; Sang, 2010). It was concluded that low income of many households in CTBR was the reason for the encroached forest land areas (table 6 & 7). However, a large majority of families in the CZs and the BZ used the forest resources at a high intensive level for their basic needs as well as income generation. In short, the locals depended heavily on the forest resources because the income from the resources was much higher than the other incomes (table 9). Their dependency on the resources causes a lot of negative impacts on biodiversity conservation of CTBR through overexploitation of the resources as well as encroachment of forest land (Sang, 2006). To sum up, poverty, occupation structure, high values of the forest resources, and negative side effects of LEK brought negative impacts to biodiversity conservation in CTBR.

To overcome these situations, the following recommendations are made. Firstly, higher yielding crops with appropriate techniques including agro-forestry models should be introduced and cultivated in areas outside the CZs as well as natural forest land. In addition, fast growing species should be introduced on the bare lands for securing firewood and increasing green cover. In particular, the EMs should be encouraged to do aquaculture and raise more poultry so that they may fulfill their subsistence foodstuffs and even earn more income. Although ecotourism creates jobs for the indigenous EMs it also brings disturbance to wildlife, so it should be carried out outside the CZs; their participation in ecotourism and forest protection should be preferred and benefits from the activities should be increased. The approach of their participation should aim at involving them not only in benefit sharing from the projects but also in the process of sustainable conservation management. Irrigation systems should be developed to enable people to grow rice for three seasons per year so that their food shortage can be lightened. The poor people in the area should have more access to credit at reasonable interest rates and over long term, enabling them to use their land effectively. Secondly, co-management arrangement (Polet, 2003) among the management board of CTBR, local government, and local people should be put into practice and effect so as to aim at sustainable uses and management of the forest resources. Thirdly, the local government and the reserve should develop the potential of ethnic minorities' indigenous knowledge in craft products, and promote home industry through craft activities. This creates long term employment for local workforce, especially for the residents in the CZs and the BZ; vocational training for the poor and unemployed people, especially in the CZs and the BZ would be also introduced.

As mentioned above, LEK contains both positive and negative sides for biodiversity conservation in CTBR. To contribute to both biodiversity conservation and development, the positive points of LEK should be applied widely and integrated with scientific knowledge. The people who are good at LEK should be encouraged to participate in the programs of biodiversity conservation, environmental education, and ecotourism. Further research to discover and understand more LEK in the area for enhancing the adaptive capacity of sustainable use is needed.

6. Apendix

Fig. A1. A captured monkey in the surveyed hamlet in the CZ in Cat Loc Sector
Source: Photo by Dinh Thanh Sang, field survey, 2010

Fig. A2. A young woman carrying bamboo poles by bicycle
Source: Photo by Dinh Thanh Sang, field survey, 2010

Fig. A3. Young women carrying fuel wood from the forest
Source: Photo by Dinh Thanh Sang; Sang, 2006

Fig. A4. Timber and a car of illegal loggers confiscated by forest guards in TZ, Lam Dong Province
Source: Photo by Dinh Thanh Sang; Sang, 2006

Fig. A5. A deforested upland with some young cashew trees in the CZ in West Cat Tien
Source: Photo by Dinh Thanh Sang, field survey, 2010

Fig. A6. A surveyed hamlet with cashew plantations in the CZ in Cat Loc Sector
Source: Photo by Dinh Thanh Sang, field survey, 2010

Fig. A7. Rice field in Ta Lai Commune in the BZ, a former area of the CZ
Source: Photo by Dinh Thanh Sang; Sang, 2006

7. Acknowledgement

This work was supported in part by the Global COE Program (Center of Excellence for Asian Conservation Ecology as a basis of human-nature mutualism), MEXT, Japan; the authors are thankful for the supports. We also thank the local people, the officials in the study area for their contribution.

8. References

Brown, K. (2002). Innovations for conservation and development. *Geographical Journal*, 168 (1), 6-17.

Cat Tien Biosphere Reserve (CTBR) (2011). Report on the illegal loggers brought to trial, January 2011, Ta Lai Cultural Centre for the Indigenous Ethnic Minorities, Vietnam.

Cavendish, W. (2000). Empirical regularities in the poverty-environment relationship of rural households: evidence from Zimbabwe. *World Development*, 28 (11), 1979-2003.

Cox, R., Cools, J. W. F. & Ebregt, A. (1995). Cat Tien National Park Conservation Project. Hanoi: Forest Inventory and Planning Institute and WWF.

CTBR (2006). Report on the biodiversity in CTBR in 2006, Dong Nai, Vietnam.

CTBR (2010). Annual report in 2010, Dong Nai, Vietnam.

Dinh Thanh Sang & Dinh Quang Diep (2007). Chau Ma minority people's indigenous knowledge of forest resource use in Cat Tien National Park. Journal of Agricultural Sciences and Technology of Nong Lam University, 3, 113-117. Available from http://srmo.hcmuaf.edu.vn/data/file/tap%20chi/2007/so%203/LN-DTSANG%20da%20sua.pdf

Dinh Thanh Sang & Ogata, K. (2010). Participation of ethnic minorities in ecotourism: Case Study of Cat Tien Biosphere Reserve, Vietnam. *International Workshop and German Alumni Summer School on "Biodiversity Management and Tourism Development"*, Lombok, Indonesia, December 2010.

Dinh Thanh Sang & Ogata, K. (2011). Ethnic minorities' use of rattan species in Cat Tien Biosphere Reserve: Prospects and Constraints for Conservation, Sustainable Use and Management. *International Workshop and German Alumni Summer School on "Multidisciplinary Approach for Biodiversity Conservation and Management in the Face of Globalization"*, Subic & Luzon, Philippines, March 2011.

Dinh Thanh Sang (2006). *Interactions between local people and protected areas; a case study of Cat Tien Biosphere Reserve, Vietnam*. Master thesis, Technical University of Dresden, Germany.

Dinh Thanh Sang (2010). Report on the field surveys of doctor's course in CTBR in 2010; Binh Phuoc, Lam Dong and Dong Nai Provinces, Vietnam.

Dinh Thanh Sang, Ogata, K., & Yabe, M. (2010). Contribution of Forest Resources to Local People's Income: A Case Study in Cat Tien Biosphere Reserve, Vietnam. *Journal of Agricultural Sciences of Kyushu University*, 55 (2), 397-402, October 2010, ISSN: 0023-6152. Available from
https://qir.kyushu-u.ac.jp/dspace/bitstream/2324/18857/1/p397.pdf

Dinh Thanh Sang, Pretzsch, J., & Ogata, K. (2009). Poverty of the Local People and their Land Use Structure Change in Cat Tien Biosphere Reserve. *International Workshop on "Further Training and Technology Transfer for Traditional and Innovative Forest Uses as well as their Genesis; Modelling of Land Use Systems for Learning and Extension"*, Tam Dao National Park, Vietnam, September 2009.

Dinh Thanh Sang, Ogata, K. & Mizoue, N. (2011). Use of edible forest plants among indigenous ethnic minorities in Cat Tien Biosphere Reserve, Vietnam. *Asian Journal of Biodiversity* 3 (1), 2011, ISSN: 2094-1519. Available from http://ejournals.ph/index.php?journal=AJB&page=article&op=view&path%5B%5D=3817

Nguyen Hung (2006). *A lot of areas in Cat Tien Biosphere Reserve have been destroyed*. P1, Tuoi Tre Newspaper. 29th June, 2006. Ho Chi Minh, Vietnam.

Olsson, P. & Folke,C. (2001). Local ecological knowledge and institutional dynamics for ecosystem management: a study of Lake Racken Watershed, Sweden. *Ecosystems,* 4 (2), 85-104.

Ostrom, E. (1990). Governing the commons: the evolution of institutions for collective action. Cambridge University Press, Cambridge, UK.

Ostrom, E. (2005). Understanding institutional diversity. Princeton University Press, Princeton, New Jersey, USA.

Polet., G. (2003). Co-management in Cat Tien National Park. In: *Co-management in protected areas in Asia: a comparative perspective*, Persoon., G., Diny M. E. van Est, & Percy E. Sajise. Nordic Institute of Asian Stuties, ISBN 87-91114-13-6, Denmark.

Prance, G. T., W. Balee, Boom, B. M., & Carneiro.R. L. (1987). Quantitative Ethnobotany and the Case for Conservation in Amazonia. *Conservation Biology*, 1, 296-310.

Sills, O., Erin, Abt, Karen Lee (2003). *Forests in a market economy*, 260-281. *Kluwer Academic Publishers*, The Netherlands.

Swinkels, R. & Turk, C. (2006*). Explaining Ethnic Minority Poverty in Viet Nam: A Summary of Recent Trends and Current Challenges.* Investment meeting on Ethnic Minority Poverty, Ministry of Planning, Ha Noi.

Ticktin, T. (2004). The ecological implications of harvesting non-timber forest products. *Journal of Applied Ecology*, 41, 11-21.

Tosun, C., (2006). Expected nature of community participation in tourism development. *Journal of Tourism Management* 27 (3), 493-504.

Vietnam News (2004). *Park rangers battle wild game poachers*, July 29, P1. Hanoi, Vietnam.

Well, M. & Brandon, K. (1992). Biodiversity conservation, affluence and poverty: mismatched costs and benefits and efforts to remedy them. *Ambio*, 21, 237-43.

14

UNESCO Biosphere Reserves Towards Common Intellectual Ground

Richard C. Mitchell, Bradley May, Samantha Purdy and Crystal Vella
Brock University, Brock Environmental Sustainability Research Unit (BESRU)
Canada

1. Introduction

This chapter emerges from ongoing research and teaching at one Canadian university, and provides a review of theoretical and empirical literature, statutory frameworks and policy documents defining the United Nations Educational, Scientific and Cultural Organization's (herein UNESCO's) three interconnected, interdependent dimensions for evaluating *sustainability science*. These dimensions include conservation of biodiversity, sustainable economic and human development, and logistic support (or capacity building) for research, monitoring, education and training. As Francis (2004, p. 21) astutely observes sustainable development…

> …implies the existence of the appropriate knowledge and governance capacity to maintain economic vitality with social inclusiveness in opportunities and benefits, provide for ecological sustainability and the protection of biodiversity to guide the use of resources, and promote social equity within and across groups and generations. All three are necessary and no one of them alone is sufficient. These requirements must also hold across a range of spatial and temporal scales.

The chapter also builds upon findings from a preliminary study (Mitchell, 2011a) with a new aim of establishing common theoretical ground for cross-scale adaptation of this research by stakeholders within UNESCO's current constellation of 564 Biosphere Reserves. The authors draw upon a transdisciplinary selection of resources including a synopsis from a two-day experts' workshop with Swedish and Canadian collaborators that analyzed socio-ecological inventories (SEIs) in two biosphere reserves (Schultz, Folke, and Olsson, 2007; Armitage and Plummer, 2010; Gafarova, May and Plummer, 2010; Mitchell, 2011b; Velaniškis, 2010), and pedagogical resources from a senior undergraduate assignment. These resources have been organized utilizing principles from international frameworks that underpin the UN Decade of Education for Sustainable Development (see also Lundholm and Plummer, 2010; Krasney, Lundholm and Plummer, 2010; Plummer, 2010; Schultz and Lundholm, 2010), and UNESCO's Man and the Biosphere Program (UNESCO, 2010).

The following sections move to an overview of contemporary thinking within sustainability research alongside teaching and service initiatives being taken up at this mid-sized southern Ontario institution. Beginning with theoretical and institutional developments that supported the emergence of a new research program founded on the same evaluative dimensions in all biosphere reserves, it includes findings and recommendations from a first-ever carbon footprint measure (Mitchell and Parmar, 2010). In addition, recommendations from a senior undergraduate report looking at how the campus might further operationalize UNESCO's framework to enhance its strategic planning and main academic functions are included. The sections are organized under the following themes to include a theoretical framework, an overview of social-ecological inventories (or SEIs), international frameworks currently being utilized to understand and evaluate resilience within complex adaptive ecosystems, and some additional resources emerging from the research program of the Brock Environmental Sustainability Research Unit (BESRU) aimed at cross-scale adaptation.

2. Theorising sustainability through transdisciplinarity

Building on Taylor's (2004) arguments for theorizing resilience and building capacity through biosphere reserve research, a common framework is presented – both a conceptual and political approach that offers the broadest possible understanding of UNESCO's dimension of sustainable human and socio-economic development. Theorising the "logistics function" from four Canadian biosphere reserves was also the focus of Whitelaw, Craig, Jamieson and Hamel (2004, p. 65) through a "'place based' framework" for ongoing research and periodic evaluations (see also Pollock's [2004] similarly conceived framework in the same volume).

It is clear that comprehensive, multi-systemic and inclusive approaches that reach beyond traditional hierarchical university structures are evolving to include the development of permanent partnerships with community-based stakeholders, Indigenous communities, governmental and non-governmental organizations, student and union leaders, faculty, administrative and alumni champions. This conceptual and methodological approach is increasingly being mirrored in academic literature of the social sciences, humanities, healthcare and scientific journals as "transdisciplinarity" (Holmes and Gastaldo, 2004; Giroux and Searls Giroux, 2004; Koizumi, 2001; Mitchell, 2010; Moore and Mitchell, 2008; Nicolescu, 2002; Robinson, 2008).

In their comprehensive literature review, Choi and Pak (2006, p. 351) note that "[t]he terms multidisciplinary, interdisciplinary and transdisciplinary are increasingly used in the literature, but are ambiguously defined and interchangeably used". Moreover, Mitchell (2011a, pp. 9-10) observes "[s]ustainability is not a contemporary phenomenon, but the solutions undertaken on the most successful Canadian campuses have taken an innovative turn (see also Beringer, 2006). Sustainable ways of life have been embedded within and across many cultures throughout history especially through North American Indigenous ways of knowing (Malott, 2008; Ralston Saul, 2008)". Human societies around the globe have organized their worldviews based upon rich experiences relating to the environments wherein they find themselves. Recently, these other knowledge systems are being referred to as traditional ecological knowledge, Indigenous or local knowledges declare Nakashima, et al. (2000).

From the University of British Columbia, Canadian geographer John Robinson (2008, pp. 72-3) declares that "[i]ssue-driven interdisciplinarity" is required for sustainability initiatives because of their "inherently complex, multi-faceted and problem-based focus" and this field of sustainability represents the "paradigm case" for applying the concept. He asserts that this intellectual and political project of forging new types of sustainability coalitions is one of being "undisciplined" in the sense that "practitioners of this style of interdisciplinarity do not find themselves at the margins between disciplines, but in the sometimes uncomfortable borderlands between the academy and the larger world". He further contends that such "transdisciplinarity" has less to do with new theoretical frameworks and the unity of knowledge "than with the emergence of problem- and solution-oriented research incorporating participatory approaches to address societal problems" (Robinson, 2008, p. 71; also Mitchell, 2011a). Visser (1999) and Nicolescu (2002) consider the contours of **transdisciplinary education** in the early twenty-first century:

- Learning is an underdeveloped concept, but is necessary for all humans to be able to adapt to continuous and ever-faster change in an increasingly complex world. Fundamental changes are urgently required in the way school systems throughout the world are organized that must include more holistic conceptualizations of schools themselves as only one part of a comprehensive learning environment.
- Learning has to do with the capacity to interact creatively and constructively with problems. In most current pedagogical practices such problems are often concealed or ignored altogether. In a manner similar to Brazilian educator Freire (1970, 1999), learning therefore needs to be re-focused on problems, including their historical and epistemological contexts.
- Learning is a transdisciplinary concept related to overarching concerns such as change and growth; community-based processes and development; complex, diverse, and emerging adaptive expressions; new designs for systems of knowledge construction interacting with, and building upon, existing knowledge bases; lifelong learning at different levels of organizational complexity; neuroscience and lifespan cognitive development; the interconnections and distinctions between and among data, information, knowledge and wisdom; and new technologies for learning, languages, cognition, and meta-cognition.

Perhaps not surprisingly, the institutionalization of transdisciplinarity within universities also has UNESCO antecedents beginning in 1987 through the creation of the International Centre of Transdisciplinary Research and Studies (Centre International de Recherches et Études Transdisciplinaires, or CIRET) in Paris. In 1995, Rumanian physicist Basarab Nicolescu co-founded the Reflection Group on Transdisciplinarity with UNESCO - a project initially involving 16 scientific and cultural personalities in the implementation of transdisciplinary methodologies in various fields of international research. Similar to Visser (1999), one of its main aims is the implementation of these principles in education, and slowly but decisively, transdisciplinarity has gained an international impact especially in superior educational settings as universities the world over have opened themselves to experimenting with transdisciplinary curricula, research activities, and conferences (Dincă, 2011).

3. Brock University applies global principles

The year 2011 is marked by important global challenges that have affected humanity as never before. While globalisation has had a positive effect on millions of people by helping them rise out of poverty, a global crisis of unusual proportions - economic, financial, social, and environmental - endangers fulfillment of the most important agenda of present-day multilateralism, the United Nations Millennium Development Goals[1]. These events coincide with a review of UNESCO's Madrid Action Plan for Biosphere Reserves (2008-2013) as well as the 40th anniversary of the Man and the Biosphere Program (MAB). MAB invites stakeholders within the World Network of Biosphere Reserves (WNBR) to engage in fostering more harmonious integration of people and nature for sustainable development through participatory dialogue, knowledge sharing and improvement of human well-being. The Madrid Action Plan notes a "commitment to innovative time-bound socio-ecological and policy actions integrating the three biosphere reserve functions and the willingness to share data, information, experience and knowledge are vital…for biosphere reserves to be learning sites during the Education for Sustainable Development (ESD) 2005-2014" (p. 10).

Lundholm and Plummer (2010) emphasize how a growing interest in environmental education has contributed to greater literacy in sustainability "dating from the 1977 UNESCO conference in Tblisi to the current Decade of Education for Sustainable Development" (2005-2014) which reached mid-term in 2009 (p. 475). Pigozzi (2010) further observes that through ESD, UNESCO seeks to integrate principles, values, and practices of sustainable development into all aspects of education and learning to address social, economic, cultural and environmental problems faced by humans in the 21st century. Nevertheless, as Lundholm and Plummer (2010) astutely inquire, in terms of "the political and pedagogical aspect of resilience, is the concept working as a heuristic cognitive tool in guiding us to look critically at ourselves [positively in terms of both human resourcefulness and strengths, and our shortcomings]?" (p. 486). It is clear that UNESCO offers internationally available conceptual and political tools for mobilizing domestic public opinion, and intellectual and academic communities in pursuit of these values and priorities. In Canada, researchers, practitioners, multidisciplinary professionals, and local, governmental and NGO stakeholders have begun to engage more fully in the current national cadre of fifteen sites shown in the map below (see also CBRA, 2011).

UNESCO embraces 193 Member States and six Associate Members, and its original mandate is still highly relevant in the 21st century where building knowledge-based societies is an imperative, where culture is crucial to any meaningful debate on sustainable development, and where science and innovation mark a new research era in fields such as climate change and water. As learning sites of excellence the 564 ecosystems comprising the current WNBR constellation offer one of the premier planetary frameworks for the development and building of international capacity to manage complex socio-ecological systems. This is achieved through greater dialogue at the science-policy interface, through environmental education and through multi-media outreach to wider communities interested in more sustainable development (adapted from UNESCO's 2008-2013 Madrid Action Plan for Biosphere Reserves). In this context it is also useful to note, as Dutch feminist Sevenhuijsen

---

[1] At the United Nations Millennium Summit in New York 6 to 8 September, 2000, the largest gathering of world leaders in history adopted Eight Millennium Development Goals to be achieved by 2015: 1 – To Eradicate Extreme Poverty and Hunger; 2 – To Achieve Universal Primary Education; 3 – To promote Gender Equality and Empower Women; 4 – To Reduce Child Mortality; 5 – To Improve Maternal Health; 6 – To Combat HIV/AIDS, Malaria and Other Diseases; 7 – To Ensure Environmental Sustainability; and 8 – To Develop a Global Partnership for Development.

(1999) has argued, that documents such as these "[p]olicy texts are sites of *power*...by establishing narrative conventions, authoritative repertoires of interpretation and frameworks of argumentation and communication, they confer power upon preferred modes of speaking and judging, and upon certain ways of expressing moral and political subjectivity" (also cited in Moss and Petrie, 2002, p. 81, emphasis added).

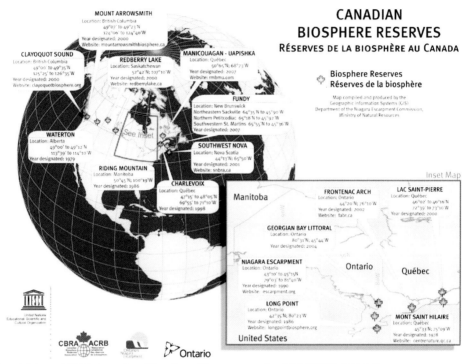

Fig. 1. Map of 2011 Canadian Biosphere Reserves under UNESCO's MAB Program (reprinted with permission from Niagara Escarpment Commission, Ontario Ministry of Natural Resources).

3.1 Brock environmental sustainability research unit

Schultz, Duit and Folke (2010, p. 663) recount how biosphere reserves were designated by UNESCO with the mission of maintaining and developing ecological and cultural diversity and securing ecosystem services for human wellbeing (see also UNESCO, 2008, p. 8) in collaboration with a suitable range of actors, often including local communities and various scientists. Since 1995, biosphere reserves have been expected to fulfill the three functions stated in the Statutory Framework and the Seville Strategy (UNESCO, 1996): (1) conserving biological and cultural diversity, (2) fostering sustainable social and economic development, and (3) supporting research, monitoring, and education. These three functions have been applied and implemented by BESRU researchers along with several of the evaluative criteria of biosphere reserves which correspond to features of adaptive co-management since they focus on monitoring and an integrated approach to conservation and development along with participation of a suitable range of actors. Based upon their case study in Sweden and the mission, functions and criteria of UNESCO's Man and the Biosphere Program (MAB), Schultz, Duit and Folke (2010) have also proposed that biosphere reserves "constitute potential sites for testing the effectiveness of participation in general and adaptive co-management in particular" (p. 663).

Nestled in the Niagara Escarpment Biosphere Reserve (designated in 1990), Brock University is one of a small but growing cadre of Canadian academic institutions located within these areas. As previously noted conservation, sustainable socio-economic development, and education are suggested as the basis for improvements in relationships between humans and their ecosystems. In the journal Environments, however, Jamieson (2004) observes that Canadian biosphere reserves have thus far tended not to function very well in achieving UNESCO's goals (see also Francis and Whitelaw, 2004) since "the average Canadian knows nothing about biosphere reserves ....Current public ignorance about biosphere reserves in Canada is partly the result of our unique situation relative to Europe - Canada has extensive areas of relatively undisturbed wilderness" (pp. 103-104). Key to any academic contribution with respect to research programs within biosphere reserves must be the commitment of a dedicated core of readily-identifiable individuals engaged in UNESCO objectives.

Responding to this gap in domestic knowledge, the Brock Environmental Sustainability Research Unit (BESRU) has begun to interrogate the scientific, pedagogical and cultural intersections of its geographical location more precisely as a proposed site of excellence. In 2011, BESRU was developed to reflect this commitment with an overall goal to:

pursue innovative and interdisciplinary research concerning the environment, sustainability and social-ecological resilience (BESRU, 2011a)

Sustainability science concepts (Kates, et al., 2001) are used to identify and address problems of interest in the areas of water, innovation and resilience, environmental governance, social-ecological systems and social justice, and the challenges faced by many of Canada's First Nations communities in a way that creates hubs and networks (BESRU, 2011a). Integral to this is the role of the university in training and educating students in sustainability science (BESRU, 2011a).

Reflecting the core values of sustainability, transdisciplinarity, and collegiality, BESRU's strategic plan involves forward-thinking research in such diverse areas as adaptation and impacts research, climate change and plant response, First Nations and source water protection, carbon pricing policy, and biosphere reserve periodic reviews (BESRU, 2011b).

3.2 Research utilising Social-Ecological Inventories (SEIs)

Social-ecological inventories (SEIs) were developed as a community-based approach for assessing resilience in Sweden's Kristianstad Vattenrike Biosphere Reserve by Lisen Schultz during her doctoral research, and as "a means to identify people with ecosystem knowledge that practice ecosystem management" (Schultz, Folke, and Olsson, 2007, p. 140). These authors highlight how SEIs are dynamic and propose their application during the preparation phase of conservation and resilience assessment projects. SEIs are a way of approaching the social landscape as carefully as the biophysical landscape with a systematic mapping of actors, their values, motives, activities, experiences over time, and networks. These authors note how the approach "complements stakeholder analyses and biological and ecological inventories, and assesses existing management systems behind the generation of ecosystem services, thus providing a starting point for participation" (Schultz, et al., 2007, p. 141).

The methodology has been further applied in the Niagara Escarpment Biosphere Reserve in Ontario, Canada since 2009. During the process "bridging organizations" are identified as those "coordinating and connecting many of the local steward groups to organizations and institutions at other levels" (Schultz, et al., 2007, pp. 140-141). The inventory complements stakeholder analyses as well as biological or ecological inventories by assessing existing management systems behind the generation of ecosystem services. As such, groups such as these "represent an undervalued and sometimes unrecognized source of knowledge and experience for ecosystem management" (Schultz, et al., 2007, p. 141).

In March, 2011 a two-day experts' workshop with Swedish and Canadian collaborators was held to compare and analyze findings from socio-ecological inventories (SEIs) undertaken in two biosphere reserves – Kristianstad Vattenrike BR in Sweden (Schultz, Folke, and Olsson, 2007), and the Niagara Escarpment BR in Canada (Armitage and Plummer, 2010; Gafarova, May and Plummer, 2010; Velaniškis, 2010). Entitled "Social-ecological inventories: Building Resilience to Environmental Change within Biosphere Reserves" (Mitchell, 2011b), the event was hosted at the Brock University main campus in southwestern Ontario by core faculty from BESRU (see http://brocku.ca/brock-environmental-sustainability-research-unit).

The workshop engaged scientists, academic researchers, practitioners and students familiar with SEIs, those in governmental and non-governmental leadership roles, those familiar with research within UNESCO Biosphere Reserves, and those familiar with similar participatory methodologies outside and beyond the boundaries of biosphere reserves. The event (as with ongoing developments within BESRU's research program) was organized utilizing principles from relevant international frameworks including those underlying the UN Decade of Education for Sustainable Development (see also Lundholm and Plummer, 2010; Krasney, Lundholm and Plummer, 2010; Plummer, 2010; Schultz and Lundholm, 2010) with the following aims:

- To advance lessons and commonalities from the application of SEIs in Sweden and Canada with those engaged in similar processes
- To develop a document outlining some common methodological and conceptual grounds

While SEIs have been framed using much of the language of conservation biologists, they involve participatory mapping of existing stewardship and monitoring of landscape management processes that facilitate baseline achievement of UNESCO's "Man and the Biosphere Program" goals (UNESCO, 2010). It is also the case that many regions of the world lie beyond the boundaries of biosphere reserves but would nonetheless benefit from adopting the SEI within their own national parks, protected areas and ecosystems under pressure from human activities. During the workshop common frameworks for interpreting and understanding SEIs emerged that could facilitate their cross-scale adaptation, and the following points reflect these participatory values:

- Since human well-being is an underlying principle and the motivation behind development of the SEI, common conceptual frameworks defining human health could facilitate identification of "bridging organizations" for those conducting the SEI within or beyond biosphere reserves.
- Due to the mission and functional criteria of biosphere reserves, accomplishing the aims of the SEI will allow greater fulfillment of their role as sites of excellence comprising the three inter-related dimensions of conservation, human and socio-economic well being, research, evaluation and education.
- Due to the globalized nature of corporate, institutional and individual *power relations*, SEIs were seen as a participatory pathway to allow these actors to be identified locally.

These same authors further observe how the means to map, analyze and facilitate stakeholder engagement in order to develop participatory conservation projects have been discussed in previous literature. They caution, however, that while 'participation' by a variety of stakeholders may be desirable from a "democratic perspective, it is not in itself a recipe for successful ecosystem management. Participation has to be connected to management practices that generate ecological knowledge, draw on experience, and learn about and respond to ecosystem dynamics". Conducting an SEI allows local actor groups "generally operating at the level below municipalities, who effect management of ecosystems and their services on the ground" to be identified (Schultz, et al., 2007, p. 141). It has also become apparent from application that the SEI prepares the methodological ground for democratic, active and meaningful stakeholder participation. "We do not claim that the SEI is complete" (Schultz, et al., 2007, p.142), and an iterative, ongoing process is envisioned that will be enhanced by those further applying SEIs in cross-scale adaptation (see particularly Shultz, Plummer and Purdy, 2011,
http://www.resalliance.org/index.php/resilience_assessment).

From an analysis of preliminary phases of the Niagara Region SEI, Velaniškis (2010) maintains "a key factor in understanding social-ecological system interactions is identifying linkages or lack of linkage among the actors who are directly involved in ecosystem and risk management" (p. 14). It was confirmed during the workshop by those engaged in further application of the Niagara SEI that the preparatory phase of identifying bridging organizations and actors rests upon trust-building and researcher transparency. Participants emphasized the importance of not being overly proscriptive and to communicate expectations from the outset highlighting the collaborative ownership of the process as well as outcomes. Clear expectations and well-defined research protocols are to be communicated since those engaged in conducting the SEI could well be understood as agents of change themselves, and as researchers even becoming a type of bridging organization (Shultz, et al., 2007) and through the exercise of reflexivity.

3.3 Measuring Brock University's carbon footprint

While there may yet remain some highly contested terrain, it is widely accepted across most disciplines that global warming and climate change are the foremost environmental challenges facing the world today (see for example the Nobel Lecture by Intergovernmental Panel on Climate Change chair Dr. Rajendra Pachauri, [2007]). It is also clear that these issues will only be tackled effectively when actors at all levels of developed and developing societies - including governments, NGO's, corporations, institutions of higher learning, communities and individuals - each take responsibility for, and attempt to minimize their own, Greenhouse Gas (GHG) Emissions.

These challenges have been taken up by Brock University through a baseline measurement of its carbon footprint (herein carbon audit) undertaken in two phases with the initial exploratory phase beginning in autumn 2008. Findings from a series of qualitatively framed interviews, participant observations, and documentary analyses were first published in the *International Journal of Sustainability in Higher Education* (Mitchell, 2011a). One notable outcome from the study included development of a transdisciplinary, campus-wide steering committee under the auspices of the University's Provost and Vice President Academic. Brock's Sustainability Coordinating Committee now consists of student representatives and senior administrative, academic, and facilities management personnel who quickly undertook a second phase of the audit from 2009-2010.

The methodology was framed by the GHG Protocol Corporate Accounting and Reporting Standard - the most widely adopted methodology for conducting corporate carbon footprints in the world. The principal investigators included the Committee's faculty co-chair and a partnership with international consultants HRCarbon to collate and analyze available data. The research was first bound by time with data only from the calendar year January 1, 2008 to December 31, 2008 and utilised the International Organization for Standardization (ISO) 14064, the emerging standard on Greenhouse Gas accounting and verification consisting of three parts:

- **Scope 1**: Direct emissions from sources that are owned or controlled by the organization/company (those from combustion of fuels in boilers, furnaces and turbines)
- **Scope 2**: Indirect emissions associated with the generation of purchased electricity consumed by the organization
- **Scope 3**: All other indirect emissions as a consequence of the activities of the organization that occur from sources neither owned nor controlled by the organization (outsourced distribution)

The research was further bound by data concerned with operations on the Main Campus only, and excluded satellite campuses, as well as university-related travel due to concerns for confidentiality although these data were recommended to be included in future audits. Findings included the following.

Scope	1	2	3
Research Year 2008	22,364 CO_2 Metric Tonnes	525 CO_2 Metric Tonnes	11,716 CO_2 Metric Tonnes

Table 1. Overall Brock University GHG (Carbon) Footprint
(Adapted from Mitchell and Parmar, 2010)

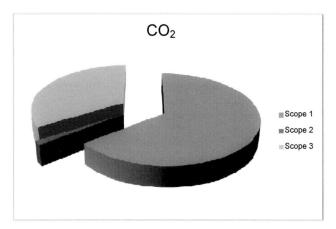

Fig. 2. Overall Brock University GHG (Carbon) Footprint Chart
(Adapted from Mitchell and Parmar, 2010)

The majority of 2008 Scope 1 and 2 emissions were generated by the burning of fossil fuels (Natural Gas) to generate electricity and a small portion to heat the Main Campus facilities. Commuting by employees, faculty and students is the second largest source of GHG emissions. The important metric of 22, 364 Mt. Tonnes of CO_2 allows for a comparative analysis against other institutions and an ability to enter into other discourses and conversations on climate change with peers around the globe. This should only occur, however, with the caveat that the research was bound by a number of factors noted previously. The final report highlighted that a more comprehensive GHG Emissions Inventory will better facilitate this comparative conversation.

The key recommendation from this initial audit (submitted to the Committee as a whole in May, 2010) included directions to "undertake a more comprehensive Carbon Audit for all facilities, functions and activities that generate emissions" (Mitchell and Parmar, 2010). The University has now acted upon this directive by joining a regional carbon monitoring initiative known as the Niagara Sustainability Initiative (see Brock University News Release, 2011). In 2012, the University will engage with NSI to continue its role in modeling carbon auditing within the Region, and as part of its ongoing commitment as one of UNESCO's 'sites of excellence' in the Niagara Escarpment Biosphere Reserve. NSI is a not-for-profit organization whose mission is to be a key interface between government, industry, community and academia, and whose aim is to promote sustainability in the Niagara Region (http://niagarasustainability.org/). In this pursuit, its main vehicle is the *Carbon Project* whose focus is in corporate greenhouse gas emission reduction.

3.4 Pedagogy

In 2011, Brock University students enrolled in a third year required Environmental Management course were asked to develop a "Brock in the Biosphere Project" (the Project). The Project was undertaken for a number of reasons. First, as noted previously, Brock University is situated on the Niagara Escarpment Biosphere Reserve. Second, the ongoing significance of this is that academic institutions have as part of their mandate a responsibility to "incorporate sustainability projects associated with these initiatives into appropriate curricula" (Mitchell, 2011a, p. 18). Third, a recently developed Sustainability Committee had been tasked with preparing a carbon audit for the university campus. Finally, a gap was identified between the UNESCO Man in the Biosphere guiding principles outlined above and Brock University's recently approved strategic plan (Brock University, 2010). This afforded students a window of opportunity to analyze the university's stated strategic direction and make recommendations as an important constituency on how to take up a more direct leadership role in biosphere reserves.

The Project required students to determine what key pillars could be used in order to operationalize Brock's strategic plan and strive to become the leading environmental steward of the Niagara community. They were then tasked with developing a 5-10 year plan of various initiatives. The Project involved the use of the Resilience Assessment workbook (Resilience Alliance Workbook, 2011) to structure information gathering and analysis. Teams met weekly with the course instructor to chart progress with work plan elements. Results from the Project were summarized by presentations to the faculty client and consultant reports. There were nine teams of between three and four members each. The following four recurring pillars were identified:

1. Education and Research
2. Community Involvement, Awareness and Partnerships
3. Protecting Biodiversity
4. Conserving Natural Resources

3.4.1 Education and research

All of the student projects identified education and research as a strategic pillar for the University to become an environmental steward and leader. Each group described various initiatives that stem from Brock University's unique geographic location. There was an opportunity identified to take advantage of this to become more "biospheric" in focus. There was also an opportunity identified to educate students on the significance of the ecosystem in which the campus is located. This could lead to greater awareness of how action taken can influence sustainability. Recommendations from the student reports encouraged departments to increase the number of environmentally related courses offered. The development of cross-referenced courses in different programs to take an interdisciplinary approach to higher education would expose students to environmental issues and sustainability. In keeping with the UN's Decade of Education for Sustainable Development (Lundholm and Plummer, 2010; Krasney, Lundholm and Plummer, 2010), the University could also work with local elementary and secondary schools in developing programs that allow students at all levels to interact and learn about the diverse Niagara ecosystem.

The broader community was also encouraged to play a role in research in the Niagara Escarpment Biosphere Reserve. Community involvement can be accomplished through environmental education, participation, plans, policies and practices about the biosphere. Research and monitoring is needed to determine the health of the Escarpment's forests and wetland habitats. Since community activities potentially impact the biosphere, research is needed to determine and understand how these human activities impact the biodiversity of the Escarpment. A network of educators and researchers needs to be built in order to exchange the necessary information to educate the community about their role in the region regarding the biosphere; this tool will help achieve sustainability.

In addition, education and research are needed to implement sustainability plans in the Niagara Region. The broader community needs to be made aware of innovations and measures in place so that they remain competitive and are able to use the new technology. An informed community promotes, and is capable of, change. Guest lecturers speaking at public events will help stimulate research at a community and post-secondary level. Annual sustainability reports can be made and be available to the public, further expanding the Region's information exchange. Organizations like the Ontario Public Interest Research Group (OPIRG) and Sierra Youth Club already in existence at Brock, welcome members to learn and inform communities about issues in their region.

3.4.2 Community involvement, awareness and partnerships

Community involvement and awareness were closely linked with education and research within the student reports. Brock students need to play a critical role in engaging the Biosphere and broader Niagara communities. Volunteers can be engaged to educate younger school children about the biosphere through different programs. The faculty can join these student efforts in on-campus naturalist groups to spread awareness through activities like trail walks, creating monthly newsletters to distribute or posters to promote sustainability within the community.

Community involvement is needed in order to positively impact the Niagara Region and allow various conservation and sustainability initiatives to be successfully executed. Members of the community can be included in meetings allowing them to give input in strategic plans and engage in meaningful discussion. The university can play a role in ensuring accurate and reliable scientific information is available so that the community can be confident when making decisions. Community support can be used to leverage funding in the implementation of initiatives, research and monitoring of the biosphere.

Collaboration and cooperation among stakeholders are critical to satisfy all parties involved in decision-making and increase potential for the best possible outcome. Brock is an emergent leader of sustainability in Niagara and can provide outreach and educational opportunities for the community by way of guest lectures, committees, meetings, interest groups and annual sustainability reports. The formation of partnerships among stakeholders is a powerful tool to achieve sustainability; different opinions and inputs are used to form the most effective plans.

3.4.3 Protecting biodiversity

Protecting biodiversity and the conservation of natural resources, discussed next, are interrelated. Identifying, promoting and regulating conservation are important first steps. Buffer zones could be established with accompanying bylaws and policies to further protect the unique nature of the Niagara Escarpment as well as immediate surroundings. The introduction of these zones would restrict development in biologically unique areas and allow an increase of green space and corridors of uninterrupted land for species of plants and animals.

Rehabilitation and restoration projects are needed to help protect biodiversity. Improvements made to landscapes that have been altered to previously accommodate human activity can help stop any further or potential negative impacts. Restoration projects provide habitat connectivity for wildlife so they may continue to spread to areas where they were once eliminated or previously uninhabited. The protection of land and biodiversity allows endangered species to re-establish their presence within the biosphere to a level that eliminates their threatened or endangered status.

3.4.4 Conserving natural resources

Conserving natural resources was identified as a related factor in aligning with biosphere objectives, as well as promoting sustainability in the Niagara Region. The Project plans observed that the university can play a leadership role in conserving the diversity of plants, animals and microorganisms so that the natural environment can thrive within significant ecosystems.

Concepts such as natural capital and ecosystem services should be incorporated into the University's daily operations. Institutional policies and practices that reflect these ideas need to be introduced or further implemented. These initiatives include but are not limited to resource conservation, recycling, composting, energy reduction and reusing materials. In order to monitor the success rate of conserving natural resources, the development of annual reports is needed to demonstrate the University's progress. As can be expected from the diversity of student backgrounds there were a number of secondary ideas developed. These are summarized in the following Table 2.

It was a novel experience for instructional staff, students and members of the Brock community who acted as consultants. Student learners were enthusiastic about applying the environmental management techniques they were learning about to a concrete problem-solving situation (Purdy, 2011). While at first the freedom afforded them was daunting, they became increasingly more comfortable with the consultant role and the challenges of working as a member of a collaborative team (Purdy, 2011). Instructional staff was intrigued by the opportunity to target a core environmental management course to a specific issue of relevance to the Niagara Region and Biosphere Reserves more generally. They were impressed with the novelty and sophistication of the research projects, some of which included multi-media in presentations. The Project helped to strengthen understanding not only of how to better make use of the university's unique position in the Biosphere Reserve but to align its policies with MAB program principles. The faculty consultant for the Project was gratified to see that a complex theoretical concept like transdisciplinarity could be practically applied for collaboration in sustainability; in addition efforts as a result of the Project and the previously discussed carbon audit has led to the University becoming involved in the broader regional NSI initiative (Mitchell, 2011c). In summary, research projects such as the "Brock in the Biosphere" Project can be used to make the application of learning relevant, proactive and useful. Not only were environmental management concepts applied by students, but the results can be used to further inform institutions within biosphere reserves in meeting their international obligations.

MONITORING	Allow individuals to undertake various initiatives and measure progress. The health of the natural environment can be measured in addition to any changes within the Niagara Escarpment and provide input into the local community
BROCK COMMUNITY	Educate and involve all students, faculty and employees in sustainability initiatives that benefit the Niagara Escarpment. Continue to invite stakeholders to the campus though hosting events (e.g. Ontario Fruit and Vegetable Growers Conference) and educational programs. Strengthen Brock's reputation as an environmental steward in influencing and shaping the Niagara community and developing sustainable initiatives.
INFRASTRUCTURE	Examine strategies to improve transportation, renewable energy production and building energy efficiency and model energy envelopes; investigate green roof technologies; For new development prevent damage to the surrounding environment and ecosystems; implement "waste to landfill" reduction plans that include recycling and composting; effectively enforce waste reduction programs and require enforcement to ensure success.
SUSTAINABLE DEVELOPMENT	Promote businesses in Niagara that practice sustainability; monitor with standards to ensure green washing is avoided; ensure all three pillars of sustainable development are considered.
ECONOMIC/ HUMAN DEVELOPMENT	Involve Chambers of Commerce and levels of government to promote economic vitality, industry diversity, partnerships and cooperation in the development of a sustainable community; develop a sustainability directory.
CULTURE	Use the biosphere to symbolize identity and heritage. Develop the concept of biosphere as a branding tool to demonstrate stewardship and sustainability. Integrate the Arts, environment, science and technology to link together topics of interest.

Table 2. Secondary Strategic Pillars Identified by Teams for the Brock in the Biosphere Project

4. Conclusion

Post-secondary institutions who find themselves situated in one of UNESCO's current (and ever-expanding) 564 Biosphere Reserves have an important obligation to ensure they reflect the goals and principles set out in international agreements and conventions. Within their strategic plans, for example, the core mission of innovative research, pedagogy and service can be adapted with little challenge to reflect UNESCO's three key functions of conservation of biodiversity, sustainable economic and human development, and logistic support (or capacity building) for research, monitoring, education and training. Entities such as the Brock Environmental Sustainability Research Unit (BESRU) can play an important bridging role in connecting with other researchers, the local community and the University itself (students, faculty, and administration).

More specifically, the chapter identified how Brock University through its support in the development of an inclusive new Environmental Sustainability Research Unit, and the application of participatory tools such as the Social-Ecological Inventory, is making progress at applying internationally accepted conservation metrics. Through their carbon audits, and proactive striving to offer socially- and academically-relevant experiential learning opportunities in the classroom and the community, the University is also taking aim towards fulfilment of UNESCO's mandate for all biosphere reserves as sites of sustainability excellence.

The emerging discourse on transdisciplinarity, as identified and defined herein, provides additional guidance on how to approach the important notion of biosphere connectivity and how to gain traction in local initiatives with UNESCO's meta-framework for sustainability science. In a case study of an Indigenous Peruvian project similarly framed by an application of transdisciplinary thinking, Apgar, Argumedo and Allen (2009, 255) declare:

> *The most critical problems humanity faces today are complex problems, characterized by high levels of uncertainty, multiple perspectives and multiple interlinked processes from local to global scales....Traditional research inquiries with specialized experts are unable to make the connections required to manage complexity. Transdisciplinary approaches can help different stakeholder groups to share and use their knowledge and experience for problem-focused inquiry. Facilitating transdisciplinarity requires good dialogue processes and the development of holistic frameworks.*

The sections above have highlighted a variety of empirical and theoretical dialogues occurring within two Biosphere Reserves – one in Sweden and one in Canada – and a concomitant transfer of knowledges that has taken place as a result. These efforts may be further summed up by a quote from Thomas Homer-Dixon who describes the need for a prospective mind, "… an attitude toward the world, ourselves within it, and our future that's grounded in the knowledge that constant change and surprise are now inevitable [and] … aggressively engages with this new world of uncertainty and risk. A prospective mind recognizes how little we understand, and how we control even less" (2007, 29). In this sense, it is all the more important that academic institutions undertake the kind of integrative work presented here, and make use of the resources and capabilities offered them to enhance evaluation, education and capacity building within and beyond Biosphere Reserves.

5. Acknowledgements

The inspiration for this chapter developed from a number of different sources. The section on social-ecological inventories was part of a study supported by Environment Canada and the subsequent research report (see Mitchell, 2011b). In addition, our thinking on sustainability in Biosphere Reserves and beyond grew out of the University's carbon audit (Mitchell, 2011a) and Executive Summary by Mitchell and Parmar (2010). Thanks go also to the students and learners in TREN 3P15, Environmental Management during 2011 for their commitment to the "Brock in the Biosphere" experiential learning research project. The authors would like to sincerely thank all members of the Brock Environmental Sustainability Research Unit (BESRU) for their support and encouragement, and in particular, current Director, Ryan Plummer.

6. References

Apgar, M.J., Argumedo, A. & Allen, W. (2009). Building Transdisciplinarity for Managing Complexity: Lessons from Indigenous Practice. *International Journal of Interdisciplinary Social Sciences* 4(5): 255-270.

Armitage, D. & Plummer, R. (2010). *Adaptation, Learning and Transformation in Theory and Practice: A State-of-the-Art Literature Review and Strategy for Application in the Niagara Region. Final Report*. Prepared for Adaptation and Impacts Research Section - Environment Canada. Toronto, Ontario. (47 pp.).

Beringer, A. (2006). "Campus sustainability audit research in Atlantic Canada: pioneering the campus sustainability assessment framework", *International Journal of Sustainability in Higher Education* 7 (4): 437-55.

Brock Environmental Sustainability Research Unit (BESRU). (2011a). *Annual Report to Environment Canada: Grants and Contributions Agreement, 2010-11*, September.

Brock Environmental Sustainability Research Unit (BESRU). (2011b). *2011-2016 Strategic Plan*, September.

Brock News. (22 August, 2011). Brock joins regional carbon monitoring initiative. Available on-line: http://www.brocku.ca/news/17175.

Brock University. (2010). Brock University Integrated Plan, Brock 2020: Taking Our Own Tack, November.

Canadian Biosphere Reserve Association/ Association canadienne des réserves de la biosphère (CBRA 2011). Available on-line: http://biospherecanada.ca/en/.

Choi, B.C.K. & Pak, A.W.P. (2006). Multidisciplinarity, interdisciplinarity and transdisciplinarity in health research, services, education, and policy: 1. Definitions, objectives and evidence of effectiveness. *Clinical & Investigative Medicine* 29(6): 351-364.

Dincă, I. (2011). *Stages in the Configuration of the Transdisciplinary Project of Basarab Nicolescu*. Available on-line:
http://basarab.nicolescu.perso.sfr.fr/Basarab/Docs_Notice/Irina_Dinca.pdf.

Francis, G. (2004). Biosphere Reserves in Canada: Ideals and some experience. *Environments* 32(3): 3-26.

Francis, G. and Whitelaw, G. (2004). Biosphere Reserves in Canada: An introduction. *Environments* 32(3): 1-2.

Friere, (1970). *Pedagogy of the Oppressed*. London: Penguin Books Ltd.

Gafarova, S., May, B. & Plummer, R. (2010). *Adaptive Collaborative Risk Management and Climate Change in the Niagara Region: A Participatory Integrated Assessment Approach for Sustainable Solutions and Transformative Change – General Overview*. Prepared for Adaptation and Impacts Research Section - Environment Canada. Toronto, Ontario. (6 pp.)

Giroux, H.A. & Searls Giroux, S. (2004). *Take Back Higher Education: Race, Youth and the Crisis of Democracy in the Post-civil Rights Era*. New York: Palgrave Macmillan.

Holmes, D. & Gastaldo, D. (2004). Rhizomatic thought in nursing: an alternative path for the development of the discipline. *Nursing Philosophy* 5: 258-67.

Homer-Dixon, T. (2007). *The Upside of Down: Catastrophe, Creativity, and the Renewal of Civilization*, Resources & Conflict Analysis Toronto: Vintage Canada, Inc.

Jamieson, G. (2004). Commentary. *Environments* 32(3): 103-106.

Kates, R.W., Clark, W.C., Corell, R., Hall, J.M., Jaeger, CC., Lowe, I., McCarthy, J.J., Schellnhuber, H.J., Bolin, B., Dickson, N.M., Faucheux, S., Gallopin, G.C., Grubler, A., Huntley,B., Jaeger, J., Jodha, N.S., Kasperson, R.E., Mabogunje,A., Matson, P., Mooney, H., Moore III, B., O'Riordan, T., & Svedin, U. (2001). Sustainability Science, *Science* 292: 641-2.

Koizumi, H. (2001). Trans-disciplinarity. *Neuroendocrinology Letters*. 22: 219-21.

Krasney, M.E., Lundholm, C. & Plummer, R. (2010). Environmental education, resilience, and learning: reflection and moving forward. *Environmental Education Research* 16(5-6): 665-672.

Lundholm, C. & Plummer, R. (2010). Resilience and learning: a conspectus for environmental education. *Environmental Education Research* 16(5-6): 475-491.

Malott, C.S. (2008). *A Call to Action – An Introduction to Education, Philosophy, and Native North America*. New York, NY: Peter Lang Publishing.

Mitchell, R.C. (2011a). Sustaining Change on a Canadian campus: Preparing Brock University for a sustainability audit. *International Journal of Sustainability in Higher Education* 12(1): 7-21.

Mitchell, R.C. (2011b). *Social-ecological Inventories: Building Resilience to Environmental Change within Biosphere Reserves*. Experts Workshop with Swedish and Canadian scientists/social scientists/students 7-8 March, 2011 in partnership with Environment Canada Adaptation and Impacts Research Section (AIRS), Stockholm Environmental Institute, and Brock Environmental Sustainability Research Unit. Available on-line: http://www.resalliance.org/cdirs/raprojects/index.php/110. (See also http://www.brocku.ca/social-sciences/undergraduate-programs/tourism environment/BESRU).

Mitchell, R.C. (2011c). Personal communication, 25 September, 2011. St. Catharines, Ontario, Canada.

Mitchell R.C. & Parmar, J. (2010) Research Report for Brock University GHG (Carbon) Footprint – Executive Summary 2008. Available on-line: http://www.brocku.ca/social-sciences/undergraduate-programs/tourism-environment/BESRU

Moore, S.A. & Mitchell, R.C. (Eds) (2008). *Power, Pedagogy and Praxis: Social Justice in the Globalized Classroom*. Rotterdam, Boston and Taipei: Sense Publishers. Available on-line: www.sensepublishers.com/files/9789087904920.pdf

Moss & Petrie (2002). From Children's Services to Children's Spaces – Public Policy, Children and Childhood. London: Routledge/Falmer.

Nakashima, D., Prott, L. & Bridgewater, P. (2000). *Tapping into the world's wisdom*. United Nations Educational, Scientific and Cultural Organization: *Sources* Vol. 125: 11-12.

Nicolescu, B. (2002). *Manifesto of Transdisciplinarity*. New York, NY: State University of New York Press.

Orr, D.W. (2004). *The Last Refuge: Patriotism, Politics, and the Environment in an Age of Terror*. Washington, DC: Island Press.

Pachauri, R. (10 December, 2007). *Intergovernmental Panel on Climate Change Nobel Lecture*. Available on-line: http://www.nobelprize.org/mediaplayer/index.php?id=795.

Pigozzi, M.J. (2010). Implementing the UN Decade for Education in Sustainable Development (DESD): achievements, open questions and strategies for the way forward. *International Review of Education* 56(2-3): 255-269.

Plummer, R. (2010). Social-ecological resilience and environmental education: synopsis, application, implications. *Environmental Education Research* 16(5-6): 493-509.

Pollock, R. M. (2004). Identifying Principles for Place-Based Governance in Biosphere Reserves. *Environments* 32(3): 27-41.

Purdy, S. (2011). Personal communication, September, 2011. St. Catharines, Ontario, Canada.

Ralston Saul, J. (2008). *A Fair Country – Telling Truths about Canada*. Toronto: Viking Canada and Penguin Group (Canada).

Resilience Alliance (RA). (2011). *Social-ecological Inventory Workbook*. Available on-line: http://www.resalliance.org/index.php/resilience_assessment

Robinson, J. (2008). Being undisciplined: transgressions and intersections in academia and beyond. *Futures* 40: 70-86.

Schultz, L., Duit, A. & Folke, C. (2010). Participation, Adaptive Co-management, and Management Performance in the World Network of Biosphere Reserves. *World Development* 39(4): 662-671.

Schultz, L., Folke, C. & Olsson, P. (2007). Enhancing ecosystem management through social-ecological inventories: lessons from Kristianstads Vattenrike, Sweden. *Environmental Conservation* 34(2): 140-152.

Schultz, L. & Lundholm, C. (2010). Learning for Resilience? Exploring learning opportunities in biosphere reserves. *Environmental Education Research* 16(5-6): 645-664.

Shultz, L., Plummer, R. & Purdy, S. (2011). *Social-ecological Inventory Workbook*. Available on-line: http://www.resalliance.org/index.php/resilience_assessment

Sevenhuijsen, S. (1999). *Citizenship and the Ethics of Care: Feminist Considerations on Justice, Morality and Politics*. London: Norton.

Taylor, P. (2004). Resilience and Biosphere Reserves. *Environments* 32(3): 79-90.

United Nations Educational, Scientific and Cultural Organization. (2010). *Man and the Biosphere Programme*. Available on-line:http://www.unesco.org/new/en/natural-sciences/environment/ecological-sciences/man-and-biosphere-programme/.

United Nations Educational, Scientific and Cultural Organization. (2008). Madrid Action Plan for Biosphere Reserves (2008-2013). Paris: UNESCO. Available on-line: http://unesdoc.unesco.org/images/0016/001633/163301e.pdf.

United Nations Educational, Scientific and Cultural Organization. (1996). *Biosphere reserves: The Seville strategy & the statutory framework of the World Network*. Paris: UNESCO. Available on-line: http://unesdoc.unesco.org/images/0010/001038/103849Eb.pdf.

Velaniškis, J. (2010). *Report on the Social-Ecological Network Characteristics and Stakeholder Perceptions about Climate Change in the Niagara Region. Final Report*. Prepared for Adaptation and Impacts Research Section - Environment Canada. Toronto, Ontario. (34 pp.)

Visser, J. (1999). *Overcoming the Underdevelopment of Learning: A Transdisciplinary View*. Paper presented at the Symposium on Overcoming the Underdevelopment of Learning held at the annual meeting of the American Educational Research Association Montreal, Canada 19-23 April, 1999.

Whitelaw, G., Craig, B., Jamieson, G. & Hamel, B. (2004). Research, Monitoring and Education: Exploring the "logistics function" of four Canadian biosphere reserves. *Environments* 32(3): 61-78.